IEE CONTROL ENGINEERING SERIES 65

Series Editors: Professor D. Atherton
Professor G. W. Irwin
Professor S. Spurgeon

Modelling and Parameter Estimation of Dynamic Systems

Other volumes in print in this series:

Volume 2 **Elevator traffic analysis, design and control** G. C. Barney and S. M. dos Santos
Volume 8 **A history of control engineering, 1800–1930** S. Bennett
Volume 14 **Optimal relay and saturating control system synthesis** E. P. Ryan
Volume 15 **Self-tuning and adaptive control: theory and application** C. J. Harris and S. A. Billings (Editors)
Volume 16 **Systems modelling and optimisation** P. Nash
Volume 18 **Applied control theory** J. R. Leigh
Volume 20 **Design of modern control systems** D. J. Bell, P. A. Cook and N. Munro (Editors)
Volume 21 **Computer control of industrial processes** S. Bennett and D. A. Linkens (Editors)
Volume 23 **Robotic technology** A. Pugh (Editor)
Volume 26 **Measurement and instrumentation for control** M. G. Mylroi and G. Calvert (Editors)
Volume 27 **Process dynamics estimation and control** A. Johnson
Volume 28 **Robots and automated manufacture** J. Billingsley (Editor)
Volume 30 **Electromagnetic suspension–dynamics and control** P. K. Slnha
Volume 31 **Modelling and control of fermentation processes** J. R. Leigh (Editor)
Volume 32 **Multivariable control for industrial applications** J. O'Reilly (Editor)
Volume 34 **Singular perturbation methodology in control systems** D. S. Naidu
Volume 35 **Implementation of self-tuning controllers** K. Warwick (Editor)
Volume 36 **Robot control** K. Warwick and A. Pugh (Editors)
Volume 37 **Industrial digital control systems (revised edition)** K. Warwick and D. Rees (Editors)
Volume 38 **Parallel processing in control** P. J. Fleming (Editor)
Volume 39 **Continuous time controller design** R. Balasubramanian
Volume 40 **Deterministic control of uncertain systems** A. S. I. Zinober (Editor)
Volume 41 **Computer control of real-time processes** S. Bennett and G. S. Virk (Editors)
Volume 42 **Digital signal processing: principles, devices and applications** N. B. Jones and J. D. McK. Watson (Editors)
Volume 43 **Trends in information technology** D. A. Linkens and R. I. Nicolson (Editors)
Volume 44 **Knowledge-based systems for industrial control** J. McGhee, M. J. Grimble and A. Mowforth (Editors)
Volume 46 **Neural networks for control and systems** K. Warwick, G. W. Irwin and K. J. Hunt (Editors)
Volume 47 **A history of control engineering, 1930–1956** S. Bennett
Volume 48 **MATLAB toolboxes and applications for control** A. J. Chipperfield and P. J. Fleming (Editors)
Volume 49 **Polynomial methods in optimal control and filtering** K. J. Hunt (Editor)
Volume 50 **Programming industrial control systems using IEC 1131-3** R. W. Lewis
Volume 51 **Advanced robotics and intelligent machines** J. O. Gray and D. G. Caldwell (Editors)
Volume 52 **Adaptive prediction and predictive control** P. P. Kanjilal
Volume 53 **Neural network applications in control** G. W. Irwin, K. Warwick and K. J. Hunt (Editors)
Volume 54 **Control engineering solutions: a practical approach** P. Albertos, R. Strietzel and N. Mort (Editors)
Volume 55 **Genetic algorithms in engineering systems** A. M. S. Zalzala and P. J. Fleming (Editors)
Volume 56 **Symbolic methods in control system analysis and design** N. Munro (Editor)
Volume 57 **Flight control systems** R. W. Pratt (Editor)
Volume 58 **Power-plant control and instrumentation** D. Lindsley
Volume 59 **Modelling control systems using IEC 61499** R. Lewis
Volume 60 **People in control: human factors in control room design** J. Noyes and M. Bransby (Editors)
Volume 61 **Nonlinear predictive control: theory and practice** B. Kouvaritakis and M. Cannon (Editors)
Volume 62 **Active sound and vibration control** M.O. Tokhi and S. M. Veres
Volume 63 **Stepping motors: a guide to theory and practice** P. P. Acarnley
Volume 64 **Control theory, 2nd edition** J. R. Leigh

Modelling and Parameter Estimation of Dynamic Systems

J.R. Raol, G. Girija and J. Singh

The Institution of Electrical Engineers

Published by: The Institution of Electrical Engineers, London, United Kingdom

The Institution of Electrical Engineers,
Michael Faraday House,
Six Hills Way, Stevenage,
Herts., SG1 2AY, United Kingdom

www.iee.org

British Library Cataloguing in Publication Data

Raol, J. R.
Modelling and parameter estimation of dynamic systems
(IEE control engineering series ; 65)
1. Parameter estimation 2. Mathematical models
I. Title II. Girija, G. III. Singh, J. IV. Institution of Electrical Engineers
519.5

ISBN 0 86341 363 3

Typeset in India by Newgen Imaging Systems (P) Ltd., Chennai
Printed in England by MPG Books Ltd., Bodmin, Cornwall

The book is dedicated, in loving memory, to:

Rinky – (Jatinder Singh)
Shree M. G. Narayanaswamy – (G. Girija)
Shree Ratansinh Motisinh Raol – (J. R. Raol)

Contents

Preface

Parameter estimation is the process of using observations from a dynamic system to develop mathematical models that adequately represent the system characteristics. The assumed model consists of a finite set of parameters, the values of which are estimated using estimation techniques. Fundamentally, the approach is based on least squares minimisation of error between the model response and actual system's response. With the advent of high-speed digital computers, more complex and sophisticated techniques like filter error method and innovative methods based on artificial neural networks find increasing use in parameter estimation problems. The idea behind modelling an engineering system or a process is to improve its performance or design a control system. This book offers an examination of various parameter estimation techniques. The treatment is fairly general and valid for any dynamic system, with possible applications to aerospace systems. The theoretical treatment, where possible, is supported by numerically simulated results. However, the theoretical issues pertaining to mathematical representation and convergence properties of the methods are kept to a minimum. Rather, a practical application point-of-view is adopted. The emphasis in the present book is on description of the essential features of the methods, mathematical models, algorithmic steps, numerical simulation details and results to illustrate the efficiency and efficacy of the application of these methods to practical systems. The survey of parameter estimation literature is not included in the present book. The book is by no means exhaustive; that would, perhaps, require another volume.

There are a number of books that treat the problem of system identification wherein the coefficients of transfer function (numerator polynomial/denominator polynomial) are determined from the input-output data of a system. In the present book, we are generally concerned with the estimation of parameters of dynamic systems. The present book aims at explicit determination of the numerical values of the elements of system matrices and evaluation of the approaches adapted for parameter estimation. The main aim of the present book is to highlight the computational solutions based on several parameter estimation methods as applicable to practical problems. The evaluation can be carried out by programming the algorithms in PC MATLAB (MATLAB is a registered trademark of the MathWorks, Inc.) and using them for data analysis. PC MATLAB has now become a standard software tool for analysis and design of control systems and evaluation of dynamic systems, including data analysis and signal processing. Hence, most of the parameter estimation algorithms are written in MATLAB based (.m) files. The programs (all of non-proprietary nature) can be downloaded from the authors' website (through the IEE). What one needs is to have access to MATLAB, control-, signal processing- and system identification-toolboxes.

Some of the work presented in this book is influenced by the authors' published work in the area of application of parameter/state estimation methods. Although some numerical examples are from aerospace applications, all the techniques discussed herein are applicable to any general dynamic system that can be described by state space equations (based on a set of difference/differential equations). Where possible, an attempt to unify certain approaches is made: i) categorisation and classification of several model selection criteria; ii) stabilised output error method is shown to be an asymptotic convergence of output error method, wherein the measured states are used (for systems operating in closed loop); iii) total least squares method is further generalised to equation decoupling-stabilised output error method; iv) utilisation of equation error formulation within recurrent neural networks; and v) similarities and contradistinctions of various recurrent neural network structures. The parameter estimation using artificial neural networks and genetic algorithms is one more novel feature of the book. Results on convergence, uniqueness, and robustness of these newer approaches need to be explored. Perhaps, such analytical results could be obtained by using the tenets of the solid foundation of the estimation and statistical theories. Theoretical limit theorems are needed to have more confidence in these approaches based on the so-called 'soft' computing technology.

Thus, the book should be useful to any general reader, undergraduate final year, postgraduate and doctoral students in science and engineering. Also, it should be useful to practising scientists, engineers and teachers pursuing parameter estimation activity in non-aero or aerospace fields. For aerospace applications of parameter estimation, a basic background in flight mechanics is required. Although great care has been taken in the preparation of the book and working out the examples, the readers should verify the results before applying the algorithms to real-life practical problems. The practical application should be at their risk. Several aspects that will have bearing on practical utility and application of parameter estimation methods, but could not be dealt with in the present book, are: i) inclusion of bounds on parameters – leading to constraint parameter estimation; ii) interval estimation; and iii) formal robust approaches for parameter estimation.

Acknowledgements

Numerous researchers all over the world have made contributions to this specialised field, which has emerged as an independent discipline in the last few years. However, its major use has been in aerospace and certain industrial systems.

We are grateful to Dr. S. Balakrishna, Dr. S. Srinathkumar, Dr. R.V. Jategaonkar (Sr. Scientist, Institute for Flight Systems (IFS), DLR, Germany), and Dr. E. Plaetschke (IFS, DLR) for their unstinting support for our technical activities that prompted us to take up this project. We are thankful to Prof. R. Narasimha (Ex-Director, NAL), who, some years ago, had indicated a need to write a book on parameter estimation. Our thanks are also due to Dr. T. S. Prahlad (Distinguished Scientist, NAL) and Dr. B. R. Pai (Director, NAL) for their moral support. Thanks are also due to Prof. N. K. Sinha (Emeritus Professor, McMaster University, Canada) and Prof. R. C. Desai (M.S. University of Baroda) for their technical guidance (JRR).

We appreciate constant technical support from the colleagues of the modelling and identification discipline of the Flight Mechanics and Control division (FMCD) of NAL. We are thankful to V.P.S. Naidu and Sudesh Kumar Kashyap for their help in manuscript preparation. Thanks are also due to the colleagues of Flight Simulation and Control & Handling Quality disciplines of the FMCD for their continual support. The bilateral cooperative programme with the DLR Institute of Flight System for a number of years has been very useful to us. We are also grateful to the IEE (UK) and especially to Ms. Wendy Hiles for her patience during this book project. We are, as ever, grateful to our spouses and children for their endurance, care and affection.

Authors,
Bangalore

Chapter 1
Introduction

Dynamic systems abound in the real-life practical environment as biological, mechanical, electrical, civil, chemical, aerospace, road traffic and a variety of other systems. Understanding the dynamic behaviour of these systems is of primary interest to scientists as well as engineers. Mathematical modelling via parameter estimation is one of the ways that leads to deeper understanding of the system's characteristics. These parameters often describe the stability and control behaviour of the system. Estimation of these parameters from input-output data (signals) of the system is thus an important step in the analysis of the dynamic system.

Actually, analysis refers to the process of obtaining the system response to a specific input, given the knowledge of the model representing the system. Thus, in this process, the knowledge of the mathematical model and its parameters is of prime importance. The problem of parameter estimation belongs to the class of 'inverse problems' in which the knowledge of the dynamical system is derived from the input-output data of the system. This process is empirical in nature and often ill-posed because, in many instances, it is possible that some different model can be fitted to the same response. This opens up the issue of the uniqueness of the identified model and puts the onus of establishing the adequacy of the estimated model parameters on the analyst. Fortunately, several criteria are available for establishing the adequacy and validity of such estimated parameters and models. The problem of parameter estimation is based on minimisation of some criterion (of estimation error) and this criterion itself can serve as one of the means to establish the adequacy of the identified model.

Figure 1.1 shows a simple approach to parameter estimation. The parameters of the model are adjusted iteratively until such time as the responses of the model match closely with the measured outputs of the system under investigation in the sense specified by the minimisation criterion. It must be emphasised here that though a good match is necessary, it is not the sufficient condition for achieving good estimates. An expanded version of Fig. 1.1 appears in Fig. B.6 (see Appendix B) that is specifically useful for understanding aircraft parameter estimation.

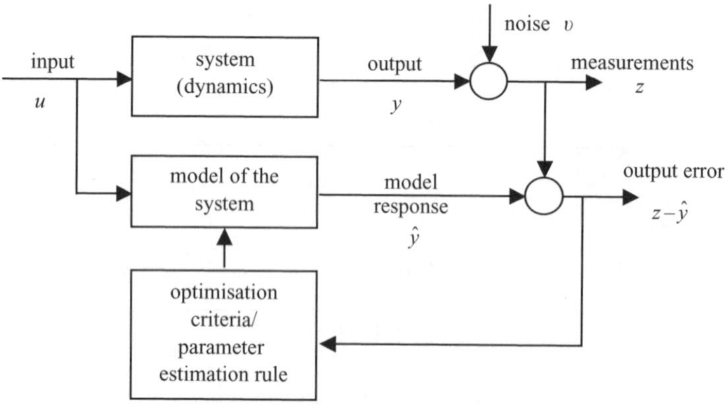

Figure 1.1 Simplified block diagram of the estimation procedure

As early as 1795, Gauss made pioneering contributions to the problem of parameter estimation of the dynamic systems [1]. He dealt with the motion of the planets and concerned himself with the prediction of their trajectories, and in the process used only a few parameters to describe these motions [2]. In the process, he invented the least squares parameter estimation method as a special case of the so-called maximum likelihood type method, though he did not name it so. Most dynamic systems can be described by a set of difference or differential equations. Often such equations are formulated in state-space form that has a certain matrix structure. The dynamic behaviour of the systems is fairly well represented by such linear or nonlinear state-space equations. The problem of parameter estimation pertains to the determination of numerical values of the elements of these matrices, which form the structure of the state-space equations, which in turn describe the behaviour of the system with certain forcing functions (input/noise signals) and the output responses.

The problem of system identification wherein the coefficients of transfer function (numerator polynomial/denominator polynomial) are determined from the input-output data of the system is treated in several books. Also included in the system identification procedure is the determination of the model structure/order of the transfer function of the system. The term modelling refers to the process of determining a mathematical model of a system. The model can be derived based on the physics or from the input-output data of the system. In general, it aims at fitting a state-space or transfer function-type model to the data structure. For the latter, several techniques are available in the literature [3].

The parameter estimation is an important step in the process of modelling based on empirical data of the system. In the present book, we are concerned with the explicit determination of some or all of the elements of the system matrices, for which a number of techniques can be applied. All these major and other newer approaches are dealt with in this book, with emphasis on the practical applications and a few real-life examples in parameter estimation.

The process of modelling covers four important aspects [2]: representation, measurement data, parameter estimation and validation of the estimated models. For estimation, some mathematical models are specified. These models could be static or dynamic, linear or nonlinear, deterministic or stochastic, continuous- or discrete-time, with constant or time-varying parameters, lumped or distributed. In the present book, we deal generally with the dynamic systems, time-invariant parameters and the lumped system. The linear and the nonlinear, as well as the continuous- and the discrete-time systems are handled appropriately. Mostly, the systems dealt with are deterministic, in the sense that the parameters of the dynamical system do not follow any stochastic model or rule. However, the parameters can be considered as random variables, since they are determined from the data, which are contaminated by the measurement noise (sensor/instrument noise) or the environmental noise (atmospheric turbulence acting on a flying aircraft or helicopter). Thus, in this book, we do not deal with the representation theory, per se, but use mathematical models, the parameters of which are to be estimated.

The measurements (data) are required for estimation purposes. Generally, the measurements would be noisy as stated earlier. Where possible, measurement characterisation is dealt with, which is generally needed for the following reasons:

1 Knowing as much as possible about the sensor/measuring instrument and the measured signals *a priori* will help in the estimation procedure, since $z = Hx + v$, i.e.,

$$measurement = (sensor\ dynamics\ or\ model) \times state\ (or\ parameter) + noise$$

2 Any knowledge of the statistics of observation matrix H (that could contain some form of the measured input-output data) and the measurement noise vector v will help the estimation process.
3 Sensor range and the measurement signal range, sensor type, scale factor and bias would provide additional information. Often these parameters need to be estimated.
4 Pre-processing of measurements/whitening would help the estimation process. Data editing would help (see Section A.12, Appendix A).
5 Removing outliers from the measurements is a good idea. For on-line applications, the removal of the outliers should be done (see Section A.35).

Often, the system test engineers describe the signals as parameters. They often consider the vibration signals like accelerations, etc. as the dynamic parameters, and some slowly varying signals as the static parameters. In the present book, we consider input-output data and the states as signals or variables. Especially, the output variables will be called observables. These signals are time histories of the dynamic system. Thus, we do not distinguish between the static and the dynamic 'parameters' as termed by the test engineers. For us, these are signals or data, and the parameters are the coefficients that express the relations between the signals of interest including the states. For the signals that cannot be measured, e.g., the noise, their statistics are assumed to be known and used in the estimation algorithms. Often, one needs to estimate these statistics.

In the present book, we are generally concerned with the estimation of the parameters of dynamic systems and the state-estimation using Kalman filtering algorithms. Often, the parameters and the states are jointly estimated using the so-called extended Kalman filtering approach.

The next and final step is the validation process. The first cut validation is the obtaining of 'good' estimates based on the assessment of several model selection criteria or methods. The use of the so-called Cramer-Rao bounds as uncertainty bounds on the estimates will provide confidence in the estimates if the bounds are very low. The final step is the process of cross validation. We partition the data sets into two: one as the estimation set and the other as the validation set. We estimate the parameters from the first set and then freeze these parameters.

Next, generate the output responses from the system by using the input signal and the parameters from the first set of data. We compare these new responses with the responses from the second set of data to determine the fit errors and judge the quality of match. This helps us in ascertaining the validity of the estimated model and its parameters. Of course, the real test of the estimated model is its use for control, simulation or prediction in a real practical environment.

In the parameter estimation process we need to define a certain error criterion [4, 5]. The optimisation of this error (criterion) cost function will lead to a set of equations, which when solved will give the estimates of the parameters of the dynamic systems. Estimation being data dependent, these equations will have some form of matrices, which will be computed using the measured data. Often, one has to resort to a numerical procedure to solve this set of equations.

The 'error' is defined particularly in three ways.

1 Output error: the difference between the output of the model (to be) estimated from the input-output data. Here the input to the model is the same as the system input.
2 Equation error: define $\dot{x} = Ax + Bu$. If accurate measurements of $\dot{x}, x$ (state of the system) and u (control input) are available, then equation error is defined as $(\dot{x}_m - Ax_m - Bu_m)$.
3 Parameter error: the difference between the estimated value of a parameter and its true value.

The parameter error can be obtained if the true parameter value is known, which is not the case in a real-life scenario. However, the parameter estimation algorithms (the code) can be checked/validated with simulated data, which are generated using the true parameter values of the system. For the real data situations, statements about the error in estimated values of the parameters can be made based on some statistical properties, e.g., the estimates are unbiased, etc. Mostly, the output error approach is used and is appealing from the point of view of matching of the measured and estimated/predicted model output responses. This, of course, is a necessary but not a sufficient condition. Many of the theoretical results on parameter estimation are related to the sufficient condition aspect. Many 'goodness of fit', model selection and validation procedures often offer practical solutions to this problem. If accurate measurements of the states and the inputs are available, the equation error methods are a very good alternative to the output error methods. However, such situations will not occur so frequently.

There are books on system identification [4, 6, 7] which, in addition to the methods, discuss the theoretical aspects of the estimation/methods. Sinha and Kuszta [8] deal with explicit parameter estimation for dynamic systems, while Sorenson [5] provides a solution to the problem of parameter estimation for algebraic systems. The present book aims at explicit determination of the numerical values of the elements of system matrices and evaluation of the approaches adapted for parameter estimation. The evaluation can be carried out by coding the algorithms in PC MATLAB and using them for system data analysis. The theoretical issues pertaining to the mathematical criteria and the convergence properties of the methods are kept to minimum. The emphasis in the present book is on the description of the essential features of the methods, mathematical representation, algorithmic steps, numerical simulation details and PC MATLAB generated results to illustrate the usefulness of these methods for practical systems.

Often in literature, parameter identification and parameter estimation are used interchangeably. We consider that our problem is mainly of determining the estimates of the parameters. Parameter identification can be loosely considered to answer the question: which parameter is to be estimated? This problem can be dealt with by the so-called model selection criteria/methods, which are briefly discussed in the book.

The merits and disadvantages of the various techniques are revealed where feasible. It is presumed that the reader is familiar with basic mathematics, probability theory, statistical methods and the linear system theory. Especially, knowledge of the state-space methods and matrix algebra is essential. The knowledge of the basic linear control theory and some aspects of digital signal processing will be useful. The survey of such aspects and parameter estimation literature are not included in the present book [9, 10, 11].

It is emphasised here that the importance of parameter estimation stems from the fact that there exists a common parameter estimation basis between [12]:

a Adaptive filtering (in communications signal processing theory [13], which is closely related to the recursive parameter estimation process in estimation theory).
b System identification (as transfer function modelling in control theory [3] and as time-series modelling in signal processing theory [14]).
c Control (which needs the mathematical models of the dynamic systems to start with the process of design of control laws, and subsequent use of the models for simulation, prediction and validation of the control laws [15]).

We now provide highlights of each chapter. Chapter 2 introduces the classical method of parameter estimation, the celebrated least squares method invented by Gauss [1] and independently by Legendre [5]. It deals with generalised least squares and equation error methods. Later in Chapter 9, it is shown that the so-called total least squares method and the equation error method form some relation to the stabilised output error methods.

Chapter 3 deals with the widely used maximum likelihood based output error method. The principle of maximum likelihood and its related development are treated in sufficient details.

In Chapter 4, we discuss the filtering methods, especially the Kalman filtering algorithms and their applications. The main reason for including this approach is its use later in Chapters 5 and 7, wherein the filter error and the estimation before modelling approaches are discussed. Also, often the filtering methods can be regarded as generalisations of the parameter estimation methods and the extended Kalman filter is used for joint state and parameter estimation.

In Chapter 5, we deal with the filter error method, which is based on the output error method and the Kalman filtering approach. Essentially, the Kalman filter within the structure of the output error handles the process noise. The filter error method is the maximum likelihood method.

Chapter 6 deals with the determination of model structure for which several criteria are described. Again, the reason for including this chapter is its relation to Chapter 7 on estimation before modelling, which is a combination of the Kalman filtering algorithm and the least squares based (regression) method and utilises some model selection criteria.

Chapter 7 introduces the approach of estimation before modelling. Essentially, it is a two-step method: use of the extended Kalman filter for state estimation (before modelling step) followed by the regression method for estimation of the parameters, the coefficients of the regression equation.

In Chapter 8, we discuss another important method based on the concept of model error. It deals with using an approximate model of the system and then determining the deficiency of the model to obtain an accurate model. This method parallels the estimation before modelling approach.

In Chapter 9, the important problem of parameter estimation of inherently unstable/augmented systems is discussed. The general parameter estimation approaches described in the previous chapters are applicable in principle but with certain care. Some important theoretical asymptotic results are provided.

In Chapter 10, we discuss the approaches based on artificial neural networks, especially the one based on recurrent neural networks, which is a novel method for parameter estimation. First, the procedure for parameter estimation using feed forward neural networks is explained. Then, various schemes based on recurrent neural networks are elucidated. Also included is the description of the genetic algorithm and its usage for parameter estimation.

Chapter 11 discusses three schemes of parameter estimation for real-time applications: i) a time-domain method; ii) recurrent neural network based recursive information processing scheme; and iii) frequency-domain based methods.

It might become apparent that there are some similarities in the various approaches and one might turn out to be a special case of the other based on certain assumptions. Different researchers/practitioners use different approaches based on the availability of software, their personal preferences and the specific problem they are tackling.

The authors' published work in the area of application of parameter/state estimation methods has inspired and influenced some of the work presented in this book. Although some numerical examples are from aerospace applications, all the techniques discussed herein are applicable to any general dynamic system that can be described by a set of difference/differential/state-space equations. The book is by no means exhaustive, it only attempts to cover the main approaches starting from simpler methods like the least squares and the equation error method to the more sophisticated approaches like the filter error and the model error methods. Even these sophisticated approaches are dealt with in as simple a manner as possible. Sophisticated and complex theoretical aspects like convergence, stability of the algorithms and uniqueness are not treated here, except for the stabilised output error method. However, aspects of uncertainty bounds on the estimates and the estimation errors are discussed appropriately. A simple engineering approach is taken rather than a rigorous approach. However, it is sufficiently formal to provide workable and useful practical results despite the fact that, for dynamic (nonlinear) systems, the stochastic differential/difference equations are not used. The theoretical foundation for system identification and experiment design are covered in Reference 16 and for linear estimation in Reference 17. The rigorous approach to the parameter estimation problem is minimised in the present book. Rather, a practical application point-of-view is adopted.

The main aim of the present book is to highlight the computational solutions based on several parameter estimation methods as applicable to practical problems. PC MATLAB has now become a standard software tool for analysis and design of the control systems and evaluation of the dynamic systems, including data analysis and signal processing. Hence, most of the parameter algorithms are written in MATLAB based (.m) files. These programs can be obtained from the authors' website (through the IEE, publisher of this book). The program/filename/directory names, where appropriate, are indicated (in bold letters) in the solution part of the examples, e.g., **Ch2LSex1.m**. Many general and useful definitions often occurring in parameter estimation literature are compiled in Appendix A, and we suggest a first reading of this before reading other chapters of the book.

Many of the examples in the book are of a general nature and great care was taken in the generation and presentation of the results for these examples. Some examples for aircraft parameter estimation are included. Thus, the book should be useful to general readers, and undergraduate final year, postgraduate and doctoral students in science and engineering. It should be useful to the practising scientists, engineers and teachers pursuing parameter estimation activity in non-aero or aerospace fields. For aerospace applications of parameter estimation, a basic background on flight mechanics is required [18, 19], and the material in Appendix B should be very useful. Before studying the examples and discussions related to aircraft parameter estimation (see Sections B.5 to B.11), readers are urged to scan Appendix B. In fact, the complete treatment of aircraft parameter estimation would need a separate volume.

1.1 A brief summary

We draw some contradistinctions amongst the various parameter estimation approaches discussed in the book.

The maximum likelihood-output error method utilises output error related cost function, and the maximum likelihood principle and information matrix. The inverse of information matrix gives the covariance measure and hence the uncertainty bounds on the parameter estimates. Maximum likelihood estimation has nice theoretical properties. The maximum likelihood-output error method is a batch iterative procedure. In one shot, all the measurements are handled and parameter corrections are computed (see Chapter 3). Subsequently, a new parameter estimate is obtained. This process is again repeated with new computation of residuals, etc. The output error method has two limitations: i) it can handle only measurement noise; and ii) for unstable systems, it might diverge. The first limitation is overcome by using Kalman filter type formulation within the structure of maximum likelihood output error method to handle process noise. This leads to the filter error method. In this approach, the cost function contains filtered/predicted measurements (obtained by Kalman filter) instead of the predicted measurements based on just state integration. This makes the method more complex and computationally intensive. The filter error method can compete with the extended Kalman filter, which can handle process as well as measurement noises and also estimate parameters as additional states. One major advantage of Kalman filter/extended Kalman filter is that it is a recursive technique and very suitable for on-line real-time applications. For the latter application, a factorisation filter might be very promising. One major drawback of Kalman filter is the filter tuning, for which the adaptive approaches need to be used.

The second limitation of the output error method for unstable systems can be overcome by using the so-called stabilised output error methods, which use measured states. This stabilises the estimation process. Alternatively, the extended Kalman filter or the extended factorisation filter can be used, since it has some implicit stability property in the filtering equation. The filter error method can be efficiently used for unstable/augmented systems.

Since the output error method is an iterative process, all the predicted measurements are available and the measurement covariance matrix R can be computed in each iteration. The extended Kalman filter for parameter estimation could pose some problems since the covariance matrix part for the states and the parameters would be of quite different magnitudes. Another major limitation of the Kalman filter type approach is that it cannot determine the model error, although it can get good state estimates. The latter part is achieved by process noise tuning. This limitation can be overcome by using the model error estimation method. The approach provides estimation of the model error, i.e., model discrepancy with respect to time. However, it cannot handle process noise. In this sense, the model error estimation can compete with the output error method, and additionally, it can be a recursive method. However, it requires tuning like the Kalman filter. The model discrepancy needs to be fitted with another model, the parameters of which can be estimated using recursive least squares method.

Another approach, which parallels the model error estimation, is the estimation before modelling approach. This approach has two steps: i) the extended Kalman filter to estimate states (and scale factors and bias related parameters); and ii) a regression method to estimate the parameters of the state model or related model. The model error estimation also has two steps: i) state estimation and discrepancy estimation using the invariant embedding method; and ii) a regression method to estimate the parameters from the discrepancy time-history. Both the estimation before modelling and the model error estimation can be used for parameter estimation of a nonlinear system. The output error method and the filter error method can be used for nonlinear problems.

The feed forward neural network based approach somewhat parallels the two-step methodologies, but it is quite distinct from these: it first predicts the measurements and then the trained network is used repeatedly to obtain differential states/measurements. The parameters are determined by Delta method and averaging. The recurrent neural network based approach looks quite distinct from many approaches, but a closer look reveals that the equation error method and the output error method based formulations can be solved using the recurrent neural network based structures. In fact, the equation error method and the output error method can be so formulated without invoking recurrent neural network theory and still will look as if they are based on certain variants of the recurrent neural networks. This revealing observation is important from practical application of the recurrent neural networks for parameter estimation, especially for on-line/real-time implementation using adaptive circuits/VLSI, etc. Of course, one needs to address the problem of convergence of the recurrent neural network solutions to true parameters. Interestingly, the parameter estimation procedure using recurrent neural network differs from that based on the feed forward neural network. In the recurrent neural network, the so-called weights (weighting matrix W) are pre-computed using the correlation like expressions between $\dot{x}, x, u$, etc. The integration of a certain expression, which depends on the sigmoid nonlinearity, weight matrix and bias vector and some initial 'guesstimates' of the states of the recurrent neural network, results into the new states of the network. These states are the estimated parameters (of the intended state-space model). This quite contrasts with the procedure of estimation using the feed forward neural network, as can be seen from Chapter 10. In feed forward neural networks, the weights of the network are not the parameters of direct interest. In recurrent neural network also, the weights are not of direct interest, although they are pre-computed and not updated as in feed forward neural networks. In both the methods, we do not get to know more about the statistical properties of the estimates and their errors. Further theoretical work needs to be done in this direction.

The genetic algorithms provide yet another alternative method that is based on direct cost function minimisation and not on the gradient of the cost function. This is very useful for types of problems where the gradient could be ill-defined. However, the genetic algorithms need several iterations for convergence and stopping rules are needed. One limitation is that we cannot get parameter uncertainties, since they are related to second order gradients. In that case, some mixed approach can be used, i.e., after the convergence, the second order gradients can be evaluated.

Parameter estimation work using the artificial neural networks and the genetic algorithms is in an evolving state. New results on convergence, uniqueness, robustness and parameter error-covariance need to be explored. Perhaps, such results could be obtained by using the existing analytical results of estimation and statistical theories. Theoretical limit theorems are needed to obtain more confidence in these approaches.

The parameter estimation for inherently unstable/augmented system can be handled with several methods but certain precautions are needed as discussed in Chapter 9. The existing methods need certain modifications or extensions, the ramifications of which are straightforward to appreciate, as can be seen from Chapter 9.

On-line/real-time approaches are interesting extensions of some of the off-line methods. Useful approaches are: i) factorisation-Kalman filtering algorithm; ii) recurrent neural network; and iii) frequency domain methods.

Several aspects that will have further bearing on the practical utility and application of parameter estimation methods, but could not be dealt with in the present book, are: i) inclusion of bounds on parameters (constraint parameter estimation); ii) interval estimation; and iii) robust estimation approaches. For i) the *ad hoc* solution is that one can pre-specify the numerical limits on certain parameters based on the physical understanding of the plant dynamics and the range of allowable variation of those parameters. So, during iteration, these parameters are forced to remain within this range. For example, let the range allowed be given as β_L and β_H. Then,

$$\text{if } \hat{\beta} > \beta_H, \qquad \text{put } \hat{\beta} = \beta_H - \varepsilon \quad \text{and}$$

$$\text{if } \hat{\beta} < \beta_H, \qquad \text{put } \hat{\beta} = \beta_L + \varepsilon$$

where ε is a small number. The procedure is repeated once a new estimate is obtained. A formal approach can be found in Reference 20.

Robustness of estimation algorithm, especially for real-time applications, is very important. One aspect of robustness is related to prevention of the effect of measurement data outliers on the estimation. A formal approach can be found in Reference 21. In interval estimation, several uncertainties (due to data, noise, deterministic disturbance and modelling) that would have an effect on the final accuracy of the estimates should be incorporated during the estimation process itself.

1.2 References

1 GAUSS, K. F.: 'Theory of the motion of heavenly bodies moving about the sun in conic section' (Dover, New York, 1963)
2 MENDEL, J. M.: 'Discrete techniques of parameter estimation: equation error formulation' (Marcel Dekker, New York, 1976)
3 LJUNG, L.: 'System identification: theory for the user' (Prentice-Hall, Englewood Cliffs, 1987)
4 HSIA, T. C.: 'System identification – least squares methods' (Lexington Books, Lexington, Massachusetts, 1977)
5 SORENSON, H. W.: 'Parameter estimation – principles and problems' (Marcel Dekker, New York and Basel, 1980)
6 GRAUPE, D.: 'Identification of systems' (Van Nostrand, Reinhold, New York, 1972)

7 EYKHOFF, P.: 'System identification: parameter and state estimation' (John Wiley, London, 1972)
8 SINHA, N. K. and KUSZTA, B.: 'Modelling and identification of dynamic system' (Van Nostrand, New York, 1983)
9 OGATA, K.: 'Modern control engineering' (Pearson Education, Asia, 2002, 4th edn)
10 SINHA, N. K.: 'Control systems' (Holt, Rinehart and Winston, New York, 1988)
11 BURRUS, C. D., McCLELLAN, J. H., OPPENHEIM, A. V., PARKS, T. W., SCHAFER, R. W., and SCHUESSLER, H. W.: 'Computer-based exercises for signal processing using MATLAB®' (Prentice-Hall International, New Jersey, 1994)
12 JOHNSON, C. R.: 'The common parameter estimation basis for adaptive filtering, identification and control', *IEEE Transactions on Acoustics, Speech and Signal Processing*, 1982, **ASSP-30**, (4), pp. 587–595
13 HAYKIN, S.: 'Adaptive filtering' (Prentice-Hall, Englewood Cliffs, 1986)
14 BOX, G. E. P., and JUNKINS, J. L.: 'Time series: analysis, forecasting and controls' (Holden Day, San Francisco, 1970)
15 DORSEY, J.: 'Continuous and discrete control systems – modelling, identification, design and implementation' (McGraw Hill, New York, 2002)
16 GOODWIN, G. C., and PAYNE, R. L.: 'Dynamic system identification: experiment design and data analysis' (Academic Press, New York, 1977)
17 KAILATH, T., SAYAD, A. H., and HASSIBI, B.: 'Linear estimation' (Prentice-Hall, New Jersey, 2000)
18 McRUER, D. T., ASHKENAS, I., and GRAHAM, D.: 'Aircraft dynamics and automatic control' (Princeton University Press, Princeton, 1973)
19 NELSON, R. C.: 'Flight stability and automatic control' (McGraw-Hill, Singapore, 1998, 2nd edn)
20 JATEGAONKAR, R. V.: 'Bounded variable Gauss Newton algorithm for aircraft parameter estimation', *Journal of Aircraft*, 2000, **3**, (4), pp. 742–744
21 MASRELIEZ, C. J., and MARTIN, R. D.: 'Robust Bayesian estimation for the linear model for robustifying the Kalman filter', *IEEE Trans. Automat. Contr.*, 1977, **AC-22**, pp. 361–371

Chapter 2
Least squares methods

2.1 Introduction

To address the parameter estimation problem, we begin with the assumption that the data are contaminated by noise or measurement errors. We use these data in an identification/estimation procedure to arrive at optimal estimates of the unknown parameters that best describe the behaviour of the data/system dynamics. This process of determining the unknown parameters of a mathematical model from noisy input-output data is termed 'parameter estimation'. A closely related problem is that of 'state estimation' wherein the estimates of the so-called 'states' of the dynamic process/system (e.g., power plant or aircraft) are obtained by using the optimal linear or the nonlinear filtering theory as the case may be. This is treated in Chapter 4.

In this chapter, we discuss the least squares/equation error techniques for parameter estimation, which are used for aiding the parameter estimation of dynamic systems (including algebraic systems), in general, and the aerodynamic derivatives of aerospace vehicles from the flight data, in particular. In the first few sections, some basic concepts and techniques of the least squares approach are discussed with a view to elucidating the more involved methods and procedures in the later chapters. Since our approach is model-based, we need to define a mathematical model of the dynamic (or static) system.

The measurement equation model is assumed to have the following form:

$$z = H\beta + v, \quad y = H\beta \tag{2.1}$$

where y is $(m \times 1)$ vector of true outputs and z is $(m \times 1)$ vector that denotes the measurements (affected by noise) of the unknown parameters (through H), β is $(n \times 1)$ vector of the unknown parameters and v represents the measurement noise/errors, which are assumed to be zero mean and Gaussian. This model is called the measurement equation model, since it forms a relationship between the measurements and the parameters of a system.

It can be said that the estimation theory and the methods have (measurement) data-dependent nature, since the measurements used for estimation are invariably noisy. These noisy measurements are utilised in the estimation procedure/algorithm/software to improve upon the initial guesstimate of the parameters that characterise the signal or system. One of the objectives of the estimator is to produce the estimates of the signal (what it means is the predicted signal using the estimated parameters) with errors much less than the noise affecting the signal. In order to make this possible, the signal and the noise should have significantly differing characteristics, e.g., different frequency spectra, widely differing statistical properties (true signal being deterministic and the noise being of random nature). This means that the signal is characterised by a structure or a mathematical model (like $H\beta$), and the noise (v) often or usually is assumed as zero mean and white process. In most cases, the measurement noise is also considered Gaussian. This 'Gaussianess' assumption is supported by the central limit theorem (see Section A.4). We use discrete-time (sampled; see Section A.2) signals in carrying out analysis and generating computer-based numerical results in the examples.

2.2 Principle of least squares

The least squares (LS) estimation method was invented by Karl Gauss in 1809 and independently by Legendre in 1806. Gauss was interested in predicting the motions of the planets using measurements obtained by telescopes when he invented the least squares method. It is a well established and easy to understand method. Still, to date, many problems centre on this basic approach. In addition, the least squares method is a special case of the well-known maximum likelihood estimation method for linear systems with Gaussian noise. In general, least squares methods are applicable to both linear as well as nonlinear problems. They are applicable to multi-input multi-output dynamic systems. Least squares techniques can also be applied to the on-line identification problem discussed in Chapter 11. For this method, it is assumed that the system parameters do not rapidly change with time, thereby assuring almost stationarity of the plant or the process parameters. This may mean that the plant is assumed quasi-stationary during the measurement period. This should not be confused with the requirement of non-steady input-output data over the period for which the data is collected for parameter estimation. This means that during the measurement period there should be some activity.

The least squares method is considered a deterministic approach to the estimation problem. We choose an estimator of β that minimises the sum of the squares of the error (see Section A.32) [1, 2].

$$J \cong \frac{1}{2}\sum_{k=1}^{N} v_k^2 = \frac{1}{2}(z - H\beta)^T (z - H\beta) \tag{2.2}$$

Here J is a cost function and v, the residual errors at time k (index). Superscript T stands for the vector/matrix transposition.

The minimisation of J w.r.t. β yields

$$\frac{\partial J}{\partial \beta} = -(z - H\hat{\beta}_{LS})^T H = 0 \quad \text{or} \quad H^T(z - H\hat{\beta}_{LS}) = 0 \tag{2.3}$$

Further simplification leads to

$$H^T z - (H^T H)\hat{\beta}_{LS} = 0 \quad \text{or} \quad \hat{\beta}_{LS} = (H^T H)^{-1} H^T z \tag{2.4}$$

In eq. (2.4), the term before z is a pseudo-inverse (see Section A.37). Since, the matrix H and the vector (of measurements) z are known quantities, $\hat{\beta}_{LS}$, the least squares estimate of β, can be readily obtained. The inverse will exist only if no column of H is a linear combination of other columns of H. It must be emphasised here that, in general, the number of measurements (of the so-called observables like y) should be more than the number of parameters to be estimated. This implies at least theoretically, that

number of measurements = number of parameters + 1

This applies to almost all the parameter estimation techniques considered in this book. If this requirement were not met, then the measurement noise would not be smoothed out at all. If we ignore v in eq. (2.1), we can obtain β using pseudo-inverse of H, i.e., $(H^T H)^{-1} H^T$. This shows that the estimates can be obtained in a very simple way from the knowledge of only H. By evaluating the Hessian (see Section A.25) of the cost function J, we can assert that the cost function will be minimum for the least squares estimates.

2.2.1 *Properties of the least squares estimates [1,2]*

a $\hat{\beta}_{LS}$ is a linear function of the data vector z (see eq. (2.4)), since H is a completely known quantity. H could contain input-output data of the system.

b The error in the estimator is a linear function of the measurement errors (v_k)

$$\tilde{\beta}_{LS} = \beta - \hat{\beta}_{LS} = \beta - (H^T H)^{-1} H^T (H\beta + v) = -(H^T H)^{-1} H^T v \tag{2.5}$$

Here $\tilde{\beta}_{LS}$ is the error in the estimation of β. If the measurement errors are large, then the error in estimation is large.

c $\tilde{\beta}_{LS}$ is chosen such that the residual, defined by $r \cong (z - H\hat{\beta}_{LS})$, is perpendicular (in general orthogonal) to the columns of the observation matrix H. This is the 'principle of orthogonality'. This property has a geometrical interpretation.

d If $E\{v\}$ is zero, then the LS estimate is unbiased. Let $\tilde{\beta}_{LS}$ be defined as earlier. Then, $E\{\tilde{\beta}_{LS}\} = -(H^T H)^{-1} H^T E\{v\} = 0$, since $E\{v\} = 0$. Here $E\{.\}$ stands for mathematical expectation (see Section A.17) of the quantity in braces. If, for all practical purposes, $z = y$, then $\hat{\beta}$ is a deterministic quantity and is then exactly equal to β. If the measurement errors cannot be neglected, i.e., $z \neq y$, then $\hat{\beta}$ is random. In this case, one can get $\hat{\beta}$ as an unbiased estimate of β. The least squares method, which leads to a biased estimate in the presence of measurement noise, can be used as a start-up procedure for other estimation methods like the generalised least squares and the output error method.

e The covariance (see Section A.11) of the estimation error is given as:

$$E\{\tilde{\beta}_{LS}\tilde{\beta}_{LS}^T\} \cong P = (H^T H)^{-1} H^T R H (H^T H)^{-1} \tag{2.6}$$

where R is the covariance matrix of v. If v is uncorrelated and its components have identical variances, then $R = \sigma^2 I$, where I is an identity matrix. Thus, we have

$$\text{cov}(\tilde{\beta}_{LS}) = P = \sigma^2 (H^T H)^{-1} \tag{2.7}$$

Hence, the standard deviation of the parameter estimates can be obtained as $\sqrt{P_{ii}}$, ignoring the effect of cross terms of the matrix P. This will be true if the parameter estimation errors like $\tilde{\beta}_{ij}$ for $i \neq j$ are not highly correlated. Such a condition could prevail, if the parameters are not highly dependent on each other. If this is not true, then only ratios of certain parameters could be determined. Such difficulties arise in closed loop identification, e.g., data collinearity, and such aspects are discussed in Chapter 9.

f The residual has zero mean:

$$r \cong (z - H\hat{\beta}_{LS}) = H\beta + v - H\hat{\beta}_{LS} = H\tilde{\beta}_{LS} + v \tag{2.8}$$

$E\{r\} = HE\{\tilde{\beta}_{LS}\} + E\{v\} = 0 + 0 = 0$ for an unbiased LS estimate.

If residual is not zero mean, then the mean of the residuals can be used to detect bias in the sensor data.

2.2.1.1 Example 2.1

A transfer function of the electrical motor speed (S rad/s) with V as the input voltage to its armature is given as:

$$\frac{S(s)}{V(s)} = \frac{K}{s + \alpha} \tag{2.9}$$

Choose suitable values of K and α, and obtain step response of S. Fit a least squares (say linear) model to a suitable segment of these data of S. Comment on the accuracy of the fit. What should be the values of K and α, so that the fit error is less than say 5 per cent?

2.2.1.2 Solution

Step input response of the system is generated for a period of 5 s using a time array ($t = 0 : 0.1 : 5$ s) with sampling interval of 0.1 s. A linear model $y = mt$ is fitted to the data for values of alpha in the range 0.001 to 0.25 with $K = 1$. Since K contributes only to the gain, its value is kept fixed at $K = 1$. Figure 2.1(a) shows the step response for different values of alpha; Fig. 2.1(b) shows the linear least squares fit to the data for $\alpha = 0.1$ and $\alpha = 0.25$. Table 2.1 gives the percentage fit error (PFE) (see Chapter 6) as a function of α. It is clear that the fit error is < 5 per cent for values of $\alpha < 0.25$. In addition, the standard deviation (see Section A.44) increases as α increases. The simulation/estimation programs are in file **Ch2LSex1.m**. (See Exercise 2.4).

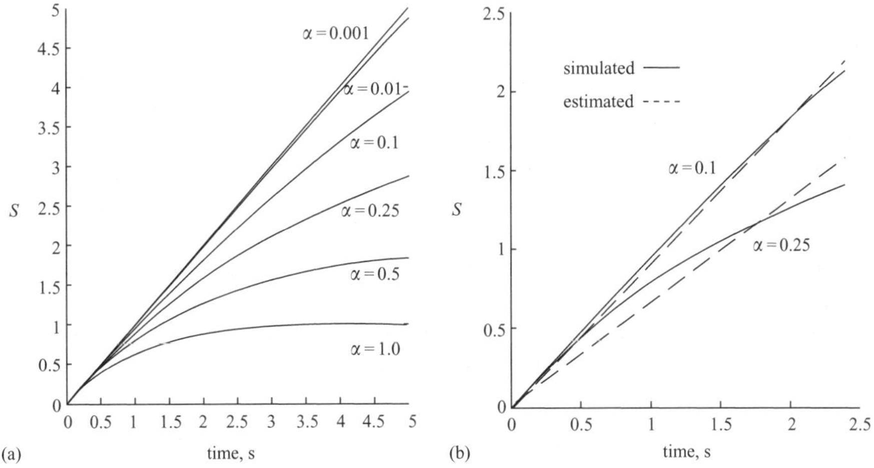

Figure 2.1 *(a) Step response for unit step input (Example 2.1); (b) linear least squares fit to the first 2.5 s of response (Example 2.1)*

Table 2.1 *LS estimates and PFE (Example 2.1)*

α	$\hat{m}$ (estimate of m)	PFE
0.001	0.999 (4.49e – 5)*	0.0237
0.01	0.9909 (0.0004)	0.2365
0.1	0.9139 (0.004)	2.3273
0.25	0.8036 (0.0086)	5.6537

* standard deviation

We see that response becomes nonlinear quickly and the nonlinear model might be required to be fitted. The example illustrates degree or extent of applicability of linear model fit.

2.2.1.3 Example 2.2

Let

$$y(k) = \beta_1 + \beta_2 k \tag{2.10}$$

Choose suitable values β_1 and β_2 and with k as the time index generate data $y(k)$. Add Gaussian noise with zero mean and known standard deviation. Fit a least squares curve to these noisy data $z(k) = y(k) + noise$ and obtain the fit error.

2.2.1.4 Solution

By varying the index k from 1 to 100, 100 data samples of $y(k)$ are generated for fixed values of $\beta_1 = 1$ and $\beta_2 = 1$. Gaussian random noise with zero mean and standard deviation (σ = square root of variance; see Section A.44) is added to the data $y(k)$ to generate three sets of noisy data samples. Using the noisy data, a linear least squares solution is obtained for the parameters β_1 and β_2. Table 2.2 shows the estimates of the parameters along with their standard deviations and the PFE of the estimated $y(k)$ w.r.t. true $y(k)$. It is clear from the Table 2.2 that the estimates of β_1 are sensitive to the noise in the data whereas the estimates of β_2 are not very sensitive. However, it is clear that the PFE for all cases are very low indicating the adequacy of the estimates. Figures 2.2(a) and (b) show the plots of true and noisy data and true and estimated output. The programs for simulation/estimation are in file **Ch2LSex2.m**.

Table 2.2 LS estimates and PFE (Example 2.2)

	β_1 (estimate) (True $\beta_1 = 1$)	β_2 (estimate) (True $\beta_2 = 1$)	PFE
Case 1 ($\sigma = 0.1$)	1.0058	0.9999	0.0056
	(0.0201)*	(0.0003)	
Case 2 ($\sigma = 1.0$)	1.0583	0.9988	0.0564
	(0.2014)	(0.0035)	

* standard deviation

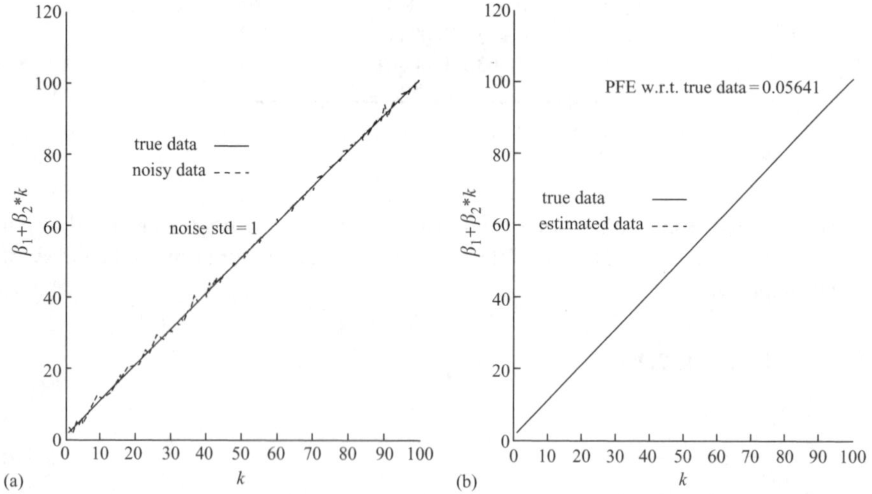

Figure 2.2 (a) Simulated data, y(k) (Example 2.2); (b) true data estimated y(k) (Example 2.2)

2.3 Generalised least squares

The generalised least squares (GLS) method is also known as weighted least squares method. The use of a weighting matrix in least squares criterion function gives the cost function for GLS method:

$$J = (z - H\beta)^T W(z - H\beta) \tag{2.11}$$

Here W is the weighting matrix, which is symmetric and positive definite and is used to control the influence of specific measurements upon the estimates of β. The solution will exist if the weighting matrix is positive definite.

Let $W = SS^T$ and $S^{-1}WS^{-T} = I$; here S being a lower triangular matrix and square root of W.

We transform the observation vector z (see eq. (2.1)) as follows:

$$z' = S^T z = S^T H\beta + S^T v = H'\beta + v' \tag{2.12}$$

Expanding the J, we get

$$\begin{aligned}(z - H\beta)^T W(z - H\beta) &= (z - H\beta)^T SS^T (z - H\beta)\\ &= (S^T z - S^T H\beta)^T (S^T z - S^T H\beta)\\ &= (z' - H'\beta)^T (z' - H'\beta)\end{aligned}$$

Due to similarity of the form of the above expression with the expression for LS, the previous results of Section 2.2 can be directly applied to the measurements z'.

We have seen that the error covariance provides a measure of the behaviour of the estimator. Thus, one can alternatively determine the estimator, which will minimise the error variances. If the weighting matrix W is equal to R^{-1}, then the GLS estimates are called Markov estimates [1].

2.3.1 *A probabilistic version of the LS [1,2]*

Define the cost function as

$$J_{ms} = E\{(\beta - \hat{\beta})^T (\beta - \hat{\beta})\} \tag{2.13}$$

where subscript ms stands for mean square.

Here E stands for the mathematical expectation, which takes, in general, probabilistic weightage of the variables.

Consider an arbitrary, linear and unbiased estimator $\hat{\beta}$ of β. Thus, we have $\hat{\beta} = Kz$, where K is matrix $(n \times m)$ that transforms the measurements (vector z) to the estimated parameters (vector β). Thus, we are seeking a linear estimator based on the measured data. Since $\hat{\beta}$ is required to be unbiased we have

$$E\{\hat{\beta}\} = E\{K(H\beta + v)\} = E\{KH\beta + Kv\} = KHE\{\beta\} + KE\{v\}$$

Since $E\{v\} = 0$, i.e., assuming zero mean noise, $E\{\hat{\beta}\} = KHE\{\beta\}$ and $KH = I$ for unbiased estimate.

This gives a constraint on K, the so-called the gain of the parameter estimator. Next, we recall that

$$\begin{aligned} J_{ms} &= E\{(\beta - \hat{\beta})^T(\beta - \hat{\beta})\} \\ &= E\{(\beta - Kz)^T(\beta - Kz)\} \\ &= E\{(\beta - KH\beta - Kv)^T(\beta - KH\beta - Kv)\} \\ &= E\{v^T K^T K v\}; \quad \text{since } KH = I \\ &= \text{Trace } E\{K v v^T K^T\} \end{aligned} \tag{2.14}$$

and defining $R = E\{vv^T\}$, we get $J_{ms} = \text{Trace}(KRK^T)$, where R is the covariance matrix of the measurement noise vector v.

Thus, the gain matrix should be chosen such that it minimises J_{ms} subject to the constraint $KH = I$. Such K matrix is found to be [2]

$$K = (H^T R^{-1} H)^{-1} H^T R^{-1} \tag{2.15}$$

With this value of K, the constraint will be satisfied. The error covariance matrix P is given by

$$P = (H^T R^{-1} H)^{-1} \tag{2.16}$$

We will see in Chapter 4 that similar development will follow in deriving KF. It is easy to establish that the generalised LS method and linear minimum mean squares method give identical results, if the weighting matrix W is chosen such that $W = R^{-1}$. Such estimates, which are unbiased, linear and minimise the mean-squares error, are called Best Linear Unbiased Estimator (BLUE) [2]. We will see in Chapter 4 that the Kalman filter is such an estimator.

The matrix H, which determines the relationship between measurements and β, will contain some variables, and these will be known or measured. One important aspect about spacing of such measured variables (also called measurements) in matrix H is that, if they are too close (due to fast sampling or so), then rows or columns (as the case may be) of the matrix H will be correlated and similar and might cause ill-conditioning in matrix inversion or computation of parameter estimates. Matrix ill-conditioning can be avoided by using the following artifice:

> Let $H^T H$ be the matrix to be inverted, then use $(H^T H + \varepsilon I)$ with ε as a small number, say 10^{-5} or 10^{-7} and I as the identity matrix of the same size $H^T H$. Alternatively, matrix factorisation and subsequent inversion can be used as is done, for example, in the UD factorisation (U = Unit upper triangular matrix, D = Diagonal matrix) of Chapter 4.

2.4 Nonlinear least squares

Most real-life static/dynamic systems have nonlinear characteristics and for accurate modelling, these characteristics cannot be ignored. If type of nonlinearity is known, then only certain unknown parameters need be estimated. If the type of nonlinearity is unknown, then some approximated model should be fitted to the data of the system. In this case, the parameters of the fitted model need to be estimated.

In general, real-life practical systems are nonlinear and hence we apply the LS method to nonlinear models. Let such a process or system be described by

$$z = h(\beta) + v \tag{2.17}$$

where h is a known, nonlinear vector valued function/model of dimension m.

With the LS criterion, we have [1, 2]:

$$J = (z - h(\beta))^T (z - h(\beta)) \tag{2.18}$$

The minimisation of J w.r.t. β results in

$$\frac{\partial J}{\partial \beta} = -2[z - h(\hat{\beta})]^T \frac{\partial h(\hat{\beta})}{\partial \beta} = 0 \tag{2.19}$$

We note that the above equation is a system of nonlinear algebraic equations. For such a system, a closed form solution may not exist. This means that we may not be able to obtain $\hat{\beta}$ explicitly in terms of observation vector without resorting to some approximation or numerical procedure. From the above equation we get

$$\left[\frac{\partial h(\hat{\beta})}{\partial \beta}\right]^T (z - h(\hat{\beta})) = 0 \tag{2.20}$$

The second term in the above equation is the residual error and the form of the equation implies that the residual vector must be orthogonal to the columns of $\partial h/\partial \beta$, the principle of orthogonality. An iterative procedure to approximately solve the above nonlinear least squares (NLS) problem is described next [2]. Assume some initial guess or estimate (called guesstimate) β^* for β. We expand $h(\beta)$ about β^* via Taylor's series to obtain

$$z = h(\beta^*) + \left\{\frac{\partial h(\beta^*)}{\partial \beta}\right\} (\beta - \beta^*) + \text{higher order terms} + v$$

Retaining terms up to first order we get

$$(z - h(\beta^*)) = \left\{\frac{\partial h(\beta^*)}{\partial \beta}\right\} (\beta - \beta^*) + v \tag{2.21}$$

Comparing this with the measurement equation studied earlier and using the results of the previous sections we obtain

$$\begin{aligned} (\hat{\beta} - \beta^*) &= (H^T H)^{-1} H^T (z - h(\beta^*)) \\ \hat{\beta} &= \beta^* + (H^T H)^{-1} H^T (z - h(\beta^*)) \end{aligned} \tag{2.22}$$

Here $H = \partial h(\beta^*)/\partial \beta$ at $\beta = \beta^*$.

Thus, the algorithm to obtain $\hat{\beta}$ from eq. (2.22) is given as follows:

(i) Choose β^*, initial guesstimate.
(ii) Linearise h about β^* and obtain H matrix.
(iii) Compute residuals $(z - h(\beta^*))$ and then compute the $\hat{\beta}$.

(iv) Check for the orthogonality condition: $H^T(z - h(\beta))|_{\beta=\hat{\beta}}$ = orthogonality condition value = 0.
(v) If the above condition is not satisfied, then replace β^* by $\hat{\beta}$ and repeat the procedure.
(vi) Terminate the iterations when the orthogonality condition is at least approximately satisfied. In addition, the residuals should be white as discussed below.

We hasten to add here that a similar iterative algorithm development will be encountered when we discuss the maximum likelihood and other methods for parameter estimation in subsequent chapters.

If the residuals $(z - h(\hat{\beta}))$ are not white, then a procedure called generalised least squares can also be adopted [1]. The main idea of the residual being white is that residual power spectral density is flat (w.r.t. frequency), and the corresponding autocorrelation is an impulse function. It means that the white process is uncorrelated at the instants of time other than $t = 0$, and hence it cannot be predicted. It means that the white process has no model or rule that can be used for its prediction. It also means that if the residuals are white, complete information has been extracted from the signals used for parameter estimation and nothing more can be extracted from the signal.

If residuals are non-white, then a model (filter) can be fitted to these residuals using the LS method and parameters of the model/filter estimated:

$$\hat{\beta}_{rLS} = (X_r^T X_r)^{-1} X_r^T$$

Here, r is the residual time history and X_r is the matrix composed of values of r, and will depend on how the residuals are modelled. Once $\hat{\beta}_r$ is obtained by the LS method, it can be used to filter the original signal/data. These filtered data are used again to obtain the new set of parameters of the system and this process is repeated until $\hat{\beta}$ and $\hat{\beta}_r$ are converged. This is also called GLS procedure (in system identification literature) and it would provide more accurate estimates when the residual errors are autocorrelated (and hence non-white) [1].

2.4.1.1 Example 2.3

Let the model be given by

$$y(k) = \beta x^2(k) \tag{2.23}$$

Add Gaussian noise with zero mean and variance such that the SNR = 2. Fit a nonlinear least squares curve to the noisy data:

$$z(k) = y(k) + noise \tag{2.24}$$

2.4.1.2 Solution

100 samples of data $y(k)$ are generated using eq. (2.23) with $\beta = 1$. Gaussian noise (generated using the function **randn**) with SNR = 2 is added to the samples $y(k)$ to

generate $z(k)$. A nonlinear least squares model is fitted to the data and β is estimated, using the procedure outlined in (i) to (vi) of Section 2.4. In a true sense, the eq. (2.23) is linear-in-parameter and nonlinear in x. *The SNR for the purpose of this book is defined as the ratio of variance of signal to variance of noise.*

The estimate $\hat{\beta} = 0.9872$ was obtained with a standard deviation of 0.0472 and PFE $= 1.1$ per cent. The algorithm converges in three iterations. The orthogonal condition value converges from 0.3792 to 1.167e − 5 in three iterations.

Figure 2.3(a) shows the true and noisy data and Fig. 2.3(b) shows the true and estimated data. Figure 2.3(c) shows the residuals and the autocorrelation of residuals with bounds. We clearly see that the residuals are white (see Section A.1). Even though the SNR is very low, the fit error is acceptably good. The simulation/estimation programs are in file **Ch2NLSex3.m**.

2.5 Equation error method

This method is based on the principle of least squares. The equation error method (EEM) minimises a quadratic cost function of the error in the (state) equations to estimate the parameters. It is assumed that states, their derivatives and control inputs are available or accurately measured. The equation error method is relatively fast and simple, and applicable to linear as well as linear-in-parameter systems [3].

If the system is described by the state equation

$$\dot{x} = Ax + Bu \qquad \text{with } x(0) = x_0 \tag{2.25}$$

the equation error can be written as

$$e(k) = \dot{x}_m - Ax_m - Bu_m \tag{2.26}$$

Here x_m is the measured state, subscript m denoting 'measured'. Parameter estimates are obtained by minimising the equation error w.r.t. β. The above equation can be written as

$$e(k) = \dot{x}_m - A_a x_{am} \tag{2.27}$$

where

$$A_a = [A \quad B] \quad \text{and} \quad x_{am} = \begin{bmatrix} x_m \\ u_m \end{bmatrix}$$

In this case, the cost function is given by

$$J(\beta) = \frac{1}{2}\sum_{k=1}^{N} [\dot{x}_m(k) - A_a x_{am}(k)]^T\ [\dot{x}_m(k) - A_a x_{am}(k)] \tag{2.28}$$

The estimator is given as

$$\hat{A}_a = \dot{x}_m \left(x_{am}^T\right) \left(x_{am} x_{am}^T\right)^{-1} \tag{2.29}$$

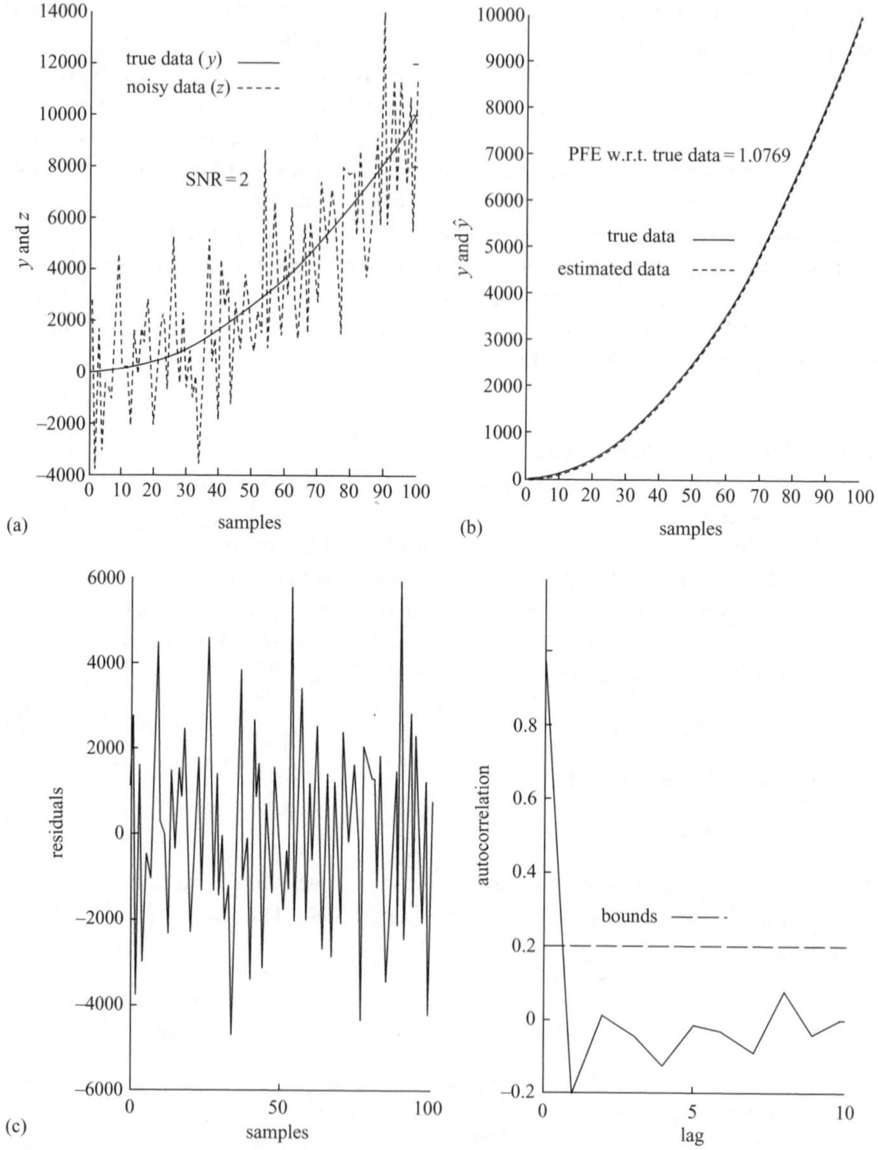

Figure 2.3 *(a) True and noisy data (Example 2.3); (b) true and estimated data (Example 2.3); (c) residuals and autocorrelation of residuals with bounds (Example 2.3)*

We illustrate the above formulation as follows:

$$\text{Let } \begin{bmatrix} \dot{x}_1 \\ \dot{x}_2 \end{bmatrix} = \begin{bmatrix} a_{11} & a_{12} \\ a_{21} & a_{22} \end{bmatrix} \begin{bmatrix} x_1 \\ x_2 \end{bmatrix} + \begin{bmatrix} b_1 \\ b_2 \end{bmatrix} u$$

Then, if there are, say, two measurements, we have:

$$x_{am} = \begin{bmatrix} x_{11m} & x_{12m} \\ x_{21m} & x_{22m} \\ u_{1m} & u_{2m} \end{bmatrix}_{3\times 2} \quad ; \quad u_m = [u_{1m} \quad u_{2m}]$$

$$\dot{x}_m = \begin{bmatrix} \dot{x}_{11m} & \dot{x}_{12m} \\ \dot{x}_{21m} & \dot{x}_{22m} \end{bmatrix}$$

Then

$$\begin{aligned} [\hat{A}_a]_{2\times 3} &= \left[[\hat{A}]_{2\times 2} \vdots [\hat{B}]_{2\times 1}\right] \\ &= [\dot{x}_m]_{2\times 2} \left[x_{am}^T\right]_{2\times 3} \left\{[x_{am}]_{3\times 2} \left[x_{am}^T\right]_{2\times 3}\right\}^{-1} \end{aligned}$$

Application of the equation error method to parameter estimation requires accurate measurements of the states and their derivatives. In addition, it can be applied to unstable systems because it does not involve any numerical integration of the dynamic system that would otherwise cause divergence. Utilisation of measured states and state-derivatives for estimation in the algorithm enables estimation of the parameters of even an unstable system directly (studied in Chapter 9). However, if the measurements are noisy, the method will give biased estimates.

We would like to mention here that equation error formulation is amenable to be programmed in the structure of a recurrent neural network as discussed in Chapter 10.

2.5.1.1 Example 2.4

Let $\dot{x} = Ax + Bu$

$$A = \begin{bmatrix} -2 & 0 & 1 \\ 1 & -2 & 0 \\ 1 & 1 & -1 \end{bmatrix} \qquad B = \begin{bmatrix} 1 \\ 0 \\ 1 \end{bmatrix}$$

Generate suitable responses with u as doublet (see Fig. B.7, Appendix B) input to the system with proper initial condition on x_0. Use equation error method to estimate the elements of the A and B matrices.

2.5.1.2 Solution

Data with sampling interval of 0.001 s is generated (using LSIM of MATLAB) by giving a doublet input to the system. Figure 2.4 shows plots of the three simulated true states of the system. The time derivatives of the states required for the estimation using the equation error method are generated by numerical differentiation (see Section A.5) of the states. The program used for simulation and estimation is **Ch2EEex4.m**. The estimated values of the elements of A and B matrices are given in Table 2.3 along with the eigenvalues, natural frequency and damping. It is clear from Table 2.3 that when there is no noise in the data, the equation error estimates closely match the true values, except for one value.

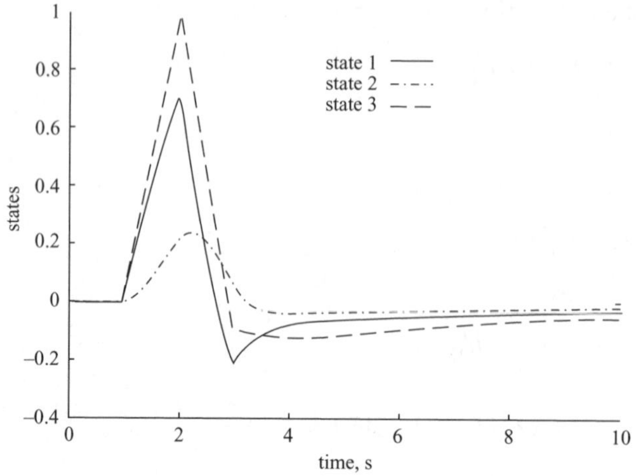

Figure 2.4 Simulated true states (Example 2.4)

Table 2.3 Estimated parameters of A and B matrices (Example 2.4)

Parameter	True values	Estimated values (data with no noise)
a_{11}	−2	−2.0527
a_{12}	0	−0.1716
a_{13}	1	1.0813
a_{21}	1	0.9996
a_{22}	−2	−1.9999
a_{23}	0	−0.00003
a_{31}	1	0.9461
a_{32}	1	0.8281
a_{33}	−1	−0.9179
b_1	1	0.9948
b_2	0	0.000001
b_3	1	0.9948
Eigenvalues	−0.1607	−0.1585
(see Section A.15)	$-2.4196 \pm j(0.6063)$	$-2.4056 \pm j(0.6495)$
Natural freq. ω (rad/s)	2.49	2.49
Damping (of the oscillatory mode)	0.97	0.965

2.5.1.3 Example 2.5

The equation error formulation for parameter estimation of an aircraft is illustrated with one such state equation here (see Sections B.1 to B.4).

Let the z-force equation be given as [4]:

$$\dot{\alpha} = Z_u u + Z_\alpha \alpha + q + Z_{\delta_e}\delta_e \tag{2.30}$$

Then the coefficients of the equation are determined from the system of linear equations given by (eq. (2.30) is multiplied in turn by u, α and δ_e)

$$\begin{aligned}
\sum \dot{\alpha} u &= Z_u \sum u^2 + Z_\alpha \sum \alpha u + \sum q u + Z_{\delta_e} \sum \delta_e u \\
\sum \dot{\alpha} \alpha &= Z_u \sum u\alpha + Z_\alpha \sum \alpha^2 + \sum q\alpha + Z_{\delta_e} \sum \delta_e \alpha \\
\sum \dot{\alpha} \delta_e &= Z_u \sum u\delta_e + Z_\alpha \sum \alpha\delta_e + \sum q\delta_e + Z_{\delta_e} \sum \delta_e^2
\end{aligned} \tag{2.31}$$

where $\sum$ is the summation over the data points ($k = 1, \ldots, N$) of u, α, q and δ_e signals. Combining the terms, we get:

$$\begin{bmatrix} \sum \dot{\alpha} u \\ \sum \dot{\alpha}\alpha \\ \sum \dot{\alpha}\delta_e \end{bmatrix} = \begin{bmatrix} \sum u^2 & \sum \alpha u & \sum qu & \sum \delta_e u \\ \sum u\alpha & \sum \alpha^2 & \sum q\alpha & \sum \delta_e u \\ \sum u\delta_e & \sum \alpha\delta_e & \sum q\delta_e & \sum \delta_e^2 \end{bmatrix} \begin{bmatrix} Z_u \\ Z_\alpha \\ 1 \\ Z_{\delta_e} \end{bmatrix}$$

The above formulation can be expressed in a compact form as

$$Y = X\beta$$

Then the equation error is formulated as

$$e = Y - X\beta$$

keeping in mind that there will be modelling and estimation errors combined in e. It is presumed that measurements of $\dot{\alpha}, u, \alpha$ and δ_e are available. If the numerical values of $\dot{\alpha}, \alpha, u, q$ and δ_e are available, then the equation error estimates of the parameters can be obtained by using the procedure outlined in eq. (2.2) to eq. (2.4).

2.6 Gaussian least squares differential correction method

In this section, the nonlinear least squares parameter estimation method is described. The method is based on the differential correction technique [5]. This algorithm can be used to estimate the initial conditions of states as well as parameters of a nonlinear dynamical model. It is a batch iterative procedure and can be regarded as complementary to other nonlinear parameter estimation procedures like the output error method. One can use this technique to obtain the start-up values of the aerodynamic parameters for other methods.

To describe the method used to estimate the parameters of a given model, let us assume a nonlinear system as

$$\dot{x} = f(x, t, C) \tag{2.32}$$

$$y = h(x, C, K) + v \tag{2.33}$$

Here x is a $n \times 1$ state vector, y is a $m \times 1$ measurement vector and v is a random white Gaussian noise process with covariance matrix R. The functions f and h are vector-valued nonlinear functions, generally assumed to be known. The unknown parameters in the state and measurement equations are represented by vectors C and K. Let x_0 be a vector of initial conditions at t_0. Then the problem is to estimate the parameter vector

$$\hat{\beta} = \left[x_0^T C^T K^T\right]^T \tag{2.34}$$

It must be noted that the vector C appears in both state and measurement equations. Such situations often arise for aircraft parameter estimation.

The iterative differential correction algorithm is applied to obtain the estimates from the noisy measured signals as [5]:

$$\hat{\beta}^{(i+1)} = \hat{\beta}^{(i)} + [(F^T W F)^{-1} F^T W \Delta y]^{(i)} \tag{2.35}$$

where

$$F = \left[\frac{\partial y}{\partial x_0} \middle| \frac{\partial y}{\partial C} \middle| \frac{\partial y}{\partial K}\right] \tag{2.36}$$

We use ∂ to denote partial differentiation here. It can be noted here that the above equations are generalised versions of eq. (2.22). W is a suitable weighting matrix and Δy is a matrix of residuals of observables

$$\Delta y = z(t_k) - y(t_k) \qquad \text{where } k = 1, 2, \ldots, N$$

The first sub matrix in F is given as

$$\frac{\partial y(t_k)}{\partial x(t_0)} = \left[\frac{\partial h(x(t_k))}{\partial x(t_k)}\right]\left[\frac{\partial x(t_k)}{\partial x(t_0)}\right] \tag{2.37}$$

with

$$\frac{d}{dt}\left[\frac{\partial x(t)}{\partial x(t_0)}\right] = \left[\frac{\partial f(t, x(t))}{\partial x(t)}\right]\left[\frac{\partial x(t)}{x(t_0)}\right] \tag{2.38}$$

The transition matrix differential eq. (2.38) can be solved with identity matrix as initial condition. The second sub matrix in F is

$$\frac{\partial y}{\partial C} = \left(\frac{\partial h}{\partial x}\right)\left(\frac{\partial x}{\partial C}\right) + \frac{\partial h}{\partial C} \tag{2.39}$$

where $(\partial x(t)/\partial C)$ is the solution of

$$\frac{d}{dt}\left[\frac{\partial x}{\partial C}\right] = \frac{\partial f}{\partial C} + \left(\frac{\partial f}{\partial x}\right)\left(\frac{\partial x}{\partial C}\right) \tag{2.40}$$

The last sub matrix in F is obtained as

$$\frac{\partial y}{\partial K} = \frac{\partial h}{\partial K} \tag{2.41}$$

Equation (2.41) is simpler than eqs (2.39) and (2.40), since K is not involved in eq. (2.32). The state integration is performed by the 4th order Runge-Kutta method.

Figure 2.5 shows the flow diagram of the Gaussian least squares differential correction algorithm. It is an iterative process. Convergence to the optimal solution/parameters (near the optimal solution – if they can be conjectured!) would help in finding the global minimum of the cost function. In this case, the least squares estimates

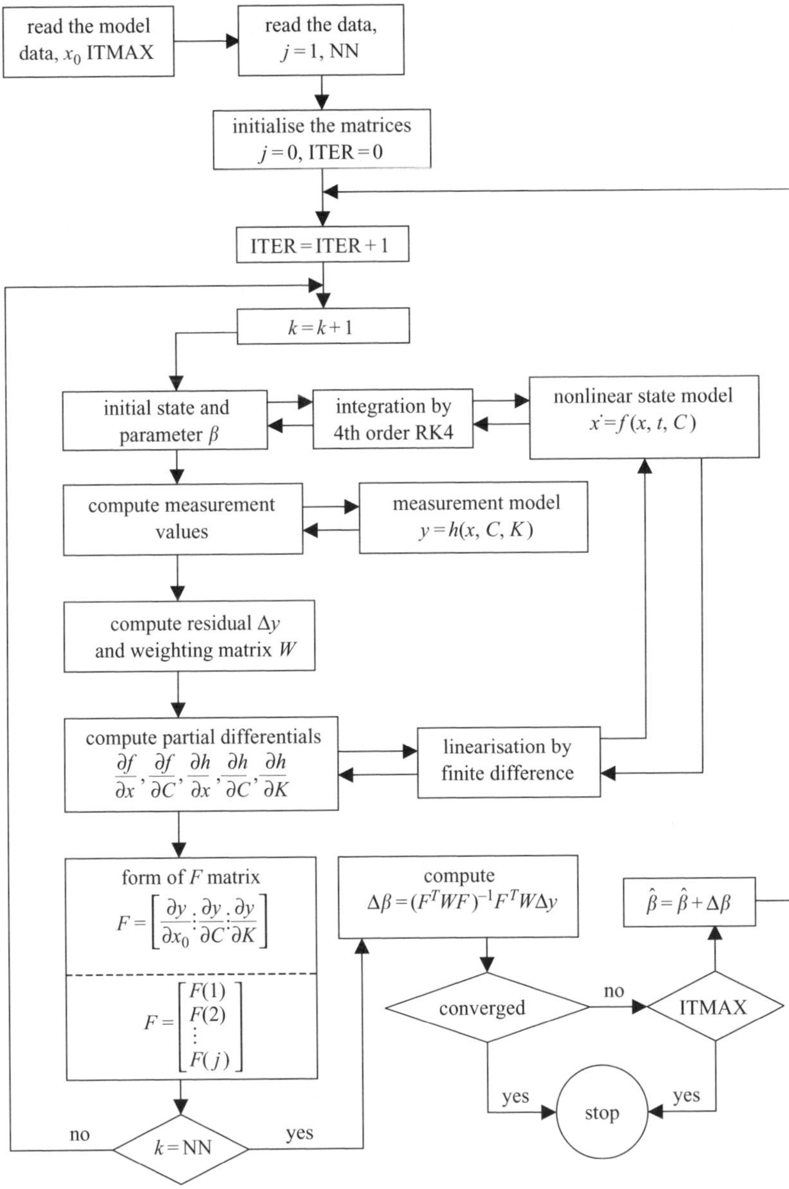

Figure 2.5 Flow diagram of GLSDC algorithm

obtained from the equation error method can be used as initial parameters for the Gaussian least squares differential correction (GLSDC) algorithm. In eq. (2.35), if matrix ill-conditioning occurs, some factorisation method can be used.

It is a well-known fact that the quality of the measurement data significantly influences the accuracy of the parameter estimates. The technique can be employed to assess quickly the quality of the measurements (aircraft manoeuvres), polarities of signals, and to estimate bias and scale factor errors in the measurements (see Section B.7).

2.6.1.1 Example 2.6

Simulated longitudinal short period (see Section B.4) data of a light transport aircraft is provided. The data consists of measurements of pitch rate q, longitudinal acceleration a_x, vertical acceleration a_z, pitch attitude θ, true air speed V and angle-of-attack α. Check the compatibility of the data (see Section B.7) using the given measurements and the kinematic equations of the aircraft longitudinal mode. Using the GLSDC algorithm, estimate the scale factor and bias errors present in the data, if any, as well as the initial conditions of the states. Show the convergence plots of the estimated parameters.

2.6.1.2 Solution

The state and measurement equations for data compatibility checking are given by:

State equations

$$\begin{aligned}\dot{u} &= (a_x - \Delta a_x) - (q - \Delta q)w - g\sin\theta \\ \dot{w} &= (a_z - \Delta a_z) - (q - \Delta q)u + g\cos\theta \\ \dot{\theta} &= (q - \Delta q)\end{aligned} \tag{2.42}$$

where $\Delta a_x, \Delta a_z, \Delta q$ are acceleration biases (in the state equations) to be estimated. The control inputs are a_x, a_z and q.

Measurement equations

$$\begin{aligned}V &= \sqrt{u^2 + w^2} \\ \alpha_m &= K_\alpha \tan^{-1}\left(\frac{w}{u}\right) + b_\alpha \\ \theta_m &= K_\theta \theta + b_\theta\end{aligned} \tag{2.43}$$

where K_α, K_θ are scale factors and b_α and b_θ are the bias errors in the measurements to be estimated.

Assuming that the a_x, a_z and q signals have biases and the measurements of V, θ and α have only scale factor errors, the Gaussian least squares differential correction algorithm is used to estimate all the bias and scale factor errors using the programs in the folder **Ch2GLSex6**. The nonlinear functions are linearised by the finite difference method. The weighting matrix is chosen as the inverse covariance matrix of the residuals. Figure 2.6(a) shows the plot of the estimated and measured V, θ and α signals at the first iteration of the estimation procedure where only integration of the states with the specified initial conditions generates the estimated responses. It is clear that there are discrepancies in the responses. Figure 2.6(b) shows the cross plot of the measured and estimated V, θ and α signals once convergence is reached. The match between the estimated and measured trajectories (which is a necessary condition for establishing the confidence in the estimated parameters) is good. The convergence of the parameter estimates is shown in Fig. 2.6(c) from which it is clear that all the parameters converge in less than eight iterations. We see that the scale factors are very close to one and the bias errors arc negligible, as seen from Table 2.4.

2.6.1.3 Example 2.7

Simulate short period (see Section B.4) data of a light transport aircraft. Adjust the static stability parameter M_w to give a system with time to double of 1 s (see Exercise 2.11). Generate data with a doublet input (see Section B.6) to pilot stick with a sampling time of 0.025 s.

State equations

$$\begin{aligned} \dot{w} &= Z_w w + (u_0 + Z_q)q + Z_{\delta_e}\delta_e \\ \dot{q} &= M_w w + M_q q + M_{\delta_e}\delta_e \end{aligned} \tag{2.44}$$

Table 2.4 Bias and scale factors (Example 2.6)

Iteration number	Δa_x	Δa_z	Δq	K_α	K_θ	u_0	w_0	θ_0
0	0	0	0	0.7000	0.8000	40.0000	9.0000	0.1800
1	0.0750	−0.0918	0.0002	0.9952	0.9984	36.0454	6.5863	0.1430
2	0.0062	−0.0116	0.0002	0.9767	0.9977	35.9427	7.4295	0.1507
3	0.0041	−0.0096	0.0002	0.9784	0.9984	35.9312	7.4169	0.1504
4	0.0043	−0.0091	0.0002	0.9778	0.9984	35.9303	7.4241	0.1504
5	0.0044	−0.0087	0.0002	0.9774	0.9984	35.9296	7.4288	0.1504
6	0.0045	−0.0085	0.0002	0.9772	0.9984	35.9292	7.4316	0.1503
7	0.0045	−0.0083	0.0002	0.9770	0.9984	35.9289	7.4333	0.1503
8	0.0046	−0.0082	0.0002	0.9769	0.9985	35.9288	7.4343	0.1503
9	0.0046	−0.0082	0.0002	0.9769	0.9985	35.9287	7.4348	0.1503
10	0.0046	−0.0082	0.0002	0.9769	0.9985	35.9287	7.4352	0.1503

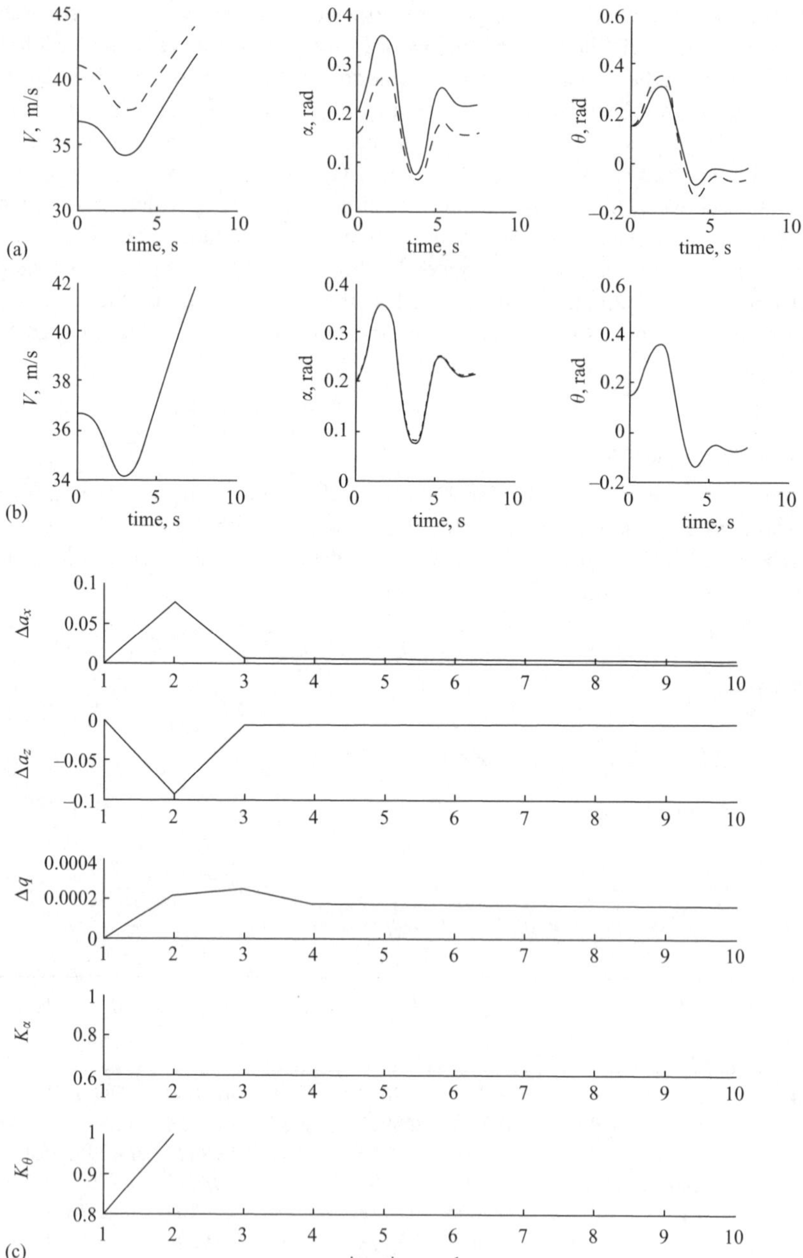

Figure 2.6 (a) Estimated and measured responses – 1st iteration GLSDC; (b) estimated and measured responses – 10th iteration GLSDC; (c) parameter convergence – GLSDC (Example 2.6)

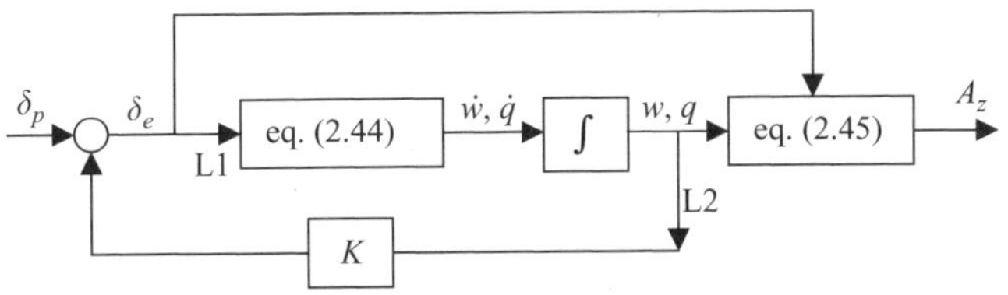

Figure 2.7 Closed loop system

Measurement equations

$$\begin{aligned} A_{z_m} &= Z_w w + Z_q q + Z_{\delta_e}\delta_e \\ w_m &= w \\ q_m &= q \end{aligned} \tag{2.45}$$

where w is vertical velocity, u_0 is stationary forward speed, q is pitch rate, A_z is vertical acceleration and δ_e is elevator deflection. Since the system is unstable, feedback the vertical velocity with a gain K to stabilise the system using

$$\delta_e = \delta_p + Kw \tag{2.46}$$

where δ_p denotes pilot input. Generate various sets of data by varying gain K. Estimate the parameters of the plant (within the closed loop (see Fig. 2.7)) using EE method described in Section 2.5. These parameters of the plant are the stability and control derivatives of an aircraft (see Sections B.2 and B.3).

2.6.1.4 Solution

Two sets of simulated data (corresponding to $K = 0.025$ and $K = 0.5$), are generated by giving a doublet input at δ_p. The equation error solution requires the derivatives of the states. Since the data are generated by numerical integration of the state equations, the derivatives of the states are available from the simulation. EE method is used for estimation of derivatives using the programs contained in the folder **Ch2EEex7**. Figure 2.8 shows the states (w, q), the derivatives of states $(\dot{w}, \dot{q})$, the control input δ_e and pilot input δ_p for $K = 0.025$. Table 2.5 shows the parameter estimates compared with the true values for the two sets of data. The estimates are close to the true values when there is no noise in the data.

This example illustrates that with feedback gain variation, the estimates of the open-loop plant (operating in the closed loop) are affected. The approach illustrated here can also be used for determination of aircraft neutral point from its flight data (see Section B.15).

2.7 Epilogue

In this chapter, we have discussed various LS methods and illustrated their performance using simple examples. A more involved example of data compatibility for aircraft was also illustrated.

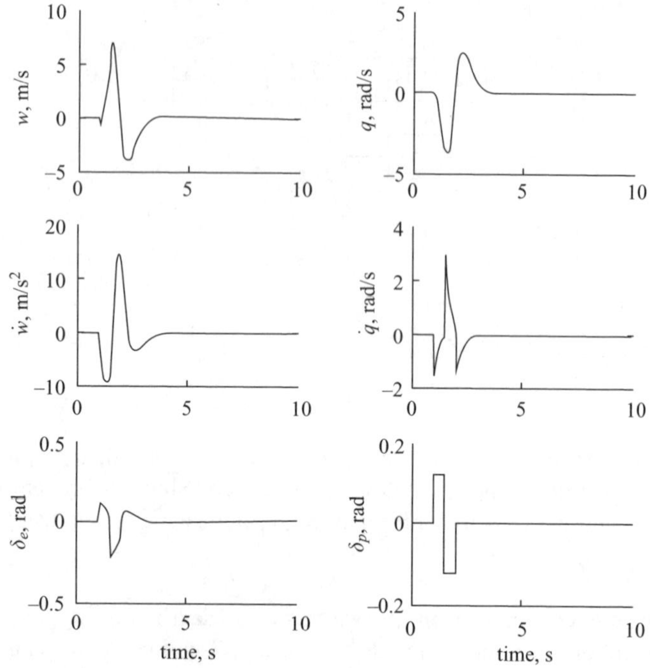

Figure 2.8 *Simulated states, state derivatives and control inputs (Example 2.7)*

Table 2.5 *Parameter estimates (Example 2.7)*

	Gain K→	0.025	0.5
Parameter	True value↓	No noise	No noise
Z_w	−1.4249	−1.4267	−1.4326
Zq	−1.4768	−1.4512	−1.3451
$Z\delta_e$	−6.2632	−6.2239	−6.0008
M_w	0.2163	0.2164	0.2040
M_q	−3.7067	−3.7080	−3.5607
$M\delta_e$	−12.784	−12.7859	−12.7173
PEEN	–	0.3164	2.2547

Mendel [3] treats the unification of the generalised LS, unbiased minimum variance, deterministic gradient and stochastic gradient approaches via equation error methods. In addition, sequential EE methods are given.

The GLS method does not consider the statistics of measurement errors. If there is a good knowledge of these statistics, then they can be used and it leads to minimum variance estimates [3]. As we will see in Chapter 4, the KF is a method to obtain minimum variance estimates of states of a dynamic system described in state-space form. It can handle noisy measurements as well as partially account for discrepancies in a state model by using the so-called process noise. Thus, there is a direct relationship between the sequential unbiased minimum variance algorithm and discrete KF [3]. Mendel also shows equivalence of an unbiased minimum variance estimation and maximum likelihood estimation under certain conditions. The LS approaches for system identification and parameter estimation are considered in Reference 6, and several important theoretical developments are treated in Reference 7. Aspects of confidence interval of estimated parameters (see Section A.8) are treated in Reference 8.

2.8 References

1 HSIA, T. C.: 'System identification – least squares methods' (Lexington Books, Lexington, Massachusetts, 1977)
2 SORENSON, H. W.: 'Parameter estimation – principles and problems' (Marcel Dekker, New York and Basel, 1980)
3 MENDEL, J. M.: 'Discrete techniques of parameter estimation: equation error formulation' (Marcel Dekker, New York, 1976)
4 PLAETSCHKE, E.: Personal Communication, 1986
5 JUNKINS, J. L.: 'Introduction to optimal estimation of dynamical systems' (Sijthoff and Noordhoff, Alphen aan den Rijn, Netherlands, 1978)
6 SINHA, N. K., and KUSZTA, B.: 'Modelling and identification of dynamic system' (Van Nostrand, New York, 1983)
7 MENDEL, J. M.: 'Lessons in digital estimation theory' (Prentice-Hall, Englewood Cliffs, 1987)
8 BENDAT, J. S., and PIERSOL, A. G.: 'Random data: analysis and measurement procedures' (John Wiley & Sons, Chichester, 1971)

2.9 Exercises

Exercise 2.1

One way of obtaining least squares estimate of (β) is shown in eqs (2.2)–(2.4). Use algebraic approach of eq. (2.1) to derive similar form. One extra term will appear. Compare this term with that of eq. (2.5).

Exercise 2.2

Represent the property of orthogonality of the least squares estimates geometrically.

Exercise 2.3

Explain the significance of the property of the covariance of the parameter estimation error (see eqs (2.6) and (2.7)). In order to keep estimation errors low, what should be done in the first place?

Exercise 2.4

Reconsider Example 2.1 and check the response of the motor speed, S beyond 1 s. Are the responses for $\alpha \geq 0.1$ linear or nonlinear for this apparently linear system? What is the fallacy?

Exercise 2.5

Consider $z = mx + v$, where v is measurement noise with covariance matrix R. Derive the formula for covariance of $(z - \hat{y})$. Here, $y = mx$.

Exercise 2.6

Consider generalised least squares problem. Derive the expression for $P = \mathrm{Cov}(\beta - \hat{\beta})$.

Exercise 2.7

Reconsider the probabilistic version of the least squares method. Can we not directly obtain K from $KH = I$? If so, what is the difference between this expression and the one in eq. (2.15)? What assumptions will you have to make on H to obtain K from $KH = I$? What assumption will you have to make on R for both the expressions to be the same?

Exercise 2.8

What are the three numerical methods to obtain partials of nonlinear function $h(\beta)$ w.r.t. β?

Exercise 2.9

Consider $z = H\beta + v$ and $v = X_v\beta_v + e$, where v is correlated noise in the above model, e is assumed to be white noise, and the second equation is the model of the correlated noise v. Combine these two equations and obtain expressions for the least squares estimates of β and β_v.

Exercise 2.10

Based on Exercise 2.9, can you tell how one can generate a correlated process using white noise as input process? (Hint: the second equation in Exercise 2.9 can be regarded as a low pass filter.)

Exercise 2.11

Derive the expression for time to double amplitude, if σ is the positive real root of a first order system. If σ is positive, then system output will tend to increase when time elapses.

Chapter 3

Output error method

3.1 Introduction

In the previous chapter, we discussed the least squares approach to parameter estimation. It is the most simple and, perhaps, most highly favoured approach to determine the system characteristics from its input and output time histories. There are several methods that can be used to estimate system parameters. These techniques differ from one another based on the optimal criterion used and the presence of process and measurement noise in the data. The output error concept was described in Chapter 1 (see Fig. 1.1). The maximum likelihood process invokes the probabilistic aspect of random variables (e.g., measurement/errors, etc.) and defines a process by which we obtain estimates of the parameters. These parameters most likely produce the model responses, which closely match the measurements. A likelihood function (akin to probability density function) is defined when measurements are (collected and) used. This likelihood function is maximised to obtain the maximum likelihood estimates of the parameters of the dynamic system. The equation error method is a special case of the maximum likelihood estimator for data containing only process noise and no measurement noise. The output error method is a maximum likelihood estimator for data containing only measurement noise and no process noise. At times, one comes across statements in literature mentioning that maximum likelihood is superior to equation error and output error methods. This falsely gives the impression that equation error and output error methods are not maximum likelihood estimators. The maximum likelihood methods have been extensively studied in the literature [1–5].

The type of (linear or nonlinear) mathematical model, and the presence of process or measurement noise in data or both mainly drive the choice of the estimation method and the intended use of results. The equation error method has a cost function that is linear in parameters. It is simple and easy to implement. The output error method is more complex and requires the nonlinear optimisation technique (Gauss-Newton method) to estimate model parameters. The iterative nature of the approach makes it a little more computer intensive. The third approach is the filter error method which is the most general approach to parameter estimation problem accounting for both process and measurement noise. Being a combination of the Kalman filter and output error method, it is the most complex of the three techniques with high computational requirements. The output error method is perhaps the most widely used approach for aircraft parameter estimation and is discussed in this chapter, after discussing the concepts of maximum likelihood. The Gaussian least squares differential correction method is also an output error method, but it is not based on the maximum likelihood principle.

3.2 Principle of maximum likelihood

Though the maximum likelihood (ML) method is accredited to Fisher [1, 2], the idea was originally given by Gauss way back in 1809. The fundamental idea is to define a function of the data and the unknown parameters [6]. This function is called the likelihood function. The parameter estimates are then obtained as those values which maximise the function. In fact, the likelihood function is the probability density of the observations (given the parameters!).

Let $\beta_1, \beta_2, \ldots, \beta_r$ be unknown physical parameters of some system and $z_1, z_2, \ldots, z_n$ the measurements of the true (data) values $y_1, y_2, \ldots, y_n$. It is assumed that the true values are a function of the unknown parameters, that is

$$y_i = f_i(\beta_1, \beta_2, \ldots, \beta_r)$$

Let z be a random variable whose probability density $p(z, \beta)$ depends on unknown parameter β. To estimate β from measurements z, choose the value of β which maximises the likelihood function $L(z, \beta) = p(z, \beta)$ [6]. The method of maximum likelihood thus reduces the problem of parameter estimation to the maximisation of a real function called the likelihood function. It is a function of the parameter β and the experimental data z. The value of the likelihood function at β and z is the probability density function of the measurement evaluated at the given observations z and the parameter β. This is to say that p becomes L when the measurements have been actually obtained and used in p. Hence, the parameter β, which makes this function most probable to have yielded these measurements, is called the maximum likelihood estimate. Next, presume that the true value y_i lies within very small interval around measurement z_i and evaluate the related probability:

$$\text{probability that } y_i \in \left[z_i - \tfrac{1}{2}\delta z_i, z_i + \tfrac{1}{2}\delta z_i\right]$$

$$\text{is given as: } \delta P_i = \int_{z_i-(1/2)\delta z_i}^{z_i+(1/2)\delta z_i} p(t)\cdot dt \approx p(z_i)\delta z_i; \qquad \text{for small } \delta z_i \tag{3.1}$$

The measurement errors are normally distributed and the probability is given by (see Section A.23):

$$\delta P_i = \frac{1}{\sqrt{2\pi}\sigma_i} \exp\left[-\frac{1}{2}\frac{(z_i - y_i)^2}{\sigma_i^2}\right]\delta z_i \tag{3.2}$$

where σ_i^2 is the variance.

The likelihood function is calculated for the statistically independent measurements, and this allows the joint probability density to be simply the product of the probabilities of the individual measurements, and is given by

$$\delta P = \prod_{i=1}^{n} p(z_i)\delta z_i = \frac{1}{(2\pi)^{n/2}\sigma_1 \cdots \sigma_n} \exp\left[\sum_{i=1}^{n} -\frac{1}{2}\frac{(z_i - y_i)^2}{\sigma_i^2}\delta z_1 \cdots \delta z_n\right] \tag{3.3}$$

The likelihood function is then given as

$$\begin{aligned} p(z \mid \beta) &= p(z_1, \ldots, z_n \mid \beta_1, \ldots, \beta_r) \\ &= \frac{1}{(2\pi)^{n/2}\sigma_1 \cdots \sigma_n} \exp\left[\sum_{i=1}^{n} -\frac{1}{2}\frac{(z_i - y_i(\beta))^2}{\sigma_i^2}\right] \end{aligned} \tag{3.4}$$

The parameter $\hat{\beta}$ that maximises this likelihood function is called the maximum likelihood parameter estimate of β (see Section A.30).

3.3 Cramer-Rao lower bound

In this section, we derive certain theoretical properties of the maximum likelihood estimator (MLE). The main point in any estimator is the error made in the estimates relative to the true parameters. However, these true parameters are unknown in the real case. Therefore, we only get some statistical indicators for the errors made. The Cramer-Rao lower bound is one such useful and, perhaps, the best measure for such errors.

The likelihood function can also be defined as:

$$L(z \mid \beta) = \log p(z \mid \beta) \tag{3.5}$$

since the function and its logarithm will have a maximum at the same argument. The maximisation yields the likelihood differential equation [6]:

$$\frac{\partial}{\partial \beta} L(z \mid \beta) = L'(z \mid \hat{\beta}) = \frac{p'}{p}(z \mid \hat{\beta}) = 0 \tag{3.6}$$

This equation is nonlinear in $\hat{\beta}$ and a first order approximation by Taylor's series expansion, can be used to obtain the estimate $\hat{\beta}$:

$$L'(z \mid \beta_0 + \Delta\beta) = L'(z \mid \beta_0) + L''(z \mid \beta_0)\Delta\beta = 0 \tag{3.7}$$

which gives increment in β as:

$$\Delta\beta = \frac{L'(z \mid \beta_0)}{-L''(z \mid \beta_0)} = -(L''(z \mid \beta_0))^{-1} L'(z \mid \beta_0) \tag{3.8}$$

The above equation tells us that if we get the right hand side term computed, then we already have obtained $\Delta\beta$, the increment/change in parameter vector. This expression is based on computation of likelihood related partials, which can be evaluated when the details of the dynamical systems are known, as will be seen later on in the chapter.

The expected value of the denominator in eq. (3.8) is defined as the Information Matrix (in general sense):

$$I_m(\beta) = E\{-L''(z \mid \beta)\} \tag{3.9}$$

The other form of $I_m(\beta)$ is derived next. Since, by the definition of the probability of a random variable

$$\int_{-\infty}^{\infty} p(z \mid \beta)\, dz = 1$$

we take first differentiation on both sides to obtain

$$\int_{-\infty}^{\infty} p'(z \mid \beta)\, dz = \int_{-\infty}^{\infty} L'(z \mid \beta)\, p(z \mid \beta)\, dz = 0 \tag{3.10}$$

using eq. (3.6).

The second differentiation yields

$$\int_{-\infty}^{\infty} p''(z \mid \beta)\, dz = \int_{-\infty}^{\infty} [L''(z \mid \beta)\, p(z \mid \beta) + L'(z \mid \beta)^2\, p(z \mid \beta)]\, dz = 0 \tag{3.11}$$

From the above equation we get

$$I_m(\beta) = E\{-L''(z \mid \beta)\} = E\{L'(z \mid \beta)^2\} \tag{3.12}$$

From the definition of information matrix, we can say that if there is large information content in the data, then $|L''|$ tends to be large, and the uncertainty in estimate $\hat{\beta}$ is small. The so-called Cramer-Rao Inequality (Information Inequality) provides a lower bound to the variance of an unbiased estimator, as will be seen in the sequel.

Let $\beta_e(z)$ be any estimator of β based on the measurement z, and then $\bar{\beta}_e(z) = E\{\beta_e(z)\}$ is the expectation of the estimate (since it depends on the random signal z). Its variance is given as

$$\sigma_{\beta e}^2 = E\{(\beta_e(z) - \bar{\beta}_e)^2\} \tag{3.13}$$

The bias in the estimator is defined as

$$E\{\beta_e - \beta\} = \int_{-\infty}^{\infty} \beta_e(z)p(z \mid \beta)\, dz - \beta = b(\beta) \tag{3.14}$$

If $b(\beta) = 0$, then it is called an unbiased estimator (see Section A.3). We have thus

$$\int_{-\infty}^{\infty} \beta_e(z)p(z \mid \beta)\, dz = \beta + b(\beta) \tag{3.15}$$

Differentiating both the sides w.r.t. β we get

$$\int_{-\infty}^{\infty} \beta_e(z)p'(z \mid \beta)\, dz = \int_{-\infty}^{\infty} \beta_e(z)L'(z \mid \beta)p(z \mid \beta)\, dz = 1 + b'(\beta) \tag{3.16}$$

since β_e is a function of only z.

In addition, we have $\int_{-\infty}^{\infty} p(z \mid \beta)\, dz = 1$ and differentiating both sides we get [6]:

$$\int_{-\infty}^{\infty} p'(z \mid \beta)\, dz = \int_{-\infty}^{\infty} L'(z \mid \beta)p(z \mid \beta)\, dz = 0 \tag{3.17}$$

Multiplying the above equation by $(-\bar{\beta}_e)$ and adding to the previous eq. (3.16) we get

$$\begin{aligned} &\int_{-\infty}^{\infty} [\beta_e(z) - \bar{\beta}_e]L'(z \mid \beta)p(z \mid \beta)\, dz = 1 + b'(\beta) \\ &\int_{-\infty}^{\infty} [\beta_e(z) - \bar{\beta}_e]\sqrt{p(z \mid \beta)} \cdot L'(z \mid \beta)\sqrt{p(z \mid \beta)}\, dz = 1 + b'(\beta) \end{aligned} \tag{3.18}$$

Now we apply the following well-known Schwarz inequality to eq. (3.18)

$$\left[\int f(z) \cdot g(z)\, dz\right]^2 \leq \int f^2(z)\, dz \cdot \int g^2(z)\, dz$$

to get (the equality applies if $f(z) = kg(z)$):

$$[1 + b'(\beta)]^2 \leq \int_{-\infty}^{\infty} [\beta_e(z) - \bar{\beta}_e]^2 p(z \mid \beta)\, dz \cdot \int_{-\infty}^{\infty} L'(z \mid \beta)^2 p(z \mid \beta)\, dz \tag{3.19}$$

using eqs (3.12) and (3.13) in the above equation, i.e., using the definition of $I_m(\beta)$ and $\sigma^2_{\beta e}$, we get

$$[1 + b'(\beta)]^2 \leq \sigma^2_{\beta e} I_m(\beta) \quad \text{or} \quad \sigma^2_{\beta e} \geq [1 + b'(\beta)]^2 (I_m(\beta))^{-1} \tag{3.20}$$

This is called the Cramer-Rao inequality. For unbiased estimator, $b'(\beta) = 0$, and hence

$$\sigma_{\beta e}^2 \geq I_m^{-1}(\beta)$$

The equality sign holds if

$$\beta_e(z) - \bar{\beta}_e = kL'(z \mid \beta)$$

For unbiased, efficient estimator we thus have:

$$\sigma_{\beta e}^2 = I_m^{-1}(\beta) \tag{3.21}$$

We emphasise here that the inverse of the information matrix is the covariance matrix and hence in eq. (3.21), we have the theoretical expression for the variance of the estimator. The information matrix can be computed from the likelihood function or related data.

The above development signifies that the variance in the estimator, for an efficient estimator, would be at least equal to the predicted variance, whereas for other cases, it could be greater but not lesser than the predicted value. Hence, the predicted value provides the lower bound. Thus, the ML estimate is also the minimum variance estimator.

3.3.1 The maximum likelihood estimate is efficient [4, 5]

We assume that it is unbiased, then for efficiency (see Section A.14) we have to show that

$$\beta_e(z) - \bar{\beta}_e \stackrel{?}{=} kL'(z \mid \beta) \tag{3.22}$$

The likelihood equation is

$$L'(z \mid \beta)|_{\beta=\hat{\beta}(z)} = 0 \tag{3.23}$$

Substituting for ML estimate: $\beta_e(z) = \hat{\beta}(z)$ and since it is unbiased ($\bar{\beta}_e = 0$), we get

$$\beta_e(z) - \bar{\beta}_e|_{\beta=\hat{\beta}(z)} = \hat{\beta}(z) - \beta|_{\beta=\hat{\beta}(z)} = 0 \tag{3.24}$$

Thus $0 = kL'(z \mid \beta)|_{\beta=\hat{\beta}(z)} = k \times 0$.

Hence, the equality is established and the ML estimator is proved efficient. This is a very important property of the ML estimator. As such, these results are quite general since we have yet not dwelt on the details of the dynamical system.

3.4 Maximum likelihood estimation for dynamic system

A linear dynamical system can be described as:

$$\dot{x}(t) = Ax(t) + Bu(t) \tag{3.25}$$

$$y(t) = Hx(t) \tag{3.26}$$

$$z(k) = y(k) + v(k) \tag{3.27}$$

We emphasise here that in many applications, the actual systems are of continuous-time. However, the measurements obtained are discrete-time, as represented by eq. (3.27).

The following assumptions are made on the measurement noise $v(k)$:

$$E\{v(k)\} = 0; \quad E\{v(k)v^T(l)\} = R\delta_{kl} \tag{3.28}$$

In the above, it is assumed that the measurement noise is zero-mean and white Gaussian with R as the covariance matrix of this noise. This assumption allows us to use the Gaussian probability concept for deriving the maximum likelihood estimator. The assumption of whiteness of the measurement noise is quite standard and very useful in engineering practice. Strictly speaking, the assumption would not hold well. However, as long as the bandwidth of the noise spectrum is much larger than the system's bandwidth, the noise can be seen as practically 'white'.

3.4.1 Derivation of the likelihood function

If z is some real valued Gaussian random variable then its probability density is given by

$$p(z) = \frac{1}{\sqrt{2\pi}\sigma} \exp\left[-\frac{1}{2}\frac{(z-m)^2}{\sigma^2}\right] \tag{3.29}$$

where $m = E(z)$ and $\sigma^2 = E\{(z-m)^2\}$.

For n – random variables $z_1, z_2, \ldots, z_n$ we have

$$p(z_1, z_2, \ldots, z_n) = \frac{1}{(2\pi)^{n/2}\sqrt{|R|}} \exp\left[-\frac{1}{2}(z-m)^T R^{-1}(z-m)\right] \tag{3.30}$$

Here $z^T = (z_1, z_2, \ldots, z_n)$; $m^T = (m_1, m_2, \ldots, m_n)$, this is a vector of mean values, and

$$R = \begin{bmatrix} r_{11} & \cdot & \cdot & r_{1n} \\ \cdot & \cdot & \cdot & \cdot \\ \cdot & \cdot & \cdot & \cdot \\ r_{1n} & \cdot & \cdot & r_{nn} \end{bmatrix} \tag{3.31}$$

is the covariance matrix with $r_{ij} = E\{(z_i - m_i)(z_j - m_j)\} = \sigma_i \sigma_j \rho_{ij}$ and $\rho_{ij} =$ correlation coefficients ($\rho_{ii} = 1$).

Applying the above development to measurements $z(k)$, assuming that the measurement errors are Gaussian, we obtain

$$p(z(k) \mid \beta, r) = \frac{1}{(2\pi)^{m/2}\sqrt{|R|}} \exp\left[-\frac{1}{2}[z(k) - y(k)]^T R^{-1}[z(k) - y(k)]\right] \tag{3.32}$$

since in this case $m = E\{z\} = E\{v + y\} = E\{v\} + E\{y\}$ and $E\{v\} = 0$.

Using eq. (3.28), we have the likelihood function as:

$$p(z(1),\ldots,z(N) \mid \beta, R) = \prod_{k=1}^{N} p(z(k) \mid \beta, R)$$

$$= ((2\pi)^m |R|)^{-N/2} \exp\left[-\frac{1}{2}\sum_{k=1}^{N}[z(k)-y(k)]^T R^{-1}[z(k)-y(k)]\right] \quad (3.33)$$

The parameter vector β is obtained by maximising the above likelihood function with respect to β by minimising the negative (log) likelihood function [4–7]:

$$L = -\log p(z \mid \beta, R)$$

$$= \frac{1}{2}\sum_{k=1}^{N}[z(k)-y(k)]^T R^{-1}[z(k)-y(k)] + \frac{N}{2}\log|R| + const \quad (3.34)$$

Based on the above two cases of minimisation arises [6]:

(i) If R is known then the cost function

$$CF = \sum_{k=1}^{N}[z(k)-y(k)]^T R^{-1}[z(k)-y(k)] \rightarrow \text{minimum} \quad (3.35)$$

since the second term in eq. (3.34) is constant.

(ii) If R is unknown then we can minimise the function with respect to R and obtain

$$\frac{\partial L}{\partial (R^{-1})} = 0$$

to get

$$\hat{R} = \frac{1}{N}\sum_{k=1}^{N}[z(k)-\hat{y}(k)][z(k)-\hat{y}(k)]^T \quad (3.36)$$

When R is substituted in the likelihood function the first term becomes $mN/2 =$ constant, and we get $CF = |R| \rightarrow$ minimum.

Minimisation of CF in (i) w.r.t. β results in

$$\frac{\partial L}{\partial \beta} = -\sum_k \left(\frac{\partial y(\beta)}{\partial \beta}\right)^T R^{-1}(z - y(\beta)) = 0 \quad (3.37)$$

This set is again a system of nonlinear equations and calls for an iterative solution. In the present case we obtain an iterative solution by the so-called Quasi-Linearisation method (also known as the Modified Newton-Raphson or Gauss-Newton method), i.e., we expand

$$y(\beta) = y(\beta_0 + \Delta\beta) \quad (3.38)$$

as

$$y(\beta) = y(\beta_0) + \frac{\partial y(\beta)}{\partial \beta}\Delta\beta \tag{3.39}$$

The quasi-linearisation is an approximation method for obtaining solutions to non-linear differential or difference equations with multipoint boundary conditions. A version of the quasi-linearisation is used in obtaining a practical workable solution in output error method [8, 9].

Substituting this approximation in eq. (3.37) we get

$$-\sum_k \left[\frac{\partial y(\beta)}{\partial \beta}\right]^T R^{-1}\left[(z - y(\beta_0)) - \frac{\partial y(\beta)}{\partial \beta_0}\Delta\beta\right] = 0 \tag{3.40}$$

$$\left[\sum_k \left[\frac{\partial y(\beta)}{\partial \beta}\right]^T R^{-1}\frac{\partial y(\beta)}{\partial \beta}\right]\Delta\beta = \sum_k \left[\frac{\partial y(\beta)}{\partial \beta}\right]^T R^{-1}(z - y) \tag{3.41}$$

Next we have

$$\Delta\beta = \left\{\sum_k \left[\frac{\partial y(\beta)}{\partial \beta}\right]^T R^{-1}\frac{\partial y(\beta)}{\partial \beta}\right\}^{-1}\left\{\sum_k \left[\frac{\partial y(\beta)}{\partial \beta}\right]^T R^{-1}(z - y)\right\} \tag{3.42}$$

The ML estimate is obtained as:

$$\hat{\beta}_{new} = \hat{\beta}_{old} + \Delta\beta \tag{3.43}$$

3.5 Accuracy aspects

Determining accuracy of the estimated parameters is an essential part of the parameter estimation process. The absence of true parameter values for comparison makes the task of determining the accuracy very difficult. The Cramer-Rao bound is one of the primary criteria for evaluating accuracy of the estimated parameters. The maximum likelihood estimator gives the measure of parameter accuracy without any extra computation, as can be seen from the following development.

For a single parameter case we have for unbiased estimate $\hat{\beta}(z)$ of β

$$\sigma^2_{\hat{\beta}} \geq I_m^{-1}(\beta)$$

where the information matrix is

$$I_m(\beta) = E\left\{-\frac{\partial^2 \log p(z \mid \beta)}{\partial \beta^2}\right\} = E\left\{\left(\frac{\partial \log p(z \mid \beta)}{\partial \beta}\right)^2\right\} \tag{3.44}$$

For several parameters, the Cramer-Rao inequality is given as

$$\sigma^2_{\hat{\beta}_i} \geq (I_m^{-1})_{ii}$$

where the information matrix is

$$(I_m)_{ij} = E\left\{-\frac{\partial^2 \log p(z \mid \beta)}{\partial\beta_i \partial\beta_j}\right\} = \left\{\left(\frac{\partial \log p(z \mid \beta)}{\partial\beta_i}\right) \cdot \left(\frac{\partial \log p(z \mid \beta)}{\partial\beta_j}\right)\right\} \tag{3.45}$$

For efficient estimation, the equality holds and we have the covariance matrix of the estimation errors:

$$P = I_m^{-1}$$

The standard deviation of the individual parameters is given by

$$\sigma_{\hat{\beta}_i} = \sqrt{P_{ii}} = \sqrt{P(i,i)}$$

and correlation coefficients are

$$\rho_{\hat{\beta}_i,\hat{\beta}_j} = \frac{P_{ij}}{\sqrt{P_{ii}\,P_{jj}}} \tag{3.46}$$

For the maximum likelihood method, we have

$$\log p(z \mid \beta) = -\frac{1}{2}\sum_{k=1}^{N}[z(k) - y(k)]^T R^{-1}[z(k) - y(k)] + const \tag{3.47}$$

The information matrix can now be obtained as follows. Differentiate both sides w.r.t. β_i to get

$$\left(\frac{\partial \log p(z \mid \beta)}{\partial\beta_i}\right) = \sum_k \left(\frac{\partial y}{\partial\beta_i}\right)^T R^{-1}(z - y) \tag{3.48}$$

Again, differentiate both sides w.r.t. β_j to get

$$\left(\frac{\partial^2 \log p(z \mid \beta)}{\partial\beta_i \partial\beta_j}\right) = \sum_k \left(\frac{\partial^2 y}{\partial\beta_i \partial\beta_j}\right)^T R^{-1}(z - y) - \sum_k \left(\frac{\partial y}{\partial\beta_i}\right)^T R^{-1}\frac{\partial y}{\partial\beta_j} \tag{3.49}$$

Taking expectation of the above equation, we get

$$(I_m)_{ij} = E\left\{-\frac{\partial^2 \log p(z \mid \beta)}{\partial\beta_i \partial\beta_j}\right\} = \sum_{k=1}^{N}\left(\frac{\partial y(k)}{\partial\beta_i}\right)^T R^{-1}\frac{\partial y(k)}{\partial\beta_j} \tag{3.50}$$

Since $E\{z - y\} = 0$, the measurement error has zero-mean. We recall here from the previous section that the increment in parameter estimate $\Delta\beta$ is given by

$$\Delta\beta = \left[\sum_k \left(\frac{\partial y}{\partial\beta}\right)^T R^{-1}\frac{\partial y}{\partial\beta}\right]^{-1} \sum_k \left(\frac{\partial y}{\partial\beta}\right)^T R^{-1}(z - y) \tag{3.51}$$

Comparing with the expression for the information matrix in eq. (3.50), we conclude that the maximum likelihood estimator gives measure of accuracy without any extra computation.

Several criteria are used to judge the 'goodness' of the estimator/estimates: Cramer-Rao bounds of the estimates, correlation coefficients among the estimates, determinant of the covariance matrix of the residuals, plausibility of the estimates based on physical understanding of the dynamical system, comparison of the estimates with those of nearly similar systems or estimates independently obtained by other methods (analytical or other parameter estimation methods), and model predictive capability. The MLE is a consistent estimator (see Section A.9).

3.6 Output error method

The output error approach is based on the assumption that only the observations contain measurement noise and there is no noise in the state equations. The mathematical model of a linear system, described in eq. (3.25) to eq. (3.27), consists of the vector x representing the system states, vector y representing the computed system response (model output), vector z representing the measured variables and u representing the control input vector. The matrices A, B and H contain the parameters to be estimated. The output error method assumes that the measurement vector z is corrupted with noise which is zero-mean and has a Gaussian distribution with covariance R, i.e., $v \sim N(0, R)$.

The aim is to minimise the error between the measured and model outputs by adjusting the unknown parameters contained in matrices A, B and H.

Let the parameter vector to be estimated be represented by Θ where $\Theta =$ [elements of A, B, H, initial condition of x].

Then, the estimate of Θ is obtained by minimising the cost function

$$J = \frac{1}{2}\sum_{k=1}^{N}[z(k) - y(k)]^T R^{-1}[z(k) - y(k)] + \frac{N}{2}\ln|R| \tag{3.52}$$

where R is the measurement noise covariance matrix. The above cost function is similar to the weighted least squares criterion with weighting matrix as W and with one extra term. The estimate of R can be obtained from

$$\hat{R} = \frac{1}{N}\sum_{k=1}^{N}[z(k) - \hat{y}(k)][z(k) - \hat{y}(k)]^T \tag{3.53}$$

once the predicted measurements are computed.

Following the development of the previous Section 3.4, the estimate of Θ at the $(i+1)$th iteration is obtained as

$$\Theta(i+1) = \Theta(i) + [\nabla_{\Theta}^2 J(\Theta)]^{-1}[\nabla_{\Theta} J(\Theta)] \tag{3.54}$$

where the first and the second gradients are defined as

$$\nabla_{\Theta} J(\Theta) = \sum_{k=1}^{N} \left[\frac{\partial y}{\partial \Theta}(k) \right]^T R^{-1} [z(k) - y(k)] \tag{3.55}$$

$$\nabla_{\Theta}^2 J(\Theta) = \sum_{k=1}^{N} \left[\frac{\partial y}{\partial \Theta}(k) \right]^T R^{-1} \left[\frac{\partial y}{\partial \Theta}(k) \right] \tag{3.56}$$

Equation (3.56) is a Gauss-Newton approximation of the second gradient. This approximation helps to speed up the convergence without causing significant error in the estimate of Θ. The development leading to the eq. (3.54) has been given in Section 3.4.

Figure 1.1 in Chapter 1 explains the output error concept. Starting with a set of suitable initial parameter values, the model response is computed with the input used for obtaining measurement data. The estimated response and the measured response are compared and the response error is used to compute the cost function. Equations (3.55) and (3.56) are used to obtain the first and second gradients of the cost function and then eq. (3.54) is used to update the model parameter values. The updated parameter values are once again used in the mathematical model to compute the new estimated response and the new response error. This updating procedure continues until convergence is achieved.

The Gauss-Newton approximation for the second gradient in eq. (3.56), also called the Fisher Information Matrix, provides a measure of relative accuracy of the estimated parameters. The diagonal elements of the inverse of the information matrix give the individual covariances, and the square root of these elements is a measure of the standard deviations called the Cramer-Rao bounds (*CRB*):

$$\text{Fisher Information Matrix} = \nabla_{\Theta}^2 J(\Theta) \tag{3.57}$$

standard deviation of estimated parameters

$$= CRB(\Theta) = \operatorname{diag}\left[\sqrt{[\nabla_{\Theta}^2 J(\Theta)]^{-1}} \right] \tag{3.58}$$

The output error method (OEM) also can be applied with equal ease to any nonlinear system, in principle:

$$\dot{x}(t) = f[x(t), u(t), \Theta] \qquad \text{with initial } x(0) = x_0 \tag{3.59}$$

$$y(t) = h[x(t), u(t), \Theta] \tag{3.60}$$

$$z(k) = y(k) + v(k) \tag{3.61}$$

In the above equations f and h are general nonlinear functions, and the initial values x_0 of the state variables need to be estimated along with the parameter vector Θ. It is evident that estimation of parameters with output error approach would require computation of the state vector x (obtained by integrating eq. (3.59)), model output vector y and sensitivity coefficients $\partial y/\partial \Theta$. The sensitivity coefficients for a linear system can be obtained analytically by partial differentiation of the system equations

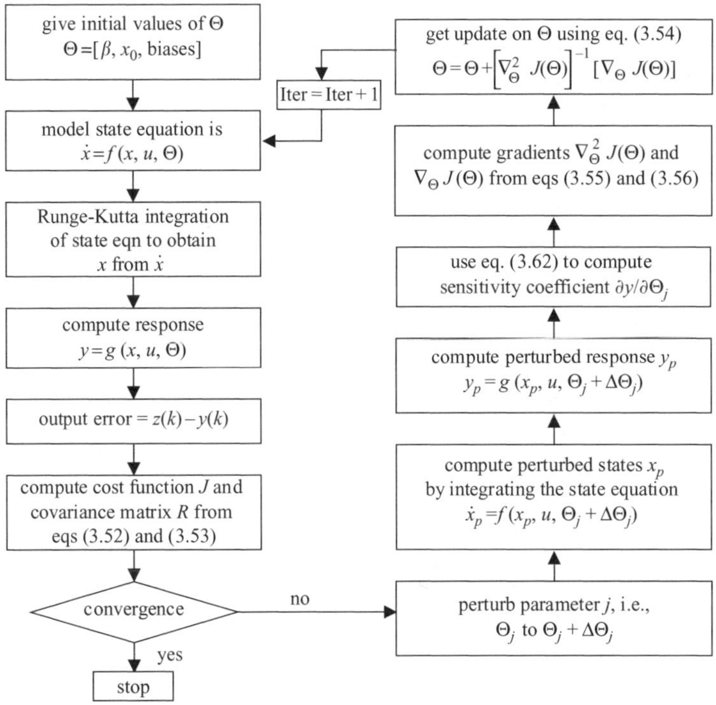

Figure 3.1 Flow chart of parameter estimation with OEM

(compare GLSDC of Chapter 2). However, for a nonlinear system, each time the model structure changes, partial differentiation of the system equations needs to be carried out to obtain $\partial y/\partial\Theta$. A better approach is to approximate the sensitivity coefficients by finite differences. In this procedure, the parameters in Θ in eqs (3.59) and (3.60) are perturbed one at a time and the corresponding perturbed model response y_p is computed. The sensitivity coefficient is then given by [8]:

$$\frac{\partial y}{\partial\Theta_j} = \frac{(y_p - y)}{\Delta\Theta_j} \tag{3.62}$$

The use of finite differencing in calculating $\partial y/\partial\Theta$ results in a program code that is more flexible and user friendly. The flow diagram of the output error computational procedure is given in Fig. 3.1.

3.7 Features and numerical aspects

The maximum likelihood method is very popular because of its several interesting features [1–12]:

- Maximum likelihood estimates are consistent, asymptotically unbiased and efficient.

- It is more general and can handle both measurement and process noise (of course, it then incorporates a Kalman filter into it, leading to the filter error method).
- If process noise is absent and measurement noise covariance is known, it reduces to the output error method.
- In case measurement noise is absent, it reduces to the equation error method, if all the states are measured.
- It is found to yield realistic values of the variances of the parameter estimates.
- It can be used to estimate the covariance of the measurement noise. In fact, it gives the covariance of residuals.

The computation of the coefficients of parameter vector Θ requires:

- Initial values of the coefficients in Θ.
- Current values of variables y at each discrete-time point k.
- Sensitivity matrix $(\partial y/\partial\Theta)_{ij} = \partial y_i/\partial\Theta_j$.
- Current state values are computed by numerical integration of the system state equations, which can be done by, say, 4th order Runge-Kutta method.

The Runge-Kutta method is fairly accurate and easier to use and, therefore, generally preferred. The sensitivity coefficients $(\partial y/\partial\Theta)_{ij}$ can be obtained explicitly for a given set of system equations by partially differentiating the equations with respect to each parameter. However, a change in the model structure would require the partial derivatives to be computed again. This becomes very cumbersome, as it requires frequent changes in the estimation algorithm. To avoid this, the sensitivity coefficients are approximately computed by using numerical differences. Assuming a small perturbation $\delta\Theta$ in the parameter Θ, the perturbed states x_p are computed and in turn used to obtain the perturbed output variable y_p. The sensitivity coefficient $\partial y/\partial\Theta$ is then given by eq. (3.62).

For nonlinear systems, the programming effort is reduced since, for every new nonlinear model, no sensitivity equations need be defined and the same routine, based on the above method, will do the job [8]. The choice of the step size for evaluating the numerical difference is typically given as

$$\partial\Theta_j \to 10^{-7} * \Theta_j$$

The gradient $\partial y/\partial\Theta_j$ may be computed using either central differencing or forward differencing. In the central differencing, the perturbed output y_p is computed for perturbations $\Theta_j + \delta\Theta_j$ and $\Theta_j - \delta\Theta_j$ in parameter Θ_j. Since there is no perceptible improvement in the accuracy of parameter estimates with central differencing compared to forward differencing, the latter is preferred as it saves CPU time. Further, forward differencing is only marginally slower compared to explicit estimation of sensitivity coefficients.

On comparing the optimisation methods for ML estimation, it is found that the quasi-linearisation method, which is equivalent to the modified Newton-Raphson method that neglects the computation of the second gradient of the error, is found to be 300–400 times faster than Powell's or Rosenbrock's method [8, 9]. It is also found to be about 150 times faster than the Quasi-Newton Method.

The method also provides direct information on accuracy of parameter estimates. However, it could have convergence problems with systems that have discontinuous nonlinearities.

The time history match is a necessary but not sufficient condition. It is quite possible that the response match would be good but some parameters could be unrealistic, e.g., unexpected sign behaviour. There could be one or more reasons for this kind of behaviour: deficient model used for the estimation or not all the modes of the system might have been sufficiently excited. One way to circumvent this problem is to add *a priori* information about the parameter in question. This can be done as shown in Chapter 9, or through adding a constraint equation in the cost function, with a proper sign (constraint) on the parameter. One more approach is to fix such parameters at some *a priori* value, which could have been determined by some other means or available independently from other source from the system.

The OEM/MLE method is so general that it can also be used for estimation of zero-shifts in measured input-output data.

3.7.1.1 Example 3.1 (see Example 2.4)

$$A = \begin{bmatrix} -2 & 0 & 1 \\ 1 & -2 & 0 \\ 1 & 1 & -1 \end{bmatrix}; \quad B = \begin{bmatrix} 1 \\ 0 \\ 1 \end{bmatrix}; \quad C = \begin{bmatrix} 1 & 0 & 0 \\ 0 & 1 & 0 \\ 0 & 0 & 1 \end{bmatrix}$$

Generate suitable responses with u as doublet input to the system and with proper initial condition on $x(0)$. Add a Gaussian white noise with zero-mean and known variance to the measurements y. Use OEM method to estimate the elements of the A and B matrices.

3.7.1.2 Solution

Data with sampling interval of 0.001 s and for duration of 5 s is generated by giving a doublet input to the system. The initial conditions for the three states are chosen as [0,0,0]. Two sets of data are generated – one with no noise in the data and the other where random noise with a $\sigma = 0.01$ is added to the data to generate noisy measurements.

The state and measurement models for estimation of the parameters (elements of A and B) are formulated as follows.

State model

$$\dot{x}_1 = a_{11}x_1 + a_{12}x_2 + a_{13}x_3 + b_1u_1$$
$$\dot{x}_2 = a_{21}x_1 + a_{22}x_2 + a_{23}x_3 + b_2u_1$$
$$\dot{x}_3 = a_{31}x_1 + a_{32}x_2 + a_{33}x_3 + b_3u_1$$

Measurement model

$$y_1 = x_1 + bias1$$

$$y_2 = x_2 + bias2$$

$$y_3 = x_3 + bias3$$

The elements of the A and B matrices together with the measurement bias values are estimated using OEM program (folder **Ch3OEMex1**). The estimated values of the elements of A and B matrices along with their standard deviations are given in Table 3.1. The table also shows the PEEN (percentage parameter estimation error

Table 3.1 Estimated elements of A and B matrices (Example 3.1)

Parameter	True values	Estimated values (data with no noise)	Estimated values (data with measurement noise $\sigma = 0.01$)		
			Case 1 (with $a_{23} = 0$)	Case 2 (with $a_{23} = -1$)	Case 3 (with $a_{23} = -3$)
a_{11}	−2	−2.0000 (0.0017)*	−2.0785 (0.0499)	−1.9247 (0.0647)	−1.9667 (0.0439)
a_{12}	0	−0.0000 (0.0037)	−0.1667 (0.1089)	−0.0602 (0.0537)	0.0109 (0.0116)
a_{13}	1	1.0000 (0.0021)	1.0949 (0.0614)	0.9392 (0.0504)	0.9782 (0.0294)
a_{21}	1	1.0000 (0.0001)	1.1593 (0.0475)	0.8190 (0.0656)	0.9125 (0.0527)
a_{22}	−2	−2.0000 (0.0017)	−1.6726 (0.1042)	−1.8408 (0.0542)	−2.0245 (0.0138)
a_{23}	$0/-1/-3$	−0.0000 (0.0037)	−0.1923 (0.0586)	−0.8558 (0.0511)	−2.9424 (0.0358)
a_{31}	1	1.0000 (0.0021)	0.9948 (0.0446)	1.0018 (0.0603)	1.0157 (0.0386)
a_{32}	1	1.0000 (0.0001)	1.0076 (0.0976)	0.9827 (0.0497)	1.0005 (0.0105)
a_{33}	−1	−1.0000 (0.0015)	−0.9981 (0.0549)	−1.0023 (0.0470)	−1.0132 (0.0257)
b_1	1	1.0000 (0.0034)	0.9978 (0.0024)	0.9977 (0.0023)	0.9979 (0.0025)
b_2	0	0.0000 (0.0019)	0.0030 (0.0023)	0.0043 (0.0024)	0.0046 (0.0030)
b_3	1	1.0000 (0.0001)	1.0011 (0.0022)	1.0022 (0.0008)	1.0004 (0.0023)
PEEN (%)		1.509e−6	11.9016	7.5914	2.3910

* the numbers in the brackets indicate the standard deviation of the parameters

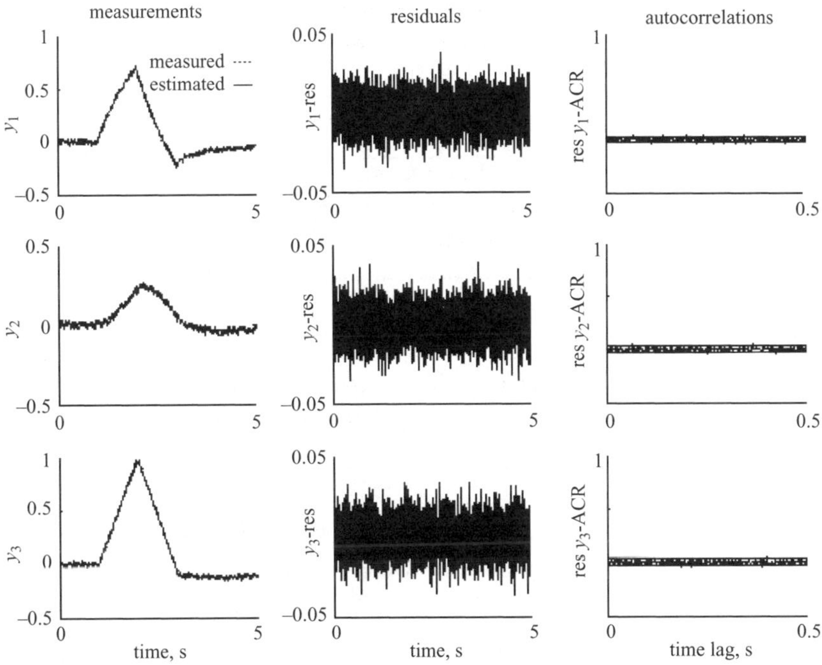

Figure 3.2 Results of estimation using OEM (Example 3.1)

norm; see Section A.36). It is clear that the estimates are very close to the true values when there is no noise in the data. When the measurements are noisy, it is seen that the estimates of those elements that are equal to zero show some deviations from the true values. The standard deviations of these derivatives are also higher compared with that of the other derivatives. This is also corroborated by the high value of the PEEN for this case. Figure 3.2 shows the comparison of the measured and estimated measurements (y_1, y_2, y_3), the residuals (y_1res, y_2res and y_3res) and the autocorrelation (ACR) of the residuals. It is clear that the residuals are white.

Since the PEEN is high when there is measurement noise in the data, it was decided to investigate this further. An observation of the estimates in Table 3.1 shows that those estimates in the dotted square in the A matrix show considerable deviation from their true values. It is to be noted that the estimates are very close to the true values when there is no noise in the data. The eigenvalues of the sub matrix $\begin{bmatrix} a_{12} & a_{13} \\ a_{22} & a_{23} \end{bmatrix}$ were evaluated and it was found that it was neutrally stable. Hence two more sets of data were generated: Case 2 with $a_{23} = -1$ and Case 3 with $a_{23} = -3$. Gaussian random noise with $\sigma = 0.01$ was added to both the sets of data. Table 3.2 lists the eigenvalues for the three cases investigated and the parameter estimates using OEM are listed in Table 3.1. It is clear that the PEEN is lower for Case 2 than for Case 1. For Case 3, the estimates are very close to the true values and the PEEN is low. This could be attributed to the stability of the system as the a_{23} is varied from 0 to -3.

Table 3.2 Eigenvalues of the sub matrix (Example 3.1)

Case number	Eigenvalues
Case 1 ($a_{23} = 0$)	$0 \pm 1.4142i$
Case 2 ($a_{23} = -1$)	$-0.5000 \pm 1.3229i$
Case 3 ($a_{23} = -3$)	$-1, -2$

When $a_{23} = 0$, the sub matrix is neutrally stable and becomes more stable for Cases 2 and 3. Thus, it is demonstrated that the interaction of the noise and stability/dynamics of the system via the sub matrix results in deficient parameter estimates from OEM.

3.7.1.3 Example 3.2

Let the dynamical system with 4 degrees of freedom (DOF) be described as

$$\begin{bmatrix} \dot{x}_1 \\ \dot{x}_2 \\ \dot{x}_3 \\ \dot{x}_4 \end{bmatrix} = \begin{bmatrix} -0.0352 & 0.107 & 0 & -32.0 \\ -0.22 & -0.44 & 3.5 & 0 \\ 1.2\text{e}{-4} & -0.0154 & -0.45 & 0 \\ 0 & 0 & 1 & 0 \end{bmatrix} \begin{bmatrix} x_1 \\ x_2 \\ x_3 \\ x_4 \end{bmatrix} + \begin{bmatrix} 0 \\ -22.12 \\ -4.66 \\ 0 \end{bmatrix} u$$

and

$$y = [I] \begin{bmatrix} x_1 \\ x_2 \\ x_3 \\ x_4 \end{bmatrix}$$

where I is the identity matrix.

Use 3211 input signal for u and generate y responses. Add Gaussian measurement noise with standard deviation $= 1.0$ and estimate the parameters of the system using output error method. Comment on the PEEN and the standard deviation of the estimates.

3.7.1.4 Solution

The above equations are of the general form $\dot{x} = Ax + Bu$ and $y = Hx$, $H = I$ in this case. Data with a sampling interval of 0.05 s, is generated by giving a 3211 input to the system. The initial conditions for the four states are chosen as [0,0,0,0]. Random noise with a $\sigma = 1.0$ is added to the data to generate noisy measurements. Data is simulated for a period of 10 s.

The state and measurement models for estimation of the parameters (elements of A and B matrices) are formulated as described in Example 3.1 with the unknown parameters in the above equations to be estimated. Measurement biases are also estimated as part of the estimation procedure. The relevant programs are contained

Table 3.3 Estimated parameters (Example 3.2)

Parameter	True values	Estimated values (data with measurement noise $\sigma = 1.0$)
a_{11}	−0.0352	−0.0287 (0.0136)*
a_{12}	0.1070	0.1331 (0.0246)
a_{14}	−32.0000	−31.8606 (0.4882)
a_{21}	−0.2200	−0.2196 (0.0009)
a_{22}	−0.4400	−0.4406 (0.0050)
a_{23}	3.5000	3.5275 (0.0897)
b_2	−22.1200	−21.9056 (0.3196)
a_{32}	−0.0154	−0.0165 (0.0007)
a_{33}	−0.4500	−0.4755 (0.0233)
b_3	−4.6600	−4.6849 (0.0890)
PEEN (%)		0.6636

* indicates the standard deviation of the parameters

in the folder **Ch3OEMex2**. The estimated parameters are listed in Table 3.3. It is to be noted that the parameters that are equal to or close to zero are kept fixed and not estimated. It is clear that the estimates are very close to the true values for all the parameters. The PEEN is also very low.

Figure 3.3(a) shows the input and the comparison of the estimated and measured data. Figure 3.3(b) shows the plot of cost function and determinant of R (starting from the 5th iteration). It is clear that the cost function converges to a value very close to 4 (which is equal to the number of observations). In addition, the $|R|$ converges to a low value, close to 0.7 for this example.

3.7.1.5 Example 3.3

Use the simulated short period data of a light transport aircraft to estimate the non-dimensional longitudinal parameters of the aircraft using OEM method. Use the 4-degree of freedom longitudinal body axis model for estimation. The relevant mass, moment of inertia and other aircraft geometry related parameters are provided below (see Section B.12):

Mass, m	$= 2280.0$ kg
Moment of inertia, I_{yy}	$= 6940.0$ kg/m^2
Mean aerodynamic chord, $\bar{c}$	$= 1.5875$ m
Wing area, S	$= 23.23$ m^2
Air density, ρ	$= 0.9077$ kg/m^3

3.7.1.6 Solution

The data are generated with a sampling interval of 0.03 s by giving a doublet input to the elevator. The measurements of $u, w, q, \theta, a_x, a_z, \dot{q}$ and δ_e are provided. Random

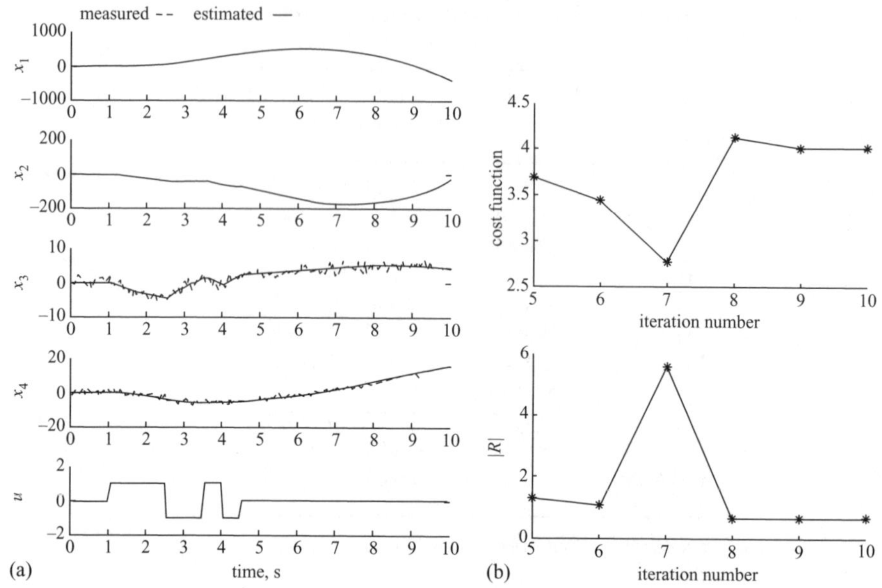

Figure 3.3 *(a) Time histories of estimated and measured data (Example 3.2); (b) cost function and $|R|$ (Example 3.2)*

noise with a standard deviation $\sigma = 0.1$ is added to the data to generate noisy measurements. The state and measurement models for estimation of the parameters in body axis (see Section B.1) are formulated as follows.

State model

$$\dot{u} = \frac{\bar{q}S}{m} C_X - qw - g\sin\theta$$

$$\dot{w} = \frac{\bar{q}S}{m} C_Z + qu + g\cos\theta$$

$$\dot{q} = \frac{\bar{q}S\bar{c}}{I_{yy}} C_m$$

$$\dot{\theta} = q$$

In the above equations we have

$$C_Z = C_{Z_0} + C_{Z_\alpha}\alpha + C_{Z_q}\frac{q\bar{c}}{2V} + C_{Z_{\delta_e}}\delta_e$$

$$C_X = C_{X_0} + C_{X_\alpha}\alpha + C_{X_{\alpha^2}}\alpha^2$$

$$C_m = C_{m_0} + C_{m_\alpha}\alpha_m + C_{m_{\alpha^2}}\alpha^2 + C_{m_q}\frac{q_m\bar{c}}{2V} + C_{m_{\delta_e}}\delta_e$$

Measurement model

$$y_1 = u + bias1$$

$$y_2 = w + bias2$$

$$y_3 = q + bias3$$

$$y_4 = \theta + bias4$$

$$y_5 = \frac{\bar{q}S}{m} C_X + bias5$$

$$y_6 = \frac{\bar{q}S}{m} C_Z + bias6$$

$$y_7 = \dot{q} + bias7$$

The parameters $C_{()}$ and measurement bias values are estimated using the output error method program (folder **Ch3OEMex3**). The estimated values of the parameters are compared with the true values of the derivatives in Table 3.4. The table also shows the PEEN. The estimates are fairly close to the true values. Figure 3.4(a) shows the time history match of the measured signals and the estimated signals. A good time history match is a necessary condition for confidence in the parameter estimates. Figure 3.4(b) shows the plot of cost function and determinant of R ($|R|$) versus the iterations. The cost function converges to a value very close to 8 (which is close to the number of observations, which is 7 in this case). In addition, the $|R|$ converges to a very low value, close to zero for this example.

Table 3.4 Estimated parameters of A and B matrices (Example 3.3)

Parameter	True values	Estimated values
C_{x0}	−0.0540	−0.0511
$C_{x\alpha}$	0.2330	0.1750
$C_{x\alpha^2}$	3.6089	3.6536
C_{z0}	−0.1200	−0.0819
$C_{z\alpha}$	−5.6800	−5.6442
$C_{z\delta}$	−0.4070	−0.3764
C_{m0}	0.0550	0.0552
$C_{m\alpha}$	−0.7290	−0.6882
$C_{m\alpha^2}$	−1.7150	−1.8265
C_{mq}	−16.3	−16.6158
$C_{m\delta}$	−1.9400	−1.9436
PEEN (%)	—	1.9641

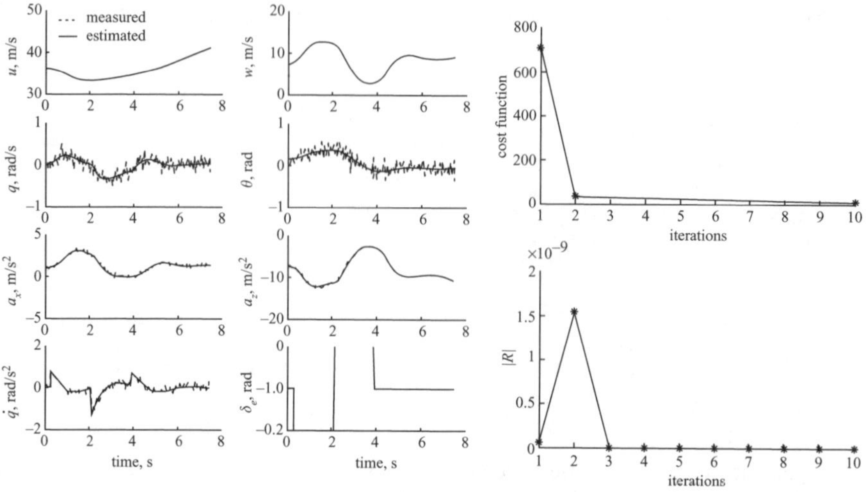

Figure 3.4 *(a) Time history match (Example 3.3); (b) cost function and $|R|$ (Example 3.3)*

3.7.1.7 Example 3.4 (Kinematic consistency checking of helicopter flight test data)

The output error program is used to perform kinematic consistency (see Section B.7) checking of helicopter flight test data. The nonlinear kinematic equations are integrated with measured rates and linear accelerations as inputs. Speed components u, v and w, attitude angles ϕ and θ and altitude h are treated as states and computed. Measurements obtained from flight data for linear accelerations, flight velocity V and sideslip angle β are defined for the c.g. location and as such need no further correction w.r.t. c.g. (see Section B.8). To correct the data for instrumentation errors, the derived time histories are compared with flight measurements and the biases (offsets) estimated.

3.7.1.8 Solution

Figure 3.5 shows the comparison of measured and model-estimated trajectories obtained by data compatibility check using standard kinematic equations. On the left hand side, the trajectory match when no bias is included is shown. It is clear that the estimated velocity V and bank angle ϕ show divergence, which could be attributed to bias errors in p (roll rate) and q (pitch rate). The trajectory match on the right hand side is obtained by estimating the biases in the measurements of p, q, ϕ and β (sideslip). The agreement, in general, has been found to be satisfactory for the measurements: altitude h, bank angle ϕ, pitch angle θ and velocity V.

For this set of helicopter data, it was observed that linear accelerations were of good quality while angular rates had small biases. Adequate agreement for the attitude angles was obtained after the measurements were corrected for biases.

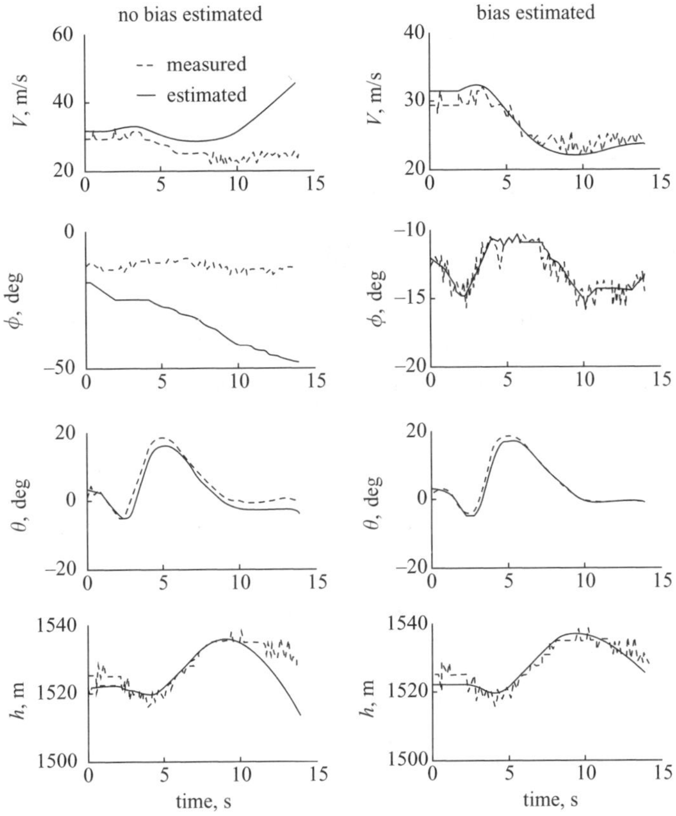

Figure 3.5 Data compatibility of measurements using kinematic equations (Example 3.4)

3.7.1.9 Example 3.5

The nuisance parameters are those assumed known even though they may not be known precisely. This is primarily done in order to reduce the number of parameters to be estimated.

In the standard maximum likelihood method, the covariance matrix is the inverse of the information matrix as mentioned in Section 3.3. However, due to the (presence of) nuisance parameters, the Fisher Information Matrix does not properly reflect the uncertainty in the primary parameter estimates of the dynamical system obtained by the ML method.

Consider the following system [13]:

$$\begin{bmatrix} \dot{u}_x \\ \dot{w}_x \\ \dot{\lambda} \\ \dot{q} \end{bmatrix} = \begin{bmatrix} x_u & x_w & -g\cos\lambda_0 & -w_0 \\ z_u & z_w & -g\sin\lambda_0 & u_0 \\ 0 & 0 & 0 & 1 \\ m_u & m_w & 0 & m_q \end{bmatrix} \begin{bmatrix} u_x \\ w_x \\ \lambda \\ q \end{bmatrix} + \begin{bmatrix} 0 \\ 0 \\ 0 \\ m_\delta \end{bmatrix} [\delta]$$

$$y = [I] \begin{bmatrix} u_x \\ w_x \\ \lambda \\ q \end{bmatrix}$$

where I is the identity matrix

Consider certain important parameters as primary parameters and assign some others to the so-called secondary parameters. Generate simulated data without state noise. Estimate Cramer-Rao Bounds (CRBs) for the parameters in turn by releasing some of the nuisance parameters as primary parameters. Comment on these estimates and CRBs. Use Gaussian random noise with zero-mean and covariance matrix R for measurements given by: $diag\{0.1^2, 0.1^2, 0.01^2, 0.006^2\}$. For nuisance parameters, assume the values (as known) with some factor of uncertainty.

3.7.1.10 Solution

The data for duration of 10 s is simulated by using a 3211 signal input for δ using sampling time = 0.05 s. The following values for the parameters are used for simulation.

$$\begin{bmatrix} \dot{u}_x \\ \dot{w}_x \\ \dot{\lambda} \\ \dot{q} \end{bmatrix} = \begin{bmatrix} -0.00335 & 0.139 & -9.8\cos(0) & -7.0 \\ -0.106 & -0.710 & -9.8\sin(0) & 36.0 \\ 0 & 0 & 0 & 1 \\ 0.00655 & -0.0293 & 0 & -2.18 \end{bmatrix} \begin{bmatrix} u_x \\ w_x \\ \lambda \\ q \end{bmatrix} + \begin{bmatrix} 0 \\ 0 \\ 0 \\ -5.29 \end{bmatrix} [\delta]$$

Random noise with standard deviations equal to 0.1, 0.1, 0.01 and 0.006 is added to the measurements.

The parameters x_u, x_w and z_u were considered as secondary parameters and the remaining five parameters namely z_w, m_u, m_w, m_q and m_δ were considered as primary parameters for estimation using OEM programs in the folder **Ch3OEMex5**. The secondary parameters were fixed at their true values to check its effect on the parameter estimates (Case 1). Figure 3.6(a) shows the time history match for this case. The parameter estimates are listed in Table 3.5 along with their standard deviations. The estimates are fairly close to the true values as is clear from the low values of PEEN.

When the nuisance parameters are known with a certain uncertainty, it is expected to have an effect on the estimated uncertainty in the parameter estimates. In order to study this effect, the secondary/nuisance parameters were assumed known with 5 per cent and 10 per cent uncertainty and used in the OEM model for parameter estimation. Table 3.5 lists the parameter estimates for these cases. It is clear that the parameter estimates are close to the true values for all these cases. However, the PEENs show an increase as the uncertainty level for the nuisance parameters

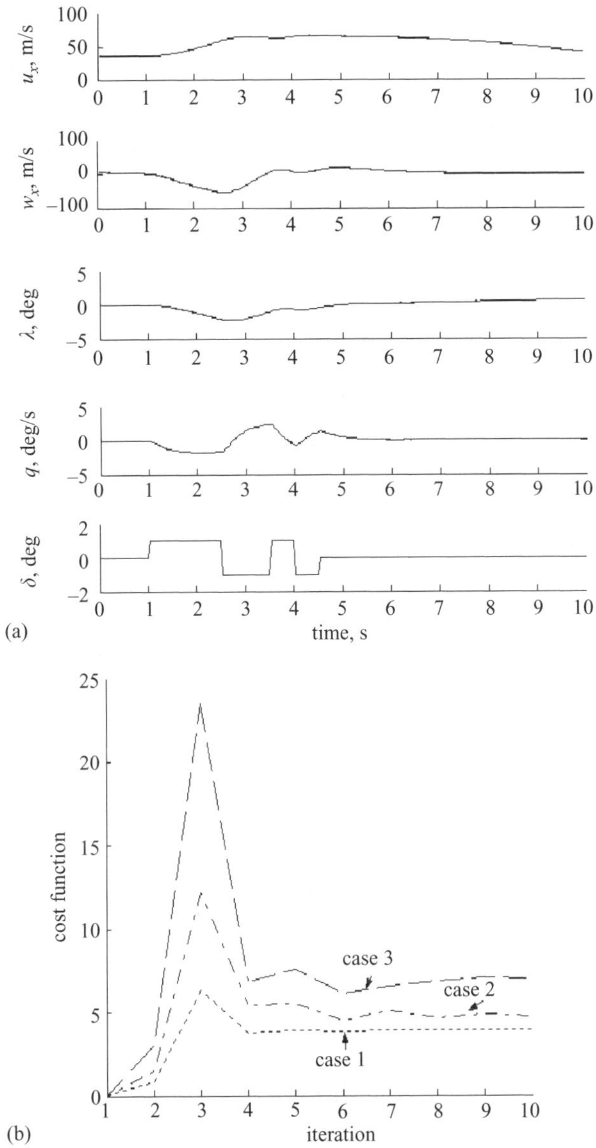

Figure 3.6 *(a) Time history match (Example 3.5) (estimated —; measured ...); (b) cost functions (Example 3.5)*

increases. There is an increase in the standard deviation of the estimates – though it is not very significant. However, it is clear from the cost function plotted in Fig. 3.6(b), that as the uncertainty in the nuisance parameters increases, there is a significant increase in the cost function.

Table 3.5 Parameter estimates (Example 3.5)

Parameter	True values	Case 1 (nuisance parameters fixed at true values)	Case 2 (nuisance parameters fixed at (true + 5%))	Case 3 (nuisance parameters fixed at (true + 10%))
z_w	−0.7100	−0.7099 (0.0007)	−0.7119 (0.0007)	−0.7116 (0.0008)
m_w	0.0066	0.0066 (0.0000)	0.0064 (0.0000)	0.0062 (0.0000)
m_u	−0.0293	−0.0292 (0.0000)	−0.0292 (0.0000)	−0.0291 (0.0000)
m_q	−2.1800	−2.1834 (0.0020)	−2.1810 (0.0021)	−2.1826 (0.0022)
m_δ	−5.2900	−5.2942 (0.0033)	−5.3013 (0.0034)	−5.3100 (0.0036)
PEEN		0.0935	0.1997	0.3512

3.8 Epilogue

Output error/maximum likelihood estimation of aircraft has been extensively treated [4–10]. Recursive MLE/adaptive filter is considered in Reference 11. The OEM/MLE based methods have found extensive applications to aircraft/rotorcraft parameter estimation. The applications are too many to be covered in this chapter. The main reason for success of the technique is that it has many nice theoretical properties and, it being an iterative process, generally gives reasonably accurate results for practical real data. The iterations refine the estimates. Another reason for its success is that it gives theoretical lower bounds on the variance of the estimates based on the Fisher information matrix, named after Fisher [1]. Thus, one can judge the accuracy of the estimates and obtain uncertainty bounds on the parameters. It can also be applied to nonlinear problems with equal ease.

3.9 References

1 FISHER, R. A.: 'On the mathematical foundations of theoretical statistics', *Philosophy Trans. Roy. Soc. London*, 1922, **222**, pp. 309–368
2 FISHER, R. A.: 'Contributions to mathematical statistics' (John Wiley & Sons, New York, 1950)
3 ASTROM, K. J.: 'Maximum likelihood and prediction error methods', *Automatica*, 1980, **16**, pp. 551–574
4 MEHRA, R. K., STEPNER, D. E., and TYLER J. S.: 'Maximum likelihood identification of aircraft stability and control derivatives', *Journal of Aircraft*, 1974, **11**, (2), pp. 81–89
5 ILIFF, K. W.: 'Parameter estimation for flight vehicles', *Journal of Guidance, Control and Dynamics*, 1989, **12**, (5), pp. 609–622

6 PLAETSCHKE, E.: 'Maximum likelihood estimation'. Lectures presented at FMCD, NAL as a part of the IFM, DLR-FMCD, NAL collaborative programme, Nov. 1987, Bangalore, India
7 MAINE, R. E., and ILIFF, K. W.: 'Application of parameter estimation to aircraft stability and control – the output error approach'. NASA report RP-1168, 1986
8 JATEGAONKAR, R. V., and PLAETSCHKE, E.: 'Maximum likelihood parameter estimation from flight test data'. DFVLR-FB 83-14, IFM/Germany, 1983
9 JATEGAONKAR, R. V., and PLAETSCHKE, E.: 'Non-linear parameter estimation from flight test data using minimum search methods'. DFVLR-FB 83-15, IFM/Germany, 1983
10 JATEGAONKAR, R. V.: 'Identification of the aerodynamic model of the DLR research aircraft ATTAS from flight test data'. DLR-FB 94-40, IFM/TUB/Germany, 1990
11 CHU, Q. P., MULDER J. A., and VAN WOERKOM, P. T. L. M.: 'Modified recursive maximum likelihood adaptive filter for nonlinear aircraft flight path reconstruction', *Journal of Guidance, Control and Dynamics*, 1996, **19**, (6), pp. 1285–1295
12 GIRIJA, G., and JATEGAONKAR, R. V.: 'Some results of ATTAS flight data analysis using maximum likelihood parameter estimation method'. DLR-FB 91-04, IFM/Germany, 1991
13 SPALL, J. C., and GARNER, J. P.: 'Parameter identification for state-space models with nuisance parameters', *IEEE Trans. on Aerospace and Electronic Systems*, 1990, **26**, (6), pp. 992–998

3.10 Exercises

Exercise 3.1

Let the spring mass system be described by $m\ddot{y} + d\dot{y} + Ky = w(t)$. Obtain the state space model in the form $\dot{x} = Ax + Bu$ and obtain $\partial\dot{x}/\partial K$, $\partial\dot{x}/\partial d$.

Exercise 3.2

The Gaussian least squares differential correction method has been discussed in Chapter 2. Comment on the differences and similarities between the Gaussian least squares differential correction method and the output error method, since both these methods use output error criterion and are applicable to dynamical systems.

Exercise 3.3

Consider the equations $\dot{x}(t) = Ax(t)+Bu(t)$ and $y(t) = Cx(t)+Du(t)$. Assume that β_1 = unknown initial values of the state variables and β_2 = unknown parameters in matrices A, B, C and D. Postulate y as a function of β_1, β_2 and u. Let $\beta = [\beta_1^T, \beta_2^T]^T$. Obtain expressions for $\partial y/\partial\beta$, $\partial\dot{x}/\partial\beta_1$ and $\partial\dot{x}/\partial\beta_2$.

(Hint: Study Gaussian least squares differential correction equations given in Chapter 2.)

Exercise 3.4

Let

$$y_1 = \beta_1 x_1 + \beta_2 x_2$$

$$y_2 = \beta_3 x_1 + \beta_4 x_2$$

$$y_3 = \beta_5 x_1 + \beta_6 x_2$$

Obtain expressions of eq. (3.56). Compare the expressions with those of eq. (10.51) and comment. The main point of this exercise is to show, on the basis of the second order gradient expression (eq. (3.56)), certain commonalities to similar developments using recurrent neural networks.

Exercise 3.5

Consider eq. (3.20) of Cramer-Rao inequality and comment on this if there is a bias in the estimate.

Exercise 3.6

Comment on the relationship between maximum likelihood and the least squares methods, by comparing eq. (3.34) for the likelihood function to eq. (2.2) for the cost function of least squares method.

Exercise 3.7

Compare and contrast eq. (3.56), the second order gradient, for maximum likelihood estimation with eq. (2.7), the covariance matix of estimation error.

Chapter 4
Filtering methods

4.1 Introduction

In the area of signal processing, we come across analogue and digital filtering concepts and methods. The real-life systems give rise to signals, which are invariably contaminated with the so-called random noise. This noise could arise due to measurement errors from the sensors, instruments, data transmission channels or human error. Some of these errors would be systematic, fixed or slowly varying with time. However, in most cases, the errors are random in nature and can be described best by a probabilistic model. A usual characteristic of such a random noise that affects the signal is Gaussian (normally distributed) noise with zero mean and some finite variance. This variance measures the power of the noise and it is often compared to the power of the signal that is influenced by the random noise. This leads to a measure called signal to noise ratio (SNR). Often the noise is assumed a white process (see Chapter 2). The aim is then to maximise SNR by filtering out the noise from the signal/data of the dynamical system. There are mainly two approaches: model free and model based. In the model free approach, no mathematical model (equations) is presumed to be fitted or used to estimate the signal from the signal plus noise. These techniques rely upon the concept of the correlation of various signals, like input-output signals and so on. In the present chapter, we use the model based approach and especially the approach based on the state-space model of a dynamical system.

Therefore, our major goal is to get the best estimate or prediction of the signal, which is buried, in the random noise. This noise could be white or time-correlated (non-white). It could be coloured noise, i.e., output of a linear lumped parameter system excited by a white noise (see Exercise 2.10). Estimation (of a signal) is a general term. One can make three distinctions in context of an estimate of a signal: filtered, predicted or smoothed estimate. We assume that the data is available up to the time 't'. Then, obtaining the estimate of a signal at the time 't' is called filtering. If we obtain an estimate, say at 't + 1', it is called prediction and if we obtain an estimate at 't − 1' by using data up to 't', it is called a smoothed estimate. In this chapter, we mainly study the problem of filtering and prediction using Kalman filtering methods [1–6].

Kalman filtering has evolved to a very high state-of-the-art method for state estimation for dynamical systems, which could be described by difference, or differential equations, especially in state-space form [1]. The impact of the Kalman filtering approach is such that it has generated worldwide extensive applications to aerospace system problems [7], and thousands of papers have been written on Kalman filtering covering: i) theoretical derivations; ii) computational aspects; iii) comparison of various versions of Kalman filtering algorithms for nonlinear systems; iv) factorisation filtering; v) asymptotic results; vi) applications to satellite orbit estimation; vii) attitude determination; viii) target tracking; ix) sensor data fusion; x) aircraft state/parameter estimation; and xi) numerous engineering and related applications. There are also more than a dozen books on Kalman filtering and closely related methods.

The main reason for its success is that it has an appealing state-space formulation and it gives algorithms that can be easily implemented on digital computers. In fact, the Kalman filter is a numerical algorithm, which also has a tremendous real-time/on-line application because of its recursive formulation as against one-shot/batch processing methods. For linear systems, it is an optimal state observer. In this chapter, Kalman filtering algorithms are discussed since they form the basis of filter error method (Chapter 5) and EBM (Chapter 7), which are used for parameter estimation of linear, nonlinear and stable/unstable dynamical systems.

4.2 Kalman filtering

It being a model based approach, we first we describe a dynamical system:

$$x(k+1) = \phi x(k) + Bu(k) + Gw(k) \tag{4.1}$$

$$z(k) = Hx(k) + Du(k) + v(k) \tag{4.2}$$

where x is an $n \times 1$ state vector; u is a $p \times 1$ deterministic control input to the system; z is an $m \times 1$ measurement vector; w is a white Gaussian noise sequence with zero mean and covariance matrix Q (also called process noise with associated matrix G); v is a white Gaussian noise sequence with zero mean and covariance matrix R (also called measurement noise); ϕ is the $n \times n$ transition matrix that takes states from k to $k+1$; B is the input gain/magnitude vector/matrix; H is the $m \times n$ measurement model/sensor dynamics matrix; and D is the $m \times p$ feedforward/direct control input matrix (often D is dropped from the Kalman filter development).

We emphasise here that, although most dynamic systems are continuous-time, the Kalman filter is an extremely popular filtering method and is best discussed using the discrete-time model. In addition, in the sequel, it will be seen that the solution of the Kalman filter requires handling of the Riccati equation, which is easier to handle in discrete form rather than in continuous-time form. One can convert the continuous-time system to a discrete-time model and use a discrete-time Kalman filtering algorithm, which can be easily implemented in a digital computer. Also, the fact that even though the continuous-time filtering algorithm would require to be implemented on a digital computer, it seems that both approaches will lead to some approximations. We feel that understanding and implementing a discrete-time Kalman filter is easier.

We observe that eq. (4.1) introduces the dynamics into the otherwise 'only measurement model' eq. (4.2), which was used in Chapter 2. Thus, the problem of state estimation using Kalman filtering can be formulated as follows: given the model of the dynamical system, statistics of the noise processes and the noisy measurement data, and the input, determine the best estimate of the state, x, of the system. Since it is assumed that the dynamical system is known, it means that the (form and) numerical values of the elements of ϕ, B and H are accurately known. If some of these elements are not known, then these can be considered as additional unknown states and appended to the otherwise state vector x yielding the extended state vector. In most circumstances this will lead to a nonlinear dynamical system for which an extended Kalman filter can be used. Life would have been much easier or even trivial if the noise processes were not present, the dynamics of the system accurately known and accurate information about the state initial values $x(0)$ available. Then simple integration (analytical or numerical) of eq. (4.1) would solve the (filtering) problem. The reality is not so simple. Initial conditions are often not known accurately, the system/plant dynamics are not always accurately known and the state and/or measurement noises are always present.

The process noise accounts for modelling errors as well as an artefact to do filter tuning for trajectory matching.

Since our aim is to obtain an estimate of the state of the dynamical system, we need to have measurements of the state. Often these are available indirectly as eq. (4.2) through the measurement model.

The mathematical models assumed are Gauss-Markov (see Section A.24), since the noise processes assumed are Gaussian and the system described in eq. (4.1) is linear. This model state is a Markov process or chain, mainly the model being the state equation of first order. This model is fairly general and is readily amenable to recursive processing of the data. In addition, it is generally assumed that the system (in fact the representation of eqs (4.1) and (4.2)) is controllable and observable (see Section A.34).

4.2.1 Covariance matrix

Consider the homogeneous state equation

$$\dot{x}(t) = A(t)x(t) \tag{4.3}$$

Then the state vector x evolves according to

$$x(t) = \phi(t, t_0)x(t_0) \tag{4.4}$$

Here, $x(t_0)$ is the initial state at time t_0. For conformity with the discrete system, we rewrite eq. (4.4) as

$$x(k+1) = \phi(k, k+1)x(k) \tag{4.5}$$

The matrix ϕ is known as the state transition matrix. It takes state from $x(k)$ at time k to $x(k+1)$ at time $k+1$ and so on. The equation for covariance matrix propagation can be easily derived based on its definition and eq. (4.5).

Let $P(k) = E\{\underline{x}(k)\underline{x}^T(k)\}$ be the covariance matrix of $\underline{x}(k)$ at time index k, where $\underline{x}(k) = \hat{x}(k) - x(k)$ at k. It reflects the errors in the estimate of x at k. We want to know how the error propagates at other times. We have from eq. (4.5):

$$\tilde{x}(k+1) = \phi\hat{x}(k) \tag{4.6}$$

Here, $\tilde{x}$ is a predicted estimate of x, considering $u = 0$, with no loss of generality. Then, we have, after adding a process noise term in eq. (4.5)

$$\begin{aligned}\tilde{P}(k+1) &= E\{(\tilde{x}(k+1) - x(k+1))(\tilde{x}(k+1) - x(k+1)^T\}\\ &= E\{(\phi\hat{x}(k) - \phi\, x(k) - Gw(k))(\phi\hat{x}(k) - \phi x(k) - Gw(k))^T\}\\ &= E\{(\phi\hat{x}(k) - \phi\hat{x}(k))(\phi\hat{x}(k) - \phi\hat{x}(k))^T\} + E\{Gw(k)w^T(k)G^T\}\end{aligned}$$

Here, we assume that state error and process noise are uncorrelated and hence the cross terms are neglected. Finally we get

$$\tilde{P}(k+1) = \phi\hat{P}(k)\phi^T + GQG^T \tag{4.7}$$

Equation (4.7) is the equation of state error covariance propagation, i.e., the state error variance at time k is modified by the process noise matrix and the new state error variance is available at time $k+1$. The transition matrix ϕ plays an important role.

4.2.2 Discrete-time filtering algorithm

For simplicity, the discrete-time algorithm is studied. We presume that the state estimate at k is evolved to $k+1$ using eq. (4.6). Now at this stage a new measurement is available. This measurement contains information regarding the state as per eq. (4.2). Therefore, intuitively, the idea is to incorporate the measurement into the data (filtering) process and obtain an improved/refined estimate of the state. We assume that the matrix H and *a priori* covariance matrix R are given or known.

4.2.2.1 Measurement/data update algorithm

Given: H, R and measurements z

Assume: $\tilde{x}(k) \rightarrow$ *a priori* estimate of state at time k, i.e., before the measurement data is incorporated.

$\hat{x}(k) \rightarrow$ updated estimate of state at time k, i.e., after the measurement data is incorporated.

$\tilde{P} \rightarrow$ *a priori* covariance matrix of state estimation error (this was derived earlier).

Then the measurement update algorithm is given as:

$$\hat{x}(k) = \tilde{x}(k) + K[z(k) - H\tilde{x}(k)] \quad \text{(state estimate/filtered estimate)} \tag{4.8}$$

$$\hat{P}(k) = (I - KH)\tilde{P}(k) \quad \text{(covariance update)} \tag{4.9}$$

The filtering eqs (4.8) and (4.9) are based on the following development. Our requirement is that we want an unbiased recursive form of estimator (filter), with minimum errors in the estimates as measured by P. Let such a recursive form be given as

$$\hat{x}(k) = K_1\tilde{x}(k) + K_2 z(k) \tag{4.10}$$

The expression in eq. (4.10) is a fair weighted combination of the *a priori* estimate (obtained) by eq. (4.6) and the new measurement. The gains K_1 and K_2 are to be optimally chosen for the above requirement of unbiased estimate.

Let $\underline{x}(k) = \hat{x}(k) - x(k)$; $\underline{x}^*(k) = \tilde{x}(k) - x(k)$ be the errors in the state estimates. Then, we have

$$\underline{x}(k) = [K_1\tilde{x} + K_2 z(k)] - x(k) = K_1\tilde{x} + K_2 Hx(k) + K_2 v(k) - x(k)$$

Using simplified measurement eq. (4.2)

$$\begin{aligned}\underline{x}(k) &= K_1[\underline{x}^*(k) + x(k)] + K_2 Hx(k) + K_2 v(k) - x(k)\\ &= [K_1 + K_2 H - I]x(k) + K_2 v(k) + K_1\underline{x}^*(k)\end{aligned}$$

Since $E\{v(k)\} = 0$ and if $E\{\underline{x}^*(k)\} = 0$ (unbiased *a priori* estimate), then

$$E\{\underline{x}(k)\} = E\{(K_1 + K_2 H - I)x(k)\}$$

Thus, in order to obtain an unbiased estimate after the measurement is incorporated, we must have $E\{\underline{x}(k)\} = 0$, and hence

$$K_1 = I - K_2 H \tag{4.11}$$

Substituting the above equation into eq. (4.10), we get

$$\begin{aligned}\hat{x}(k) &= (I - K_2 H)\tilde{x}(k) + K_2 z(k)\\ &= \tilde{x}(k) + K_2[z(k) - H\tilde{x}(k)]\end{aligned} \tag{4.12}$$

For further development, we change K_2 to K as the Kalman (filter) gain. Essentially, eq. (4.12) is the measurement data update algorithm, but we need to determine the expression for gain K. The structure of the filter has now been well defined:

Current estimate = previous estimate + gain × (error in measurement prediction)

The term $[z(k) - H\tilde{x}(k)]$ is called the measurement prediction error or the residual of the measurement. It is also called innovation. The above form is common to many recursive algorithms.

Next, we formulate P to determine the covariance of the state error after the measurement is incorporated.

$$\begin{aligned}\hat{P} &= E\{\underline{x}(k)\underline{x}^T(k)\} = E\{(\hat{x}(k) - x(k))(\hat{x}(k) - x(k))^T\}\\ &= E\{(\tilde{x}(k) - x(k) + K[Hx(k) + v(k) - H\tilde{x}(k)])(\cdot)^T\}\\ &= E\left\{[(I - KH)\underline{x}^* + Kv(k)]\left[\underline{x}^{*^T}(I - KH)^T + v^T(k)K^T\right]\right\}\end{aligned}$$

$$\hat{P} = (I - KH)\tilde{P}(I - KH)^T + KRK^T \tag{4.13}$$

In the above, '·' means that the second term within the parenthesis is the same as the first term.

Next, we optimally choose K so that the error covariance matrix $\hat{P}$ is minimised in terms of some norm. Let the cost function

$$J = E\{\underline{x}^T(k)\underline{x}(k)\}$$

be minimised with respect to the gain matrix K. This is equivalent to

$$\begin{aligned} J &= trace\{\hat{P}\} \\ &= trace\{(I - KH)\tilde{P}(I - KH)^T + KRK^T\} \end{aligned} \tag{4.14}$$

$$\frac{\partial J}{\partial K} = -2(I - KH)\tilde{P}H^T + 2KR = 0 \qquad \text{(the null matrix)}$$

$$KR = \tilde{P}H^T - KH\tilde{P}H^T$$

$$KR + KH\tilde{P}H^T = \tilde{P}H^T$$

$$K = \tilde{P}H^T(H\tilde{P}H^T + R)^{-1} \tag{4.15}$$

Substituting the expression of K into eq. (4.13) and simplifying, we get

$$\hat{P} = (I - KH)\tilde{P} \tag{4.16}$$

Finally, the Kalman filter equations are put collectively in the following form.

State propagation

State estimate: $\tilde{x}(k+1) = \phi\hat{x}(k)$ (4.17)

Covariance (*a priori*): $\tilde{P}(k+1) = \phi\hat{P}(k)\phi^T + GQG^T$ (4.18)

Measurement update

Residual: $r(k+1) = z(k+1) - H\tilde{x}(k+1)$ (4.19)

Kalman gain: $K = \tilde{P}H^T(H\tilde{P}H^T + R)^{-1}$ (4.20)

Filtered estimate: $\hat{x}(k+1) = \tilde{x}(k+1) + Kr(k+1)$ (4.21)

Covariance (*a posteriori*): $\hat{P} = (I - KH)\tilde{P}$ (4.22)

Although K and P would vary as the filter is running, the time index is dropped for simplicity. However, Q and R are assumed pre-determined and constant.

We note here that $K = \tilde{P}H^TS^{-1}$ with $S = H\tilde{P}H^T + R$. This matrix S is the covariance matrix of residuals. The actual residuals can be computed from eq. (4.19) and they can be compared with standard deviations obtained by taking the square root of the diagonal elements of S. This process of checking and tuning the filter to bring the computed residuals within the bound of at least two standard deviations is an important filter tuning exercise for the correct solution of the problem. This process of tuning in conjunction with eq. (4.18) is called the covariance-matching concept for adaptive estimation in Kalman filtering algorithm.

4.2.3 Continuous-time Kalman filter

Although the discrete-time filtering algorithm is widely preferred for digital implementation, we briefly discuss continuous-time filtering algorithm here.

Let us define the continuous-time model of the dynamical system as

$$\dot{x}(t) = Ax(t) + w(t) \tag{4.23}$$

$$z(t) = Hx(t) + v(t) \tag{4.24}$$

We have the following assumptions:

1 The noise processes $w(t)$ and $v(t)$ are uncorrelated Gaussian random processes with spectral density matrices $Q(t)$ and $R(t)$, respectively (see Section A.29).
2 $E\{x(0)\} = \hat{x}_0$; $E\{(\hat{x}_0 - x(0))(\hat{x}_0 - x(0))^T\} = P_0$
3 We have very accurate knowledge of A, H, Q and R.

Then, continuous-time KF is given as [3]:

$$\dot{\hat{x}}(t) = A\hat{x}(t) + K(t)[z(t) - H\hat{x}(t)] \qquad \text{(state evolution)} \tag{4.25}$$

$$\dot{P}(t) = AP(t) + P(t)A^T + Q(t) - KRK^T; \qquad P(0) = P_0 \tag{4.26}$$

$$K = PH^TR^{-1} \qquad \text{(Kalman gain)} \tag{4.27}$$

The eq. (4.26) is called the matrix Riccati equation, which needs to be solved for obtaining P, which is used in computation of the Kalman gain. The comparison of eqs (4.26) and (4.27) with eqs (4.18) and (4.20) shows that the computations for the continuous-time Kalman filter are more involved due to the continuous-time matrix Riccati equation. One simple route is to assume that a steady state is reached, thereby considering $\dot{P} = 0$, and solving eq. (4.26) by an appropriate method [2, 3]. In addition, another method is given in Reference 3 (see Section A.43).

4.2.4 Interpretation and features of the Kalman filter

Insight into the Kalman filter functioning can be easily obtained by considering the continuous-time Kalman filter gain eq. (4.27)

Let K for the scalar system be given as

$$K = c\left(\frac{\sigma_x^2}{\sigma_v^2}\right)$$

Here, $H = c$, $P = \sigma_{\underline{x}}^2$ and $R = \sigma_v^2$. The state eq. (4.25) simplifies to

$$\dot{\hat{x}}(t) = a\hat{x}(t) + K[z(t) - c\hat{x}(t)]$$

If the measurement uncertainty is large, represented by σ_v^2, then the Kalman gain will be low for fixed value of $\sigma_{\underline{x}}^2$. Then the filter does not put more emphasis on measurement and the state estimate will be based only on the previous estimate. Similarly, if $\sigma_{\underline{x}}^2$ is low, then K will be low as well. This is intuitively appealing for the state update. If $\sigma_{\underline{x}}^2$ is large, then K will be large and more emphasis will be put on the measurement, assuming relatively low σ_v^2. Hence, based on the relative value of the scalar ratio $\sigma_{\underline{x}}^2/\sigma_v^2$, the Kalman gain adapts to the value, which is intuitively appealing. This is just achieved by the optimisation of the cost function, without invoking this appealing feature in the first place.

For the discrete-time filter, we have the Kalman gain as

$$K = \tilde{P}H^T(H\tilde{P}H^T + R)^{-1}$$

For the scalar case, we have

$$K = \sigma_{\underline{x}}^2 c\left(c^2\sigma_{\underline{x}}^2 + \sigma_v^2\right)^{-1} = \frac{\sigma_{\underline{x}}^2 c}{\left(c^2\sigma_{\underline{x}}^2 + \sigma_v^2\right)}$$

We presume that $c = 1$, then

$$K = \frac{\sigma_{\underline{x}}^2}{\left(\sigma_{\underline{x}}^2 + \sigma_v^2\right)}$$

For constant process noise variance, increase in σ_v^2 signifies decrease in K and hence the filter puts more weightage on the previous state estimate and less on the new measurement. Similarly, for constant σ_v^2, increase in $\sigma_{\underline{x}}^2$ will cause K to increase, and more emphasis will be put on the measurement. Thus, in KF, the filter shifts its emphasis based on the information content/uncertainties in the measurement data. Ironically, this mechanisation points to a major limitation of the Kalman filter, i.e., filter tuning of the parameters Q and R. However, it can be seen from the foregoing, that it is only the ratio of Q and R that matters. For matrices, the ratio will be in the form of individual norms of matrices Q and R (see Section A.33) or any other measure can be used. The filter tuning aspect is addressed in Section 4.5 of this chapter.

We need to evaluate the performance of the filter to see if proper tuning has been achieved or not and whether the estimates make sense. Two possibilities exist:

1 to check the whiteness of the measurement residuals (see Chapters 2 and 6, and Section A.1);
2 to see if the computed covariances match the theoretical covariances obtained from the covariance equations of the filter (eqs (4.20) and (4.22)).

Test 1 signifies that as the measurement residual is white, no information is left out to be utilised in the filter. The white process is an unpredictable process. Test 2 signifies that the computed covariances from the data match the filter predictions (theoretical estimates of the covariances), and hence proper tuning has been achieved. These tests are valid for all types of Kalman filter versions, be it extended Kalman filter or factorisation filtering algorithm.

Some features of the Kalman filter are given below:

a It is a finite dimensional linear filter.
b It can be considered as a system driven by residuals and producing the state estimates.

c It obtains unbiased (by the design, see eq. (4.11)) and minimum variance (see eq. (4.14)) estimates of the state.
d It obtains theoretical estimates of the state error covariance at each instant of time.
e It is a recursive filter and incorporates the data as they are received. Uniform sampling of data is not a great need for this filter.
f It can be easily adapted to real-time estimation of states. The only restriction is the computation of P and K, which would be time consuming. Often parallel Kalman filtering equations can be used. For linear systems, Kalman gain K and covariances can be pre-computed as can be seen from eqs (4.18), (4.20) and (4.22), since these computations do not depend upon the measurement data. This will simplify the on-line implementation.
g It can be extended to nonlinear systems.
h With this modification, it can be used for joint state and parameter estimation.
i It is also applicable to continuous time, time varying linear and nonlinear systems.
j It can be modified to handle correlated process noise [2].
k It has intuitively appealing features, which using a continuous-time Kalman filter can be easily explained.

4.3 Kalman UD factorisation filtering algorithm

The Kalman filter solution could diverge due to one or more of the following reasons [8]:

(i) modelling errors (due to nonlinear system);
(ii) wrong *a priori* statistics (P, Q, R);
(iii) finite word length implementation of the filter.

For handling (i) a properly tuned extended Kalman filter should be used. If feasible, accurate mathematical models of the system should be used, since the Kalman filter utilises the mathematical model of the underlying system itself. For handling (ii) proper tuning should be done. Reliable estimates of Q and R or 'ratio' of Q and R should be determined. Adaptive tuning methods should be used. For (iii) factorisation-filtering methods should be used, or the filter should be implemented on a computer with large word length.

In the Kalman filter, eq. (4.22) is especially ill-conditioned. Due to round off errors in computation and their propagation, the covariance matrix P could be rendered non-positive definite, whereas theoretically it should be at least semi-positive definite. In addition, matrix P should be symmetric, but during computation it could lose this property. All these will lead the Kalman filter to diverge, meaning thereby that the residuals will grow in size and the filter estimate will not converge in the sense of mean square to the true state. This is not the problem with the Kalman filter but its implementation on a finite word length. These effects are circumvented or greatly reduced by implementing a Kalman filter in its factorised form. These algorithms do not process covariance matrix P in its original form, but process its square root. Such factorisation implicitly preserves the symmetry and ensures the non-negativity of the covariance matrix P. There are several such algorithms available in the literature. One such algorithm, which is widely used, called the UD factorisation filtering algorithm is given here. Here, U and D are matrix factors of the covariance matrix P of the Kalman filter, where U is a unit upper triangular matrix and D is a diagonal matrix.

The UD factorisation filter has the following merits [8]:

a It is numerically reliable, accurate and stable.
b It is a square root type algorithm, but does not involve square rooting operations.
c The algorithm is most efficiently and simply mechanised by processing vector measurements (observables), one component at a time.
d For linear systems, the UD filter (UDF) is algebraically equivalent to the Kalman filter.

The major advantage from UD comes from the fact that the square root type algorithms process square roots of the covariance matrices and hence, they essentially use half the word length normally required by the conventional Kalman filters. In the UD filter, the covariance update formulae of the conventional KF and the estimation recursion are reformulated, so that the covariance matrix does not appear explicitly. Specifically, we use recursions for U and D factors of covariance matrix $P = UDU^T$. Computing and updating with triangular matrices involve fewer arithmetic operations and thus greatly reduce the problem of round off errors, which might cause ill-conditioning and subsequent divergence of the algorithm, especially if the filter is implemented on a finite word length machine. This is more so for real-time implementation on board computers where the word length could be small, e.g., 16 or 32 bit.

The filter algorithm for linear system is given in two parts.

Time propagation
We have for the covariance update

$$\tilde{P}(k+1|k) = \phi \hat{P}(k)\phi^T + GQG^T \tag{4.28}$$

Given $\hat{P} = \hat{U}\hat{D}\hat{U}^T$ and Q as the process noise covariance matrix, the time update factors $\tilde{U}$ and $\tilde{D}$ are obtained through a modified Gram-Schmidt orthogonalisation process [8].

We define $V = [\phi\hat{U}|G]$ and $\bar{D} = \mathrm{diag}[\hat{D}, Q]$, and $V^T = [v_1, v_2, \ldots, v_n]$. P is reformulated as $\tilde{P} = \tilde{V}\tilde{D}\tilde{V}^T$. The U and D factors of $\tilde{V}\tilde{D}\tilde{V}^T$ may be computed as described below.

For $j = 1, \ldots, n$ the following equations are recursively evaluated.

$$\tilde{D}_j = \langle v_j, v_j \rangle_{\bar{D}} \tag{4.29}$$

$$\tilde{U}_{ij} = \left(\frac{1}{\tilde{D}_j}\right) \langle v_i, v_j \rangle_{\bar{D}} \qquad i = 1, \ldots, j-1 \tag{4.30}$$

$$v_i = v_i - \tilde{U}_{ij} v_j \tag{4.31}$$

Here, $\langle v_i, v_j \rangle_{\bar{D}} = v_i^T \bar{D} v_j$ is the weighted inner product between v_i and v_j.

Therefore, the time propagation algorithm directly and efficiently produces the required U, D factors, taking the effect of previous U, D factors and the process noise. Thus, it also preserves the symmetry of the (original) P matrix.

Measurement update

The measurement update in Kalman filtering combines *a priori* estimates $\tilde{x}$ and error covariance $\tilde{P}$ with a scalar observation $z = cx + v$ to construct an updated estimate and covariance given as

$$
\begin{aligned}
K &= \frac{\tilde{P}c^T}{s} \\
\hat{x} &= \tilde{x} + K(z - c\tilde{x}) \\
s &= c\hat{P}c^T + R \\
\hat{P} &= \tilde{P} - Kc\tilde{P}
\end{aligned}
\tag{4.32}
$$

Here, $\tilde{P} = \tilde{U}\tilde{D}\tilde{U}^T$; c is the measurement matrix, R is the measurement noise covariance, and z is the vector of noisy measurements.

Kalman gain K, and updated covariance factors $\hat{U}$ and $\hat{D}$ can be obtained from the following equations [8]:

$$
\begin{aligned}
g &= \tilde{U}^T c^T; \qquad g^T = (g_1, \ldots, g_n) \\
w &= \tilde{D}g; \\
\hat{d}_1 &= \frac{\tilde{d}_1 R}{s_1}, \qquad s_1 = R + w_1 g_1
\end{aligned}
\tag{4.33}
$$

For $j = 2, \ldots, n$ the following equations are evaluated:

$$
\begin{aligned}
s_j &= s_{j-1} + w_j g_j \\
\hat{d}_j &= \frac{\tilde{d}_j s_{j-1}}{s_j} \\
\hat{u}_j &= \tilde{d}_j + \lambda_j K_j, \qquad \lambda_j = -\frac{g_j}{s_{j-1}} \\
K_{j+1} &= K_j + w_j \tilde{u}_j; \qquad \tilde{U} = [\tilde{u}_1, \ldots, \tilde{u}_n]
\end{aligned}
$$

The Kalman gain is given by

$$
K = \frac{K_{n+1}}{s_n} \tag{4.34}
$$

Here, $\tilde{d}$ is the predicted diagonal element, and $\hat{d}_j$ is the updated diagonal element of the D matrix.

The time propagation and measurement update of the state vector is similar to KF and hence, not repeated here. We also note that the measurement update/data processing can be done sequentially, meaning thereby that each observable can be processed in turn, and state estimate updated. This avoids the matrix inversion in the Kalman gain formulation. Several nice properties and theoretical development of UD factorisation KF are given in Reference 8.

4.3.1.1 Example 4.1

Simulate data of a target moving with constant acceleration and acted on by an uncorrelated noise, which perturbs the constant acceleration motion. Add measurement noise with standard deviation of one to generate measurements of position and velocity. Estimate the states of the system using a UD factorisation based linear Kalman filter (UDKF) and the noisy position and velocity measurements. Evaluate the filter performance using the standard procedure.

4.3.1.2 Solution

The target data (position and velocity) is generated using the state and measurement eqs (4.1) and (4.2) by adding random process noise with $\sigma = 0.001$ and measurement noise with $\sigma = 1$.

The state vector, x consists of target position (x_p), velocity (x_v) and acceleration (x_a), $x = [x_p, x_v, x_a]$.

For this case

$$\text{the state transition matrix} \quad \phi = \begin{bmatrix} 1 & \Delta t & \Delta t^2/2 \\ 0 & 1 & \Delta t \\ 0 & 0 & 1 \end{bmatrix}$$

$$\text{process noise matrix} \quad G = \begin{bmatrix} \Delta t^2/2 \\ \Delta t \\ 1 \end{bmatrix}$$

$$\text{observation matrix} \quad H = \begin{bmatrix} 1 & 0 \\ 0 & 1 \end{bmatrix}$$

Using the program **Genmeas.m** in the folder **Ch4UDex1**, both the position and velocity measurements are generated for a duration of 100 s. The sampling time of $\Delta t = 0.25$ s is chosen for simulation.

The initial condition of the states used for the simulation: $x_0 = [200, 10, 0.5]$.

For use in UDKF, the state model is formulated with the three states and the measurement model is formulated using noisy measurements of position and velocity. The state estimation programs are contained in the folder **Ch4UDex1**. The initial conditions for the filter are chosen as $\hat{x}_0 = [190.0, 8.0, 0.4]$. The initial state error covariance is chosen to reflect the difference in the true x_0 and $\hat{x}_0$.

Figure 4.1 shows the estimated position and velocity measurements compared with the measured values. The figure also shows the position and velocity innovations along with their theoretical bounds ($2\sqrt{S_{ii}(k)}$, S = innovation covariance), the autocorrelation function (ACR) of residuals with their bounds ($\pm 1.96/\sqrt{N}$, N = number of data points, $N = 400$) and the position and velocity state errors along

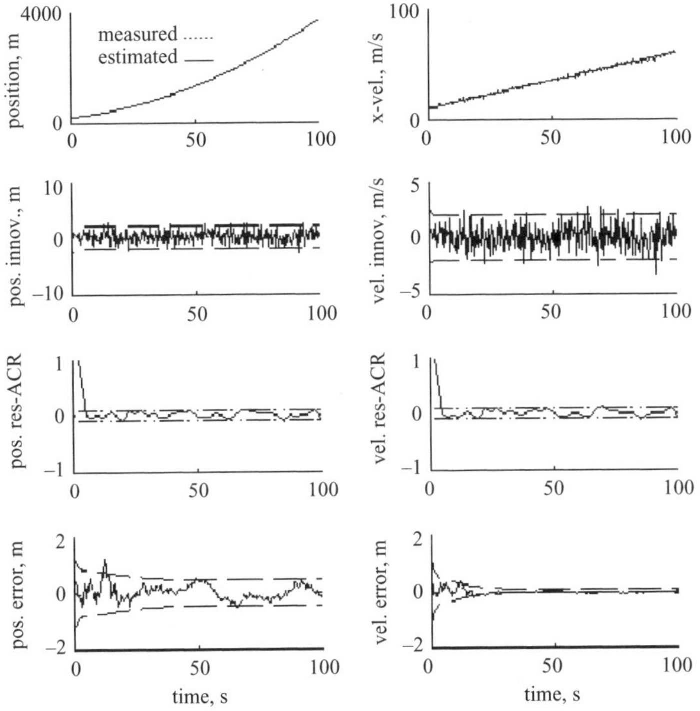

Figure 4.1 *Measurements, innovations, autocorrelation of residuals and state errors (Example 4.1). (Note: for the ACR plot the X-axis (time axis) is actually equivalent to the number of lags, e.g., 10 s = 40 lags × 0.25 s. Similar clarification holds for related examples in the book.)*

with the $\pm 2\sqrt{P_{ii}(k)}$ bounds. It is clear that the filter performance is very good as is evident from the figure where all the estimated quantities fall within their theoretical bounds. For this example, the residual mean = [0.0656 and − 0.0358] and PFE (percentage fit error) of the predicted measurements w.r.t. the true measurements = [0.0310, 0.4009].

4.4 Extended Kalman filtering

Real-life dynamical systems are nonlinear and estimation of the states of such systems is often required. The nonlinear system can be expressed with the following set of equations (see Chapter 3):

$$\dot{x}(t) = f[x(t), u(t), \Theta] \tag{4.35}$$

$$y(t) = h[x(t), u(t), \Theta] \tag{4.36}$$

$$z(k) = y(k) + v(k) \tag{4.37}$$

Here, f and h are general nonlinear vector valued functions, and Θ is the vector of unknown parameters given by

$$\Theta = [x_0, b_u, b_y, \beta] \tag{4.38}$$

Here, x_0 represents values of the state variables at time $t = 0$; b_u represents the bias in control inputs (nuisance parameters); b_y represents the bias in model response y (nuisance parameters); and β represents parameters in the mathematical model that defines the system characteristics.

Comparing eqs (4.35 and 4.36) and eqs (4.1) and (4.2), we see that the linear KF recursions eqs (4.17–4.22) cannot be directly used for state estimation of the nonlinear systems. One can, however, linearise the nonlinear functions f and h and then apply the KF recursions with proper modification to these linearised problems. The linearisation of f and h could be around the pre-supposed nominal states, e.g., in orbit estimation problem, the nominal trajectory could be the circular orbit of the satellite to be launched. When the satellite is launched, it will acquire a certain orbit, which will be the actual orbit but affected by noisy measurements. Therefore, there will be three trajectories: nominal, estimated and the true trajectory. Often, the extended Kalman filter is preferred since the linearisation will be around previous/current best state estimates, which are more likely to represent the truth, rather than the linearisation around the nominal states, leading to linearised KF (LKF). Hence, in this section, an extended Kalman filter is considered which has application to aircraft parameter estimation as well. In EKF, the estimated state would converge to the true states for relatively large initial state errors, whereas this may not be so true for the linearised Kalman filter.

An extended Kalman filter is a sub-optimal solution to a nonlinear filtering problem. The nonlinear functions f and h in eqs (4.35) and (4.36) are linearised about each new estimated/filtered state trajectory as soon as it becomes available. Simultaneous estimation of states and parameters is achieved by augmenting the state vector with unknown parameters (as additional states) and using the filtering algorithm with the augmented nonlinear model [2, 3, 5].

The new augmented state vector is

$$x_a^T = \left[x^T \Theta^T\right] \tag{4.39}$$

$$\dot{x} = \begin{bmatrix} f(x_a, u, t) \\ 0 \end{bmatrix} + \begin{bmatrix} G \\ 0 \end{bmatrix} w(t) \tag{4.40}$$

$$\dot{x} = f_a(x_a, u, t) + G_a w(t) \tag{4.41}$$

$$y(t) = h_a(x_a, u, t) \tag{4.42}$$

$$z_m(k) = y(k) + u(k), \qquad k = 1, \ldots, N \tag{4.43}$$

Here

$$f_a^T(t) = \left[f^T 0^T\right]; \quad G_a^T = \left[G^T 0^T\right] \tag{4.44}$$

The estimation algorithm is obtained by linearising eqs (4.35) and (4.36) around the prior/current best estimate of the state at each time and then applying the KF algorithm to the linearised model. The linearised system matrices are defined as

$$A(k) = \left.\frac{\delta f_a}{\delta x_a}\right|_{x_a=\hat{x}_a(k),\ u=u(k)} \tag{4.45}$$

$$H(k) = \left.\frac{\delta h_a}{\delta x_a}\right|_{x_a=\tilde{x}_a(k),\ u=u(k)} \tag{4.46}$$

and the state transition matrix is given by

$$\phi(k) = \exp\left[-A(k)\Delta t\right] \quad \text{where } \Delta t = t_{k+1} - t_k \tag{4.47}$$

For the sake of clarity and completeness, the filtering algorithm is given in two parts: (i) time propagation, and (ii) measurement update [2–4]. In the above equations, we notice the time-varying nature of A, H and ϕ, since they are evaluated at the current state estimate, which varies with time k.

4.4.1.1 Time propagation

The current estimate is used to predict the next state, so that the states are propagated from thc present state to the next time instant.

The predicted state is given by

$$\tilde{x}_a\,(k+1) = \hat{x}_a(k) + \int_{t_k}^{t_{k+1}} f_a[\hat{x}_a(t), u(k), t]\,dt \tag{4.48}$$

In the absence of knowledge of process noise, eq. (4.48) gives the predicted estimate of the state based on the initial/current estimate. The covariance matrix for state error (here state is x_a) propagates from instant k to $k+1$ as

$$\tilde{P}(k+1) = \phi\,(k)\,\hat{P}\,(k)\,\phi^T\,(k) + G_a\,(k)\,QG_a^T\,(k) \tag{4.49}$$

Here, $\tilde{P}(k+1)$ is the predicted covariance matrix for the instant $k+1$, G_a is the process noise related coefficient matrix, and Q is the process noise covariance matrix.

4.4.1.2 Measurement update

The extended Kalman filter updates the predicted estimates by incorporating the measurements as and when they become available as follows:

$$\hat{x}_a(k+1) = \tilde{x}_a(k+1) + K(k+1)\,\{z_m(k+1) - h_a[\tilde{x}_a(k+1), u(k+1), t]\} \tag{4.50}$$

Here, K is the Kalman gain matrix.

The covariance matrix is updated using the Kalman gain and the linearised measurement matrix from the predicted covariance matrix $\tilde{P}(k+1)$.

The Kalman gain expression is given as

$$K(k+1) = \tilde{P}(k+1)H^T(k+1)[H(k+1)\tilde{P}(k+1)H^T(k+1) + R]^{-1} \tag{4.51}$$

A posteriori covariance matrix expression is given as

$$\hat{P}(k+1) = [I - K(k+1)H(k+1)]\,\tilde{P}(k+1) \tag{4.52}$$

The EKF computationally is more complex than simple KF. Major cost is due to the linearisations at every instant of time. For moderately nonlinear functions, the EKF would give reasonably accurate state estimates. If nonlinearities were severe, then repeated linearisations around newly estimated states, especially during measurement update, can be made. This yields so-called iterative EKF. In addition, a procedure called forward-backward filtering can be used. In this procedure, the EKF is used, in the first pass, as forward filtering. Then the EKF is run backward from the final point t_f to the initial time t_0, utilising the same measurements. This process refines the estimates, but then it cannot be used in real-time applications.

The UD factorisation filter can also be conveniently used in the EKF mode, since eqs (4.51) and (4.52) can be put in the factorisation form and processed.

We note from eq. (4.48) that state (estimate) propagation is achieved by integration of the nonlinear function f_a between times t_k and t_{k+1}, thereby maintaining the effect of nonlinearity of f. Also, in eq. (4.50), nonlinear function h_a is used for predicting the measurements. These two features essentially give credence to the filter and hence extended KF.

The EKF can be used for parameter estimation of linear/nonlinear systems. However, since the covariance matrices are approximations, computed based on linearised nonlinear functions f and h, there is no guarantee of stability and performance, prior to experimental data analysis. However, in practice, the approach seems to work well if linearisation is accurate and proper tuning of the filter is achieved. Although EKF is a nonlinear filtering solution, the modelling errors could prevail and these might degrade the performance of the algorithm. To have good matching of the states proper tuning using the Q matrix should be done. The approach of model error discussed in Chapter 8 could minimise the effect of modelling errors on state estimation. One major demerit of EKF is that it is computationally demanding and not easily amenable to parallelisation of the algorithm, since the computations of the covariances are coupled with the filter computations. Often EKF/EUDF algorithms are used in conjunction with regression (LS) techniques leading to the so-called two-step procedure. This is discussed in Chapter 7.

4.4.1.3 Example 4.2

Simulate data of a second order system with the following state and measurement matrices:

$$\begin{bmatrix}\dot{x}_1\\ \dot{x}_2\end{bmatrix} = \begin{bmatrix}a_{11} & a_{22}\\ a_{33} & a_{44}\end{bmatrix}\begin{bmatrix}x_1\\ x_2\end{bmatrix} + \begin{bmatrix}b_1\\ b_2\end{bmatrix}u = \begin{bmatrix}0.06 & -2.0\\ 2.8 & 0.08\end{bmatrix}\begin{bmatrix}x_1\\ x_2\end{bmatrix} + \begin{bmatrix}-0.6\\ 1.5\end{bmatrix}u$$

$$\begin{bmatrix}z_1\\ z_2\end{bmatrix} = \begin{bmatrix}1 & 0\\ 0 & 1\end{bmatrix}\begin{bmatrix}x_1\\ x_2\end{bmatrix} + v$$

Use a doublet signal as input to the dynamic system (with sampling interval of 0.05 s). Use UD factorisation based EKF (EUDF) to estimate the states and parameters of the system using measurements of z_1 and z_2. Study the effect of measurement noise on the estimation results. Evaluate the performance of the filter using the standard procedure.

4.4.1.4 Solution

Simulated data of 10 s duration is generated using the above equations (folder **Ch4EUDFex2sim**) with a sampling time of 0.05 s. State noise with $\sigma = 0.001$ is added to generate the states. Measurements have SNR = 10. For state and parameter estimation, the state model is formulated with the two states x_1, x_2 and the six parameters of the A and B matrices in the above equations as augmented states in EUDF (eq. (4.39)). This results in a state model with eight states – two pertaining to the states x_1 and x_2 and six pertaining to the parameters a_{11}, a_{12}, a_{21}, a_{22}, b_1, b_2. The EUDF parameter estimation programs are contained in the folder **Ch4EUDFex2**. The initial states/parameters for the Kalman filter are assumed 50 per cent away from their true values. The initial state-error covariance matrix is chosen to reflect this uncertainty. The values of the process and measurement noise covariances are kept fixed at the values used in the simulation.

Figure 4.2(a) shows the estimated measurements compared with the noisy measurements. The figure also shows that the innovations pertaining to the two measurements fall within their theoretical bounds and that the autocorrelation of the residuals falls within their theoretical bounds as well. Figure 4.2(b) shows the convergence of the parameters. It is clear that even in the presence of noise in the

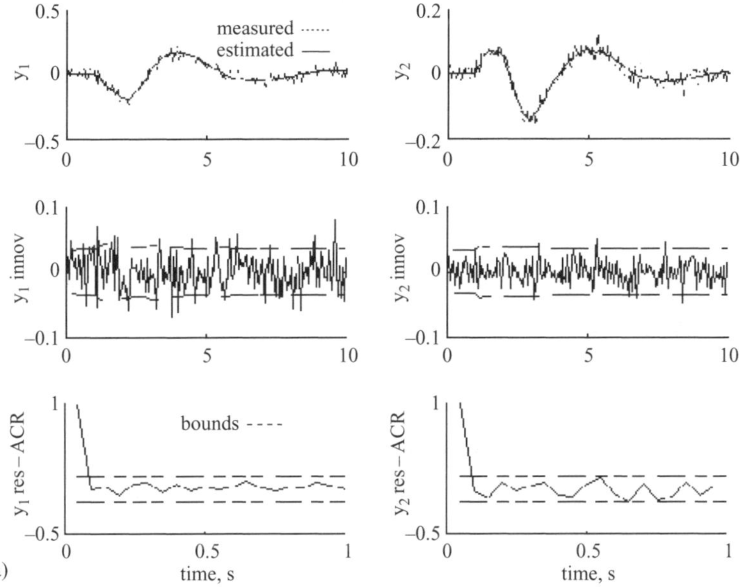

(a)

Figure 4.2 (a) Measurements, innovations and autocorrelation of residuals (Example 4.2)

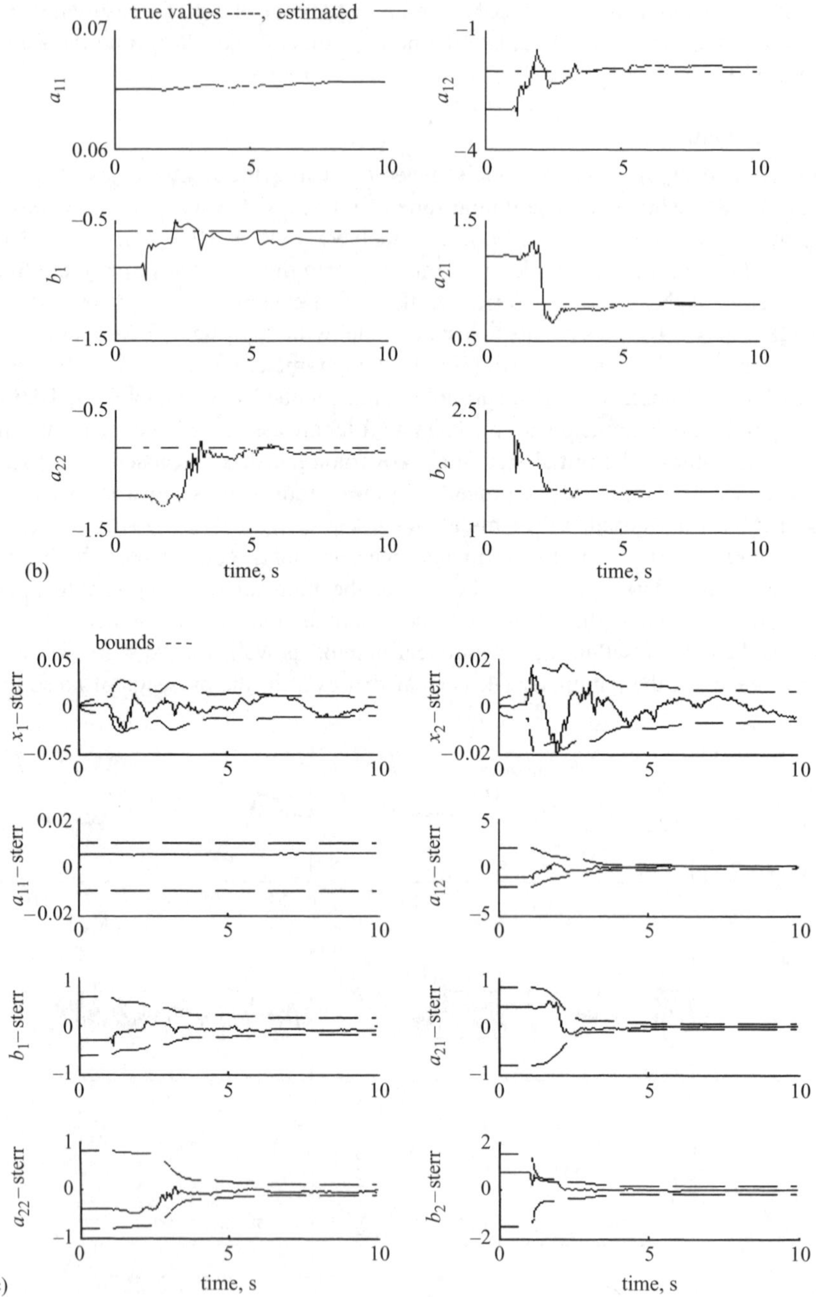

Figure 4.2 Continued. (b) Convergence of parameter estimates (Example 4.2); (c) state errors with bounds (Example 4.2)

Table 4.1 Parameter estimates (EUDF) (Example 4.2)

Parameters	True	Estimated (no noise)	Estimated (SNR = 10)
a_{11}	0.06	0.0662 (0.0149)	0.0656 (0.0050)
a_{12}	−2.0	−2.0003 (0.0450)	−1.9057 (0.0956)
a_{21}	0.8	0.8005 (0.0202)	0.8029 (0.0892)
a_{22}	−0.8	−0.8038 (0.0340)	−0.8431 (0.0345)
b_1	−0.6	−0.5986 (0.0353)	−0.6766 (0.0548)
b_2	1.5	1.5078 (0.0356)	1.5047 (0.0734)
PEEN (%)	–	0.3833	4.5952

data, the parameters converge very close to their true values. Figure 4.2(c) shows that the state errors are well within the theoretical bounds. Table 4.1 lists the estimated parameters along with their standard deviations. The standard deviations are given by the square root of the diagonal elements of the estimation error covariance matrix, $\sigma = \sqrt{P_{ii}(k)}$. The estimated parameters and the standard deviations in Table 4.1 are those at the last data point (200 for this case). The parameter estimates are very close to the true values when there is no measurement noise in the data. In this case, a very small value of R is used in the filter computation. However, it should be noted that process noise is present in the data. Some of the estimated parameters show slight deviations from the true values when there is noise in the data. However, it is clear that the PEEN is less than 5 per cent, which is acceptable when there is noise in the data.

4.4.1.5 Example 4.3

Use the simulated short period data of a light transport aircraft with process noise to estimate the non-dimensional longitudinal parameters of the aircraft using Kalman filtering method. Use the 4DOF longitudinal body axis model for estimation. The relevant mass, moment of inertia and other aircraft geometry related parameters are provided in Example 3.3.

4.4.1.6 Solution

Using the equations given in Example 3.3, the data are generated with a sampling interval of 0.03 s by giving a doublet input to the elevator. Random noise with $\sigma = 0.001$ is added to the states u, w, q, θ. The states with additive process noise are used to generate measurements (data set 1) of u, w, q, θ, a_x, a_z, $\dot{q}$. Random noise is added to these measurements to generate noisy data with SNR = 10 (data set 2). Both the sets of data are used for parameter estimation using UDKF. For estimating the parameters using UDKF, the parameters are modelled as augmented states in the state model (eq. (4.39)). For this case there are 4 states and 11 parameters so that the state model has 15 states. Seven measurements u, w, q, θ, a_x, a_z, $\dot{q}$ are used and all the 11 parameters are estimated using the programs in the folder **Ch4EUDFex3**. The process and measurement noise covariances are kept fixed at the values used

Table 4.2 Estimated parameters of a light transport aircraft (Example 4.3)

Parameter	True values	Estimated (no noise)	Estimated (SNR = 10)
C_{x0}	−0.0540	−0.05680 (0.0039)	−0.0592 (0.0085)
$C_{x\alpha}$	0.2330	0.2529 (0.0235)	0.2543 (0.0262)
$C_{x\alpha^2}$	3.6089	3.5751 (0.0619)	3.7058 (0.1131)
C_{z0}	−0.1200	−0.1206 (0.0046)	−0.1249 (0.0166)
$C_{z\alpha}$	−5.6800	−5.6759 (0.0196)	−5.7247 (0.0783)
$C_{z\delta}$	−0.4070	−0.4067 (0.0108)	−0.5049 (0.0477)
C_{m0}	0.0550	0.0581 (0.0049)	0.0576 (0.0081)
$C_{m\alpha}$	−0.7290	−0.7466 (0.0334)	−0.7092 (0.0433)
$C_{m\alpha^2}$	−1.7150	−1.6935 (0.0831)	−1.7843 (0.1097)
C_{mq}	−16.3	−16.2660 (0.3857)	−15.3075 (0.7980)
$C_{m\delta}$	−1.9400	−1.9397 (0.0110)	−1.8873 (0.0450)
PEEN (%)	–	0.4424	5.6329

in simulation of the data. The initial states and parameters for the Kalman filter are assumed 10 per cent away from their true values. The initial state-error covariance matrix is chosen to reflect this uncertainty.

The estimated values of the parameters are compared with the true values (aerodynamic derivatives) in Table 4.2. The table also shows the PEEN. The estimates are fairly close to the true values even when there is noise in the data.

Figure 4.3(a) shows the estimated measurements compared with the noisy measurements. The convergence of the pitching moment related derivatives: $C_{m\alpha}$, $C_{m\alpha^2}$, C_{mq}, $C_{m\delta}$ is shown in Fig. 4.3(b). It is clear that even in the presence of noise in the data, the parameters converge close to their true values. Some deviation is observed for C_{mq} estimate. Figure 4.3(c) shows that the state errors for the pitching moment parameters are well within their theoretical bounds.

4.5 Adaptive methods for process noise

We have seen in previous sections that the Kalman filter requires tuning for obtaining optimal solutions. The process noise covariance matrix Q and measurement noise covariance matrix govern this tuning process. In practice, the system models and the noise statistics are known with some uncertainty. This could lead to degradation in the performance of the filter. Thus, there is a need to estimate these uncertain parameters adaptively, leading to adaptive estimation algorithms [2]. The adaptive techniques generally are complex and need more computations. As far as the uncertainties in the basic model of the system are concerned, there are several approaches for model compensation and estimation [2]. One relatively simple and practical approach is based on the principle of model error discussed in Chapter 8. The estimation algorithm will determine optimal estimation of model error of the so-called (model) discrepancy time history. However, this method as such does not handle process noise. The point is that we have, say, data from a nonlinear system, the accurate model for which is not known. Then, since KF needs the system model, we end up using an approximate known model. This will cause divergence in state estimation. We can use the EKF to

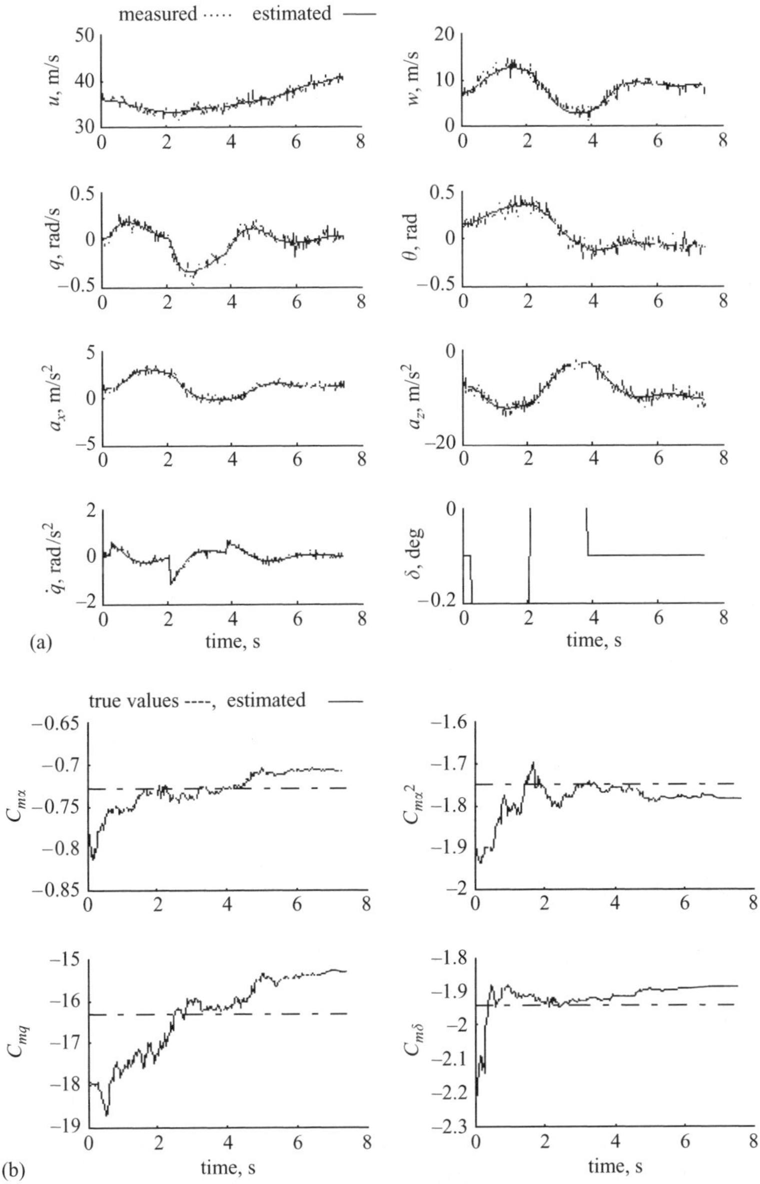

Figure 4.3 *(a) Time history match (Example 4.3); (b) parameter convergence – pitching moment derivatives (Example 4.3)*

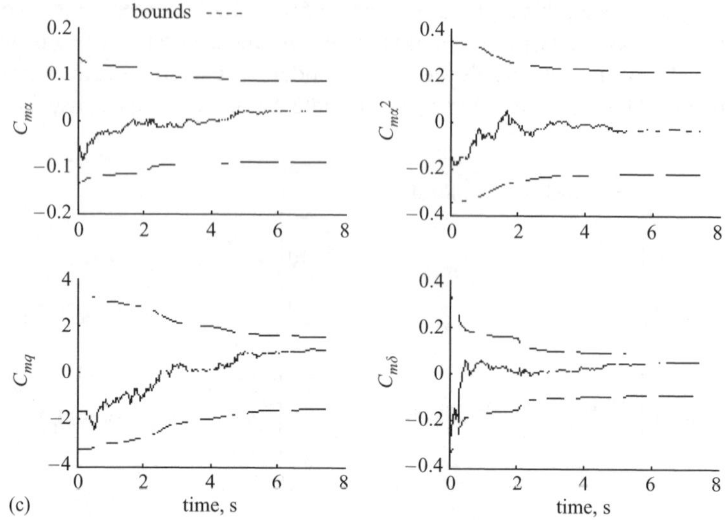

Figure 4.3 Continued. (c) State error – pitching moment derivatives (Example 4.3)

estimate uncertain/unknown parameters of the postulated nonlinear state model, as we discussed in Section 4.4.

In the present section, we discuss some approaches that can be used as adaptive filtering methods. In general, the measurement noise statistics can be obtained from the statistical characteristics of the sensors. In addition, analysis of the data from previous similar experiments for sensor (noise) characterisation can be used. However, it will be difficult to obtain *a priori* reliable information on the process noise covariance. Since the process noise covariance used in KF accounts not only for process noise affecting the states but also any model inaccuracies, it requires special attention. Here we address mainly the problem of determination/adaptation of Q.

4.5.1 Heuristic method

The method is based on the observation that the Kalman filter performance depends only on the relative strength of the process and measurement noise characteristics and not on their absolute values. This feature of the Kalman filter is of great practical value since it means that there is no need to make any absolute calibration of noise measurements, though this will greatly help in general. This aspect of the Kalman filter is used to develop a heuristic approach wherein the process noise covariance is assumed dependent on the measurement noise covariance. The implementation of the procedure involves an appropriate choice of proportionality factor/relationship. If the measurement noise covariance R is assumed constant throughout, then the process noise covariance can be approximated by

$$Q = q_1 R \tag{4.53}$$

The factor q_1 is chosen based on trial and error using measurement data collected from various experiments. One form of Q can be expressed as follows:

$$Q_k = \left[q_1\sqrt{R_k}\exp(-q_2k\Delta t)\right]^2; \qquad k = 1, 2, \ldots, N \tag{4.54}$$

The above form has been arrived at based on the engineering judgement and post-experiment data analysis. The values q_i are tuned to achieve the best performance. Thus, in this heuristic approach, the number of parameters to be tuned is reduced to only two. We see that as $k \to N$, eventually $\exp(-q_2k\Delta t) \to$ small, and hence Q is made less dominant. It is quite probable that for a given problem at hand, a different form of eq. (4.54) might bc suitable. The present form has been found to work well for target tracking applications [9].

This being a heuristic method, it requires substantial post-experimental data analysis for similar systems as the one in question, to arrive at factors q_1 and q_2. For each specific problem, one has to do this exercise. Often such data are available from previous experiments. In addition, most recent experiments can be used. Subsequently, the on-line application requires trivial effort and is computationally simple.

4.5.2 Optimal state estimate based method

The method [2] is based on the aim of adaptation to improve the state estimation performance. In the KF, the primary requirement is to have a good estimate of the filter gain even if the accuracy in estimating the process noise covariance is poor. In this method, the filter gain is obtained as a solution to the likelihood equation. Then the process noise covariance is obtained from the estimated gain. For on-line applications, a sub-optimal solution has been developed [2]. Under the assumption of steady state performance over the most recent N_w sample times (a sliding window of size N_w), a unique estimate of K and R_m can be obtained even if a unique estimate of Q cannot be obtained.

If matrix $\hat{S}$ is chosen as one of the parameters to be estimated, then an estimate of $\hat{S}$ is obtained using

$$\hat{S} = \frac{1}{N_w}\sum_{k=i-N_w+1}^{i} r(k)r^T(k) \tag{4.55}$$

Here, $r(k) = z(k) - H\hat{x}(k)$ are the residuals.

Using $\hat{R}_m$ and eqs (4.18), (4.20) and (4.22) and following the 'reverse' procedure, the estimates of $\hat{Q}$ can be obtained from the following relations [2]:

$$\begin{aligned}\tilde{P}_c &= K\hat{S}(H^T)^{-1}\\ \hat{P}_c &= (I - KH)\tilde{P}_c\\ \hat{Q} &= G^{-1}(\tilde{P}_c - \phi\hat{P}_c\phi^T)G^{-T}\end{aligned} \tag{4.56}$$

In the above equations '-1' represents pseudo inverse, and in case G is not invertible.

The basic tenet of the method is that for a small window length, the covariance of residuals is computed. One can then use eqs (4.18), (4.20) and (4.22) to do the reverse operations and compute the estimate of Q as shown earlier.

Although the method requires more computations, it could be made suitable for on-line applications.

4.5.3 Fuzzy logic based method

The method is based on the principle of covariance matching. Here, the estimates of residual covariance and the theoretical values as computed by the filter are compared and the covariance of process noise is tuned until the two agree [2]. Fuzzy logic (Section A.22) is then used to implement the covariance matching method [10] to arrive at an adaptive KF. This approach is suitable for on-line applications.

Since the residual is the difference between the actual measurements and measurement prediction based on the filter's internal model, a mismatch would indicate erroneous model formulation. This particular characteristic of the mismatch can be used to perform the required adaptation using the fuzzy logic rules. The advantages derived from the use of the fuzzy technique are the simplicity of the approach, the possibility of accommodating the heuristic knowledge about the phenomenon and the relaxation of some of the *a priori* assumptions on the process [10].

For a sufficiently accurate discretised and linearised model, the statistical properties of the innovation process are assumed similar to their theoretical estimates. Hence, the residuals (also called innovations) have the following covariance matrix (see eq. (4.20)):

$$\begin{aligned} S(k+1) &= H\tilde{P}H^T + R(k+1) \\ &= H(\phi\hat{P}(k)\phi^T + \hat{Q}(k))H^T + R(k+1) \end{aligned} \tag{4.57}$$

Here, $\hat{Q}(k) = \hat{\sigma}^2(k)\bar{Q}(k)$ where $\bar{Q}$ is some fixed known *a priori* covariance matrix. The current $Q(k)$ is altered at each instant based on: if the innovation is neither too near nor too far from zero, then leave the estimate of $Q(k)$ almost unchanged; if it is very near to zero, then reduce the estimate of $Q(k)$; if it is very far from zero, then increase the estimate of $Q(k)$. This is intuitively appealing since it achieves the covariance matching as discussed earlier.

The above adjustment mechanism can be implemented using fuzzy logic as follows. At each instant, the input variable (to fuzzy system) percentage is given by the parameter:

$$r_s(k+1) = \frac{r(k+1)}{\sqrt{s(k+1)}} \tag{4.58}$$

Here, $r(k+1)$ is the innovation component and $s(k+1)$ is the $(k+1)^{th}$ value of S. Then $r_s(k+1)$ gives the measure of actual amplitude of innovation compared to its theoretical assumed value.

The following If...Then...fuzzy rules can be used to generate output variables, based on linguistic description of the input variable $r_s(k+1)$ [10]:

If r_s is near zero, then ψ is near zero.
If r_s is small, then ψ is near one.
If r_s is medium, then ψ is a little larger than one.
If r_s is moderately large, then ψ is moderately larger than one.
If r_s is large, then ψ is large.

Subsequently, ψ is used to compute:

$$\hat{\sigma}^2(k+1) = \psi(k+1)\hat{\sigma}^2(k) \tag{4.59}$$

Here we assume some start-up value of the factor $\hat{\sigma}^2(k)$. This estimate will oscillate and it should be smoothed by using some smoothing techniques [2, 10].

Thus, the fuzzy rule based system has r_s as the input variables and ψ as the output variables. Thus, the input variables r_s define the Universe of discourse U_{rs} and the output variables ψ define Universe of discourse U_ψ.

The Universe spaces can be discretised into five (or even more) segments and thc fuzzy sets are defined by assigning triangular (or any other type of) membership functions to each of the discretised Universe. The membership functions of r_s and ψ can be denoted as m_r and m_ψ respectively. The membership function defines to what degree a member belongs to the fuzzy set. Representative fuzzy membership functions are : i) trapezoidal, ii) triangular, iii) Gaussian, or combination of these; one function is shown in Appendix A (p. 311). Finally, the adaptive estimation algorithm requires crisp values hence a defuzzification procedure based on 'centre of the area' method is used at each step (see Section A.22).

4.5.3.1 Example 4.4

Generate the target position data in the three axes of the Cartesian (XYZ) frame of reference using the state and measurement models having the general form of eqs (4.1) and (4.2). The state vector x consists of target position (p), velocity (v) and acceleration (a) in each of the axes, X, Y and Z. Use a linear Kalman filter to estimate the target states. Demonstrate the effects of the three adaptive process noise estimation methods on the target state estimation performance of the Kalman filter.

4.5.3.2 Solution

The state transition matrix and process noise matrix used for generating the simulated data in each of the three axes of the Cartesian (XYZ) frame of reference are the same as those in Example 4.1. However, in this case, the observation matrix has the following form: $H = \begin{bmatrix} 1 & 0 & 0 \end{bmatrix}$.

The state vector has nine states represented by $\boldsymbol{x} = [x_p, x_v, x_a, y_p, y_v, y_a, z_p, z_v, z_a]$. It is to be noted that (p, v, a) used as subscripts indicate the position, velocity and acceleration respectively. The acceleration increments over a sampling period are assumed discrete-time zero-mean white noise. Process noise with $\sigma = 0.001$ is added to generate the true state trajectories. A low value of process noise variance yields nearly a constant acceleration motion. The noise variances in each of the coordinate axes are assumed equal. Position measurements in all the three axes are generated by addition of measurement noise with $\sigma = 10$. Measurements are generated for a duration of 100 s with $\Delta t = 0.25$ s.

Initial condition of the states used for the simulation is

$$x_0 = [200\ \ 2\ \ 0\ \ 200\ \ 10\ \ 0.01\ \ 200\ \ -0.5\ \ 0.001]$$

Using known value of the measurement noise covariance ($R = 100$) in the Kalman filter, the three adaptive filtering methods: the heuristic method (HMQ), the optimal state estimation based method (OSQ) and the fuzzy logic based method (FLQ), outlined in the previous section, are used for adaptation of Q. Since the target motion is decoupled in the three axes, in the adaptive Kalman filters implemented in this example, the state model is formulated with the three states (p, v, a) in each of the three axes X, Y and Z. The noisy measurements of position are used for measurement update. The adaptive state estimation programs are contained in the folder **Ch4KFADex4**. The initial conditions for the filter are chosen as $\hat{x}_0 = [195.2, 1.006, 0, 195.2, 1.998, 0, 195.2, 0.6689, 0]$. The initial state error covariance is chosen to have a large value. The tuning factors used in the three filters for this case of simulated data are: $q_1 = 0.2$ for HMQ, window length $N = 10$ for OSQ and low = 0, high = 100 for FLQ.

Figure 4.4(a) shows the estimated position states X, Y and Z using all the three filters compared with the true states. The match indicates good performance of the

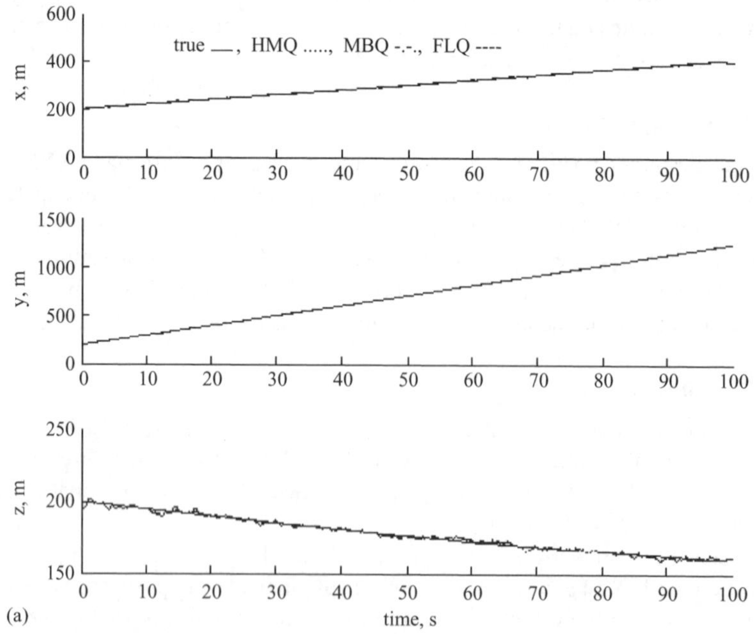

Figure 4.4 (a) Estimated position states compared with the true positions (Example 4.4)

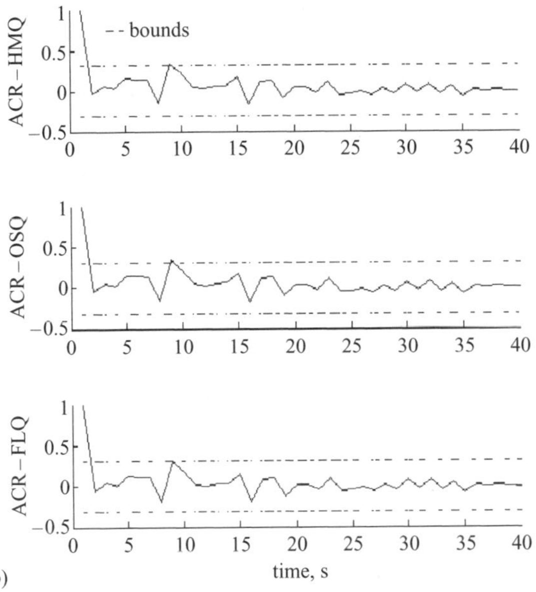

(b)

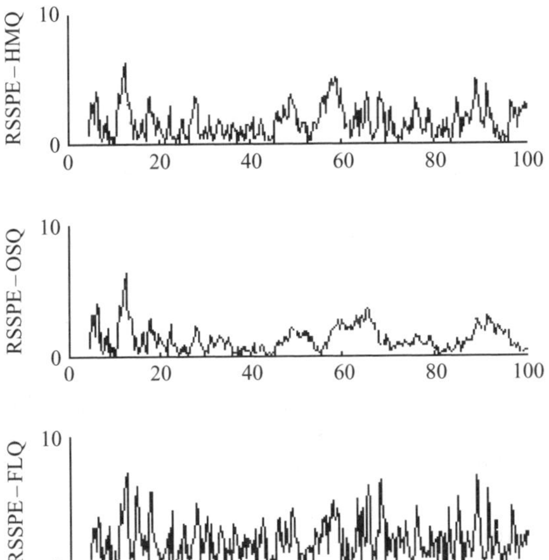

(c)

Figure 4.4 Continued. (b) Autocorrelation of innovations with bounds (Example 4.4); (c) root sum of squares position error (Example 4.4)

Table 4.3 Fit error (%) – simulated data (Example 4.4)

Q tuning method	Fit error (%)		
	X	Y	Z
HMQ	0.9256	0.3674	1.6038
OSQ	0.9749	0.3873	1.6895
FLQ	0.8460	0.3358	1.4659

three adaptive state estimation algorithms. Figure 4.4(b) shows the autocorrelation function with bounds. The autocorrelation plots indicate that the residuals satisfy the whiteness test and the values are well within the 95 per cent confidence limits as is clear from the bounds plotted in dotted lines. In Fig. 4.4(c) the root sum squares position error (RSSPE; see Sections A.38 and A.39) is plotted.

The RSSPE values are low, indicating good accuracy of the position estimates. The percentage fit errors (%FE) are given in Table 4.3. The values indicate that the performance of all the three adaptive filtering schemes is similar in terms of fit error. However, it can be seen from the table that the percentage fit errors obtained from the fuzzy logic based method are lower. When the measurement noise statistics are known fairly well, all the three methods of adaptive estimation give almost similar performances.

4.6 Sensor data fusion based on filtering algorithms

We see that eq. (4.2) defines the measurement model of the dynamical system. Thus, z represents a vector of 'm-observables', e.g., position, velocity, acceleration of a vehicle or angular orientation or temperature, pressure etc. in an industrial plant. The KF then uses these measurement variables and produces optimal states of the system. The fact that z as such is a combination of several observables (and their numerical values) the KF itself does what is called sensor data fusion. This fusion is called data level fusion. This is viable and practical if the measurement sensors are commensurate, such that the measurements can be combined in z. If the sensors are of dissimilar types, then the data level fusion may not be feasible. In addition, the data might be coming from different locations and communication channels could get overloaded. In such cases, it might be desirable to process the data at each sensor node that generates the data. The processed data then can be sent to a central station/node, where the state-vector level fusion can be easily accomplished. The state-vector level fusion here means that the state estimates arriving from different nodes can be fused using some fusion equation/algorithm to get the fused state estimates. Such aspects fall in the general discipline of multisensor data fusion (MSDF), which generalises to multisource multisensor information fusion.

Although MSDF aspects directly do not belong to the parameter estimation problem, they are included here for the following reasons:

- KF, *per se*, is a kind of data fusion algorithm.
- Many estimation principles and methods discussed in the present book can be used in MSDF discipline for state estimation, system identification, feature extraction, image processing and related studies.
- At a basic level, the processing operation in MSDF is dominated by numerical procedures, which are similar to those used in linear estimation and statistical theory of which parameter estimation can be considered as a specialised branch.

MSDF is defined as a process of combining information from multiple sensors/sources to produce the most appropriate and unified data about an object [11]. The object could be an entity, activity or event. As a technology, the MSDF integrates many disciplines: communication and decision theory, uncertainty management, numerical methods, optimisation and control theory and artificial intelligence. The applications of MSDF are varied: automated target recognition, autonomous vehicles, remote sensing, manufacturing processes, robotics, medical and environmental systems. In all these systems, data could arise from multiple sources/sensors located at different positions to provide redundancy and/or to extend the temporal or spatial coverage of the object. The data after fusion are supposed to provide improved and more reliable estimates of the state of the object and more specific inferences than could be obtained using a single sensor.

Theoretically, the measurement/data level fusion obtains optimal states with less uncertainty. But this approach may not be practicable for certain applications, since the volume of data to be transmitted to the fusion centre could exceed the capacity of existing data links among the individual channels/stations/nodes. In such cases, the state-vector fusion is preferable. Each node utilises an estimator to extract the state vector of the object's trajectory and state error covariance matrices from the sensor measurements of its own node. These estimates are transmitted to a central station/node via data links and state-vector fusion is accomplished to obtain a composite state vector and a composite state error covariance matrix. In addition, data at different nodes could be from different types of sensors: optical, infrared or electromagnetic sources.

4.6.1 Kalman filter based fusion algorithm

We assume that at each node the sensor data has been pre-processed (i.e., registration of data, synchronisation, etc.). The estimates of the states are obtained from each sensor's measurements using the KF.

State/covariance time propagation

$$\tilde{x}^m(k+1) = \phi \hat{x}^m(k) \tag{4.60}$$

$$\tilde{P}^m = \phi \hat{P}^m \phi^T + GQG^T \tag{4.61}$$

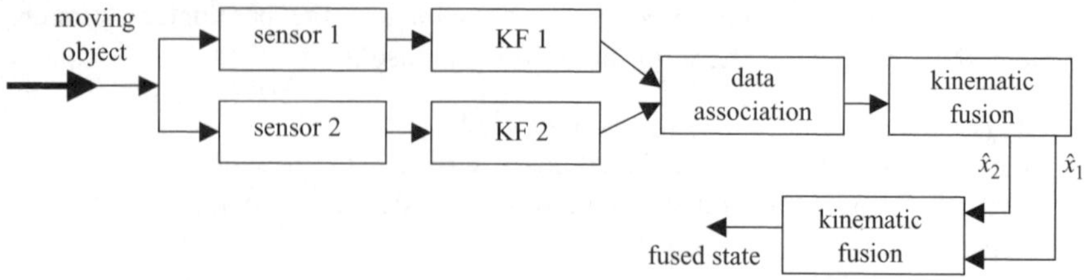

Figure 4.5 State-vector fusion strategy

State/covariance update

$$
\begin{aligned}
r(k+1) &= z^m(k+1) - H\tilde{x}^m(k+1) \\
K^m &= \tilde{P}^m H^T [H\tilde{P}^m H^T + R^m]^{-1} \\
\hat{x}^m(k+1) &= \tilde{x}^m(k+1) + K^m r^m(k+1) \\
\hat{P}^m &= (I - K^m H)\tilde{P}^m
\end{aligned}
\tag{4.62}
$$

In the above equations, m stands for number of sensors ($m = 1, 2, \ldots$). These filters use the same state dynamics. The measurement models and the measurement noise statistics could be different (i.e., $H \to H^1, H^2, \ldots,$ and $R \to R^1, R^2, \ldots,$). Then the fused states can be obtained using the following equations [12]:

$$\hat{x}^f = \hat{x}^1 + \hat{P}^1(\hat{P}^1 + \hat{P}^2)^{-1}(\hat{x}^2 - \hat{x}^1) \tag{4.63}$$

$$\hat{P}^f = \hat{P}^1 - \hat{P}^1(\hat{P}^1 + \hat{P}^2)^{-1}\hat{P}^{1^T} \tag{4.64}$$

From the above, it is observed that the fused state and covariance utilise the quantities from the individual filters only. These estimates are global fusion states/covariances. Figure 4.5 shows a typical scheme for sensor fusion.

4.6.2 Data sharing fusion algorithm

We see from the above state-vector fusion that it requires the inverse of covariance matrices. The data sharing fusion algorithm [13] does not require such a matrix inversion and it involves information feedback from the global filter to the local filters. The filtering algorithm is given by:

Time propagation of global estimates:

$$
\begin{aligned}
\tilde{x}^f(k+1) &= \phi\hat{x}^f(k) \\
\tilde{P}^f(k+1) &= \phi\hat{P}^f(k)\phi^T + GQG^T
\end{aligned}
\tag{4.65}
$$

The local filters are reset as [13]

$$
\begin{aligned}
\tilde{x}^m(k+1) &= \tilde{x}^f(k+1) \\
\tilde{P}^m(k+1) &= \tilde{P}^f(k+1)
\end{aligned}
\tag{4.66}
$$

The measurement update (of state/gain) is given by

$$\begin{aligned} K^m &= (1/\gamma^m)\tilde{P}^f(k+1)H^T[H\tilde{P}^f(k+1)H^T + (1/\gamma^m)R^m]^{-1} \\ \hat{x}^m(k+1) &= \tilde{x}^f(k+1) + K^m[z^m(k+1) - H\tilde{x}^f(k+1)] \end{aligned} \tag{4.67}$$

Then the global fusion of m local estimates is obtained from

$$\begin{aligned} \hat{x}^f(k+1) &= \sum^m \hat{x}^m(k+1) - (m-1)\hat{x}^f(k+1) \\ \hat{P}^f(k+1) &= \left[I - \sum^m K^m H\right] \hat{P}^f(k+1) \left[I - \sum^m K^m H\right]^T \\ &\quad + \sum^m K^m R^m K^m \end{aligned} \tag{4.68}$$

We see from eq. (4.67) that there is information feedback from the global filter to the local filters. In addition, it does not require measurement update of covariances at local nodes. Due to information feedback from the global filter to the local filters, there is implicit data sharing between the local filters. This feature provides some robustness to the fusion filter, especially if there is a measurement data loss in one of the local filters, then the overall performance of the fusion filter will not degrade as much as the KF based fusion filter.

4.6.3 *Square-root information sensor fusion*

The KF can be considered based on covariance matrices and their updates, and hence it is often termed the (conventional) covariance based KF, and interestingly, the state is called the 'covariance state' as against the 'information state' of the information filter. The information matrices are propagated and updated along with propagation of information states.

The state is updated based on a sensor measurement containing relevant information about the state. The observations can be modelled as usual using the linear model:

$$z = Hx + v \tag{4.69}$$

Here v is an m-vector of measurement noise with identity covariance matrix. The least squares solution of x is obtained by minimisation of J:

$$J(x) = (z - Hx)^T(z - Hx) \tag{4.70}$$

We now assume that we have an *a priori* unbiased estimate $\tilde{x}$ of x along with an *a priori* information matrix. The information matrix is the inverse of the (conventional) Kalman filter covariance matrix P. Thus, we have an *a priori* state information pair: $(\tilde{x}, P^{-1})$.

We now modify the cost function J by inclusion of the *a priori* information pair to obtain [8]:

$$J_a(x) = (z - Hx)^T(z - Hx) + (x - \tilde{x})^T\tilde{P}^{-1}(x - \tilde{x}) \tag{4.71}$$

The information matrix P^{-1} (being square of some quantity) can be factored as $\tilde{P}^{-1} = \tilde{C}^T \tilde{C}$, and inserted in eq. (4.71) to get

$$J_a(x) = (z - Hx)^T (z - Hx) + (x - \tilde{x})^T \tilde{C}^T \tilde{C}(x - \tilde{x})$$

The second term in $J_a(x)$ can be expanded and simplified as follows:

$$\begin{aligned}(x - \tilde{x})^T \tilde{C}^T \tilde{C}(x - \tilde{x}) &= (x^T \tilde{C}^T - \tilde{x}^T \tilde{C}^T)(\tilde{C}x - \tilde{C}\tilde{x}) \\ &= (\tilde{x}^T \tilde{C}^T - x^T \tilde{C}^T)(\tilde{C}\tilde{x} - \tilde{C}x) \\ &= (\tilde{C}\tilde{x} - \tilde{C}x)^T (\tilde{C}\tilde{x} - \tilde{C}x)\end{aligned}$$

Inserting back this simplified term in $J_a(x)$, we get

$$\begin{aligned}J_a(x) &= (z - Hx)^T (z - Hx) + (\tilde{C}\tilde{x} - \tilde{C}x)^T (\tilde{C}\tilde{x} - \tilde{C}x) \\ &= (z - Hx)^T (z - Hx) + (\tilde{z} - \tilde{C}x)^T (\tilde{z} - \tilde{C}x)\end{aligned} \tag{4.72}$$

We define $\tilde{z} = \tilde{C}\tilde{x}$.

The second term can be written as $\tilde{z} = \tilde{C}x + \tilde{v}$ following eq. (4.69). From the above development, we can see that the cost function J_a represents the combined system:

$$\begin{bmatrix} \tilde{z} \\ z \end{bmatrix} = \begin{bmatrix} \tilde{C} \\ H \end{bmatrix} x + \begin{bmatrix} \tilde{v} \\ v \end{bmatrix} \tag{4.73}$$

Thus, the *a priori* information artifice forms a data equation similar to the measurement eq. (4.69) and hence, can be considered as an additional measurement.

The above inference provides the basis of the square-root information filter (SRIF). The square-root information pair (as a new observation like a data equation), and the existing measurements are put in the following form and orthogonal transformation is applied to obtain the LS solution [8]:

$$T_0 \begin{bmatrix} \tilde{C}(k-1) & \tilde{z}(k-1) \\ H(k) & z(k) \end{bmatrix} = \begin{bmatrix} \hat{C}(k) & \hat{z}(k) \\ 0 & e(k) \end{bmatrix}; \qquad k = 1, \ldots, N \tag{4.74}$$

With $e(k)$ being the sequence of residuals. Here, T_0 is the Householder transformation matrix. We see that updated information pair $(\hat{z}(k), \hat{C}(k))$ is generated. The process of estimation can be continued with inclusion of next/new measurement $z(k+1)$ and so on. This obtains the recursive SRIF [8]. Next, the square-root information sensor fusion algorithm is given.

Let us assume that we have a two-sensor system with H_1 and H_2 as observation models. Then one can fuse the data at the local node [14]:

$$T_0 \begin{bmatrix} \tilde{C}(k-1) & \tilde{z}(k-1) \\ H_1(k) & z_1(k) \\ H_2(k) & z_2(k) \end{bmatrix} = \begin{bmatrix} \hat{C}(k) & \hat{z}(k) \\ 0 & e(k) \end{bmatrix}; \qquad k = 1, \ldots, N \tag{4.75}$$

If combined with state dynamics, the above process will give the state estimates as the effect of two-sensor data fusion. The process can be easily extended to more than two sensors. Alternatively, one can process the individual sensor measurement data using SRIF at each node to obtain the estimate of information state-vector. It is interesting to note that fusion of these states and information (matrix) is done trivially:

$$\hat{z}_f = \hat{z}_1 + \hat{z}_2 \quad \text{and} \quad \hat{C}_f = \hat{C}_1 + \hat{C}_2 \tag{4.76}$$

In the domain of square-root information philosophy, the state $\hat{z}$ is the information state. Finally, the fused covariance state can be obtained as:

$$\hat{x}_f = \hat{C}_f^{-1}\hat{z}_f$$

Thus, we see that the data equation concept arising out of the information pair and the Householder orthogonal matrix transformation obtain very elegant and simple expressions and solutions to the sensor data fusion problem at either sensor data level fusion or the information state-vector level fusion. These fusion solutions will have enhanced numerical reliability, stability, modularity and flexibility, which stem from the foundation of square-root information processing philosophy. One can obtain a complete filter by considering state dynamics with (correlated) process noise and bias parameters [8].

One important merit of the SRIF based fusion process is that the smaller range of numbers, arising due to propagation of square-root matrices (rather than the original full range matrices), enables the results to be represented by fewer bits. This feature could result in substantial savings in communication overheads.

4.6.3.1 Example 4.5

Generate simulated data of a target moving with constant acceleration and acted on by an uncorrelated process noise, which perturbs the constant acceleration motion. Generate measurements of position of the target from two sensors with different noise characteristics. Obtain state estimates of the target using fusion of the data from the two sensors using Kalman filter based (KFBF) and data sharing (DSF) fusion algorithms.

1 Evaluate the performance of these algorithms.
2 Assuming that there is no measurement available (data loss) during a part of the target trajectory, evaluate the performance of the filters.

4.6.3.2 Solution

The state transition matrix and process noise matrix used for generating the simulated data in each of the three axes of the Cartesian (XYZ) frame of reference are the same as in Example 4.1. Process noise with $\sigma = 0.001$ is added to generate the state trajectories. The state vector has three states represented by $x = [p, v, a]$, [position, velocity, acceleration]. The observation matrix is: $H = \begin{bmatrix} 1 & 0 & 0 \end{bmatrix}$. Position measurements from sensors S1 and S2 are generated by adding measurement noise with $\sigma = 1$ and $\sigma = 3$. Measurements are generated for a duration of 125 s with $\Delta t = 0.25$ s. Initial condition of the states used for the simulation is $x_0 = [200\ 1\ 0.1]$.

The programs for data simulation and data fusion using the KFBF and DSF algorithms are contained in the folder **Ch4KFBDSex5**. Measurement data loss for 50 s

Table 4.4 Percentage state errors (Example 4.5)

	Normal (no data loss)			Data loss in Sensor 1		
	Position	Velocity	Acceleration	Position	Velocity	Acceleration
KFB1	0.1608	1.2994	7.8860	0.6429	5.7262	41.9998
KFB2	0.2025	1.8532	9.1367	0.2025	1.8532	9.1367
KFBF	0.1610	1.3361	7.1288	0.5972	4.6024	30.6382
DS1	0.1776	1.3558	8.2898	0.2065	1.9263	13.1959
DS2	0.1759	1.3720	8.2337	0.2051	1.9431	13.1646
DSF	0.1612	1.3483	8.2517	0.1919	1.9144	13.1817

Table 4.5 H_∞ norms (fusion filter) (Example 4.5)

	Normal	Data loss in S1
KFBF	0.0888	1.2212
DSF	0.0890	0.1261

(between 25–75 s of the target trajectory) is simulated in the sensor measurement S1. The initial conditions for the filter are chosen as $\hat{x}_0 = [180\;0.6\;0.09]$ for both the filters in the KFBF fusion algorithm and for the single global filter in the DSF algorithm.

Table 4.4 gives the percentage state errors of the estimated states w.r.t. the true states. Table 4.5 gives the H_∞ norm (see Section A.26). The results clearly show the superior performance of the DSF algorithm compared with the normal KFBF algorithm when there is measurement data loss in one of the sensors. Their performance is similar when there is no data loss. Figures 4.6(a) and (b) show the state errors with bounds for KFBF and DSF algorithms. The norms of the covariances of the two fusion algorithms are shown in Fig. 4.6(c) from which it is clear that the DSF algorithm has a lower value when there is data loss. It can be concluded that the performance of the KFBF suffers when there is data loss whereas that of the DSF remains generally unaffected, except for velocity state error, which, though reduced in magnitude for DSF, occasionally, crosses the theoretical bounds.

4.7 Epilogue

The KF related algorithms have a wide variety of applications, besides state estimation: parameter estimation, sensor data fusion, sensor fault detection, etc. Numerically reliable solutions/algorithms are extensively treated in References 8

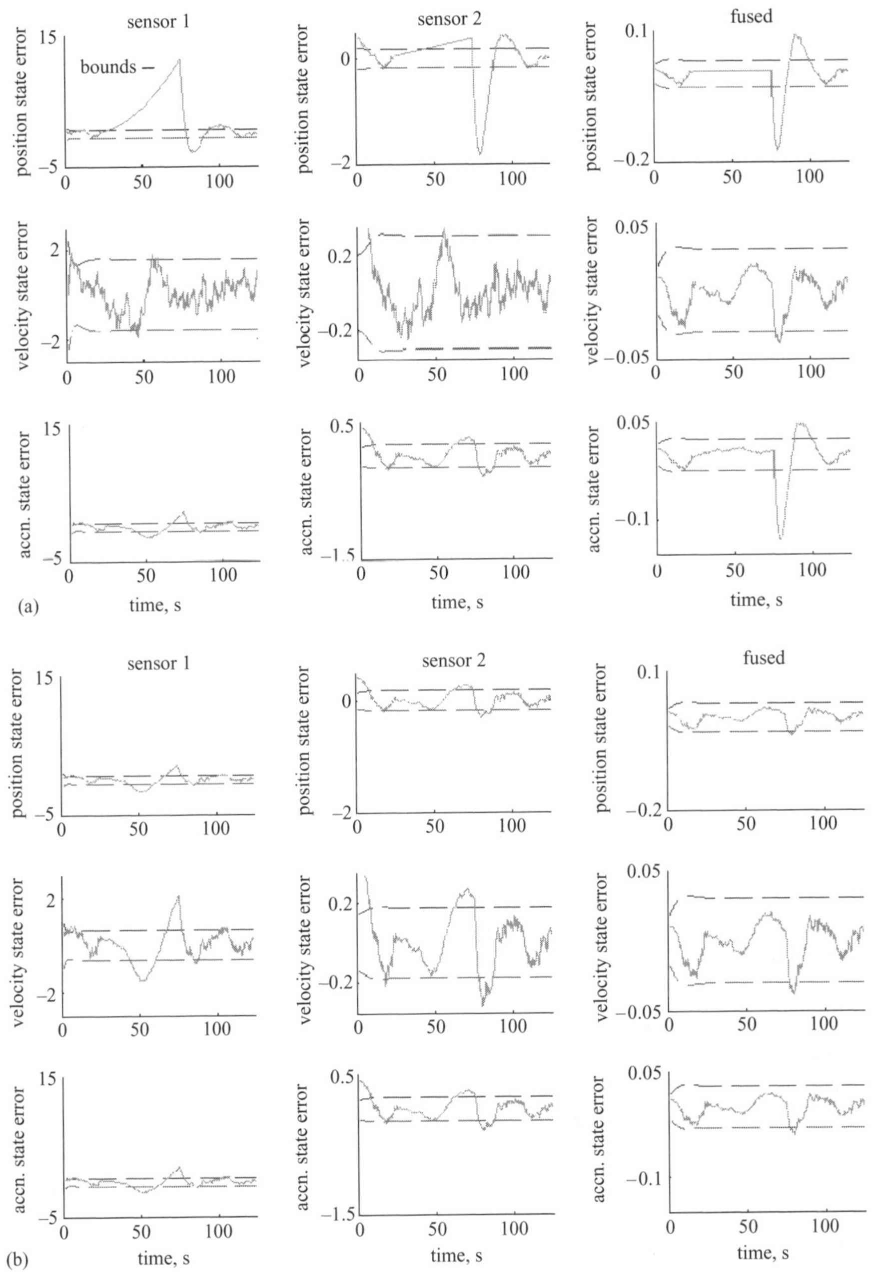

Figure 4.6 *(a) State errors with bounds for KFBF with data loss in Sensor 1 (Example 4.5); (b) state errors with bounds for DSF with data loss in Sensor 1 (Example 4.5)*

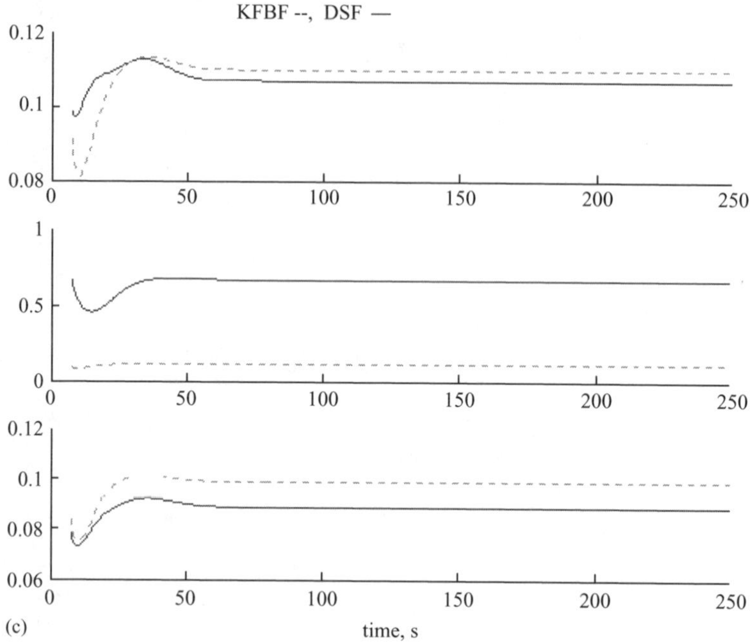

Figure 4.6 Continued. (c) Comparison of norms of covariance matrix for local and fusion filters for KFBF and DSF (Example 4.5)

and 15. The innovations-approach to LS estimation is considered in References 16 and 17. In Reference 18, the concept of modified gain EKF is presented for parameter estimation of linear systems. Reference 18 considers the design of nonlinear filters and gives the conditions under which Kalman equations may be generalised. Also, aircraft parameter/state estimation has been considered [20, 21]. Reference 22 considers H-infinity filtering (see Section A.26) algorithm, which can also be used for sensor data fusion [23]. It will be worthwhile to explore the possibility of developing the EKF type filtering algorithms based on H-infinity filtering concepts so that they can be used for joint state-parameter estimation. The main reason for the utility of the H-infinity based concept is that it does not require many statistical assumptions as needed in developing conventional filtering algorithms. One possibility is to use the H-infinity filtering algorithm in the two-step procedure discussed in Chapter 7. In Reference 24, the estimation theory for tracking and navigation problems is extensively dealt with.

4.8 References

1 KALMAN, R. E.: 'A new approach to linear filtering and prediction problems', *Trans. of ASME, Series D, Journal of Basic Engineering*, 1960, **8**, pp. 35–45

2 MAYBECK, P. S.: 'Stochastic models, estimation and control', vol. 1 (Academic Press, New York, 1979)

3 GELB, A. (Ed.): 'Applied optimal estimation' (MIT Press, Massachussetts, 1974)
4 GREWAL, M. S. and ANDREWS, M. S.: 'Kalman filtering: theory and practice' (Prentice Hall, New Jersey, 1993)
5 ANDERSON, B. D. O.: 'Optimal filtering' (Prentice-Hall, New Jersey, 1979)
6 SORENSON, H. W.: 'Kalman filtering: theory and application' (IEEE Press, New York, 1985)
7 SCHMIDT, F.: 'The Kalman filter: its recognition and development for aerospace applications', *Journal of Guidance and Control*, 1981, **4**, (1), pp. 4–7
8 BIERMAN, G. J.: 'Factorisation methods for discrete sequential estimation' (Academic Press, New York, 1977)
9 RAOL, J. R., and GIRIJA, G.: 'Evaluation of adaptive Kalman filtering methods for target tracking applications'. Paper No. AIAA-2001-4106, August 2001
10 JETTO, L., LONGHI, S., and VITALI, D.: 'Localization of a wheeled mobile robot by sensor data fusion based on fuzzy logic adapted Kalman filter', *Control Engg. Practice*, 1999, **4**, pp. 763–771
11 HALL, D. L.: 'Mathematical techniques in multisensor data fusion' (Artech House, Inc., Boston, 1992)
12 SAHA, R. K.: 'Effect of common process noise on two-track fusion', *Journal of Guidance, Control and Dynamics*, 1996, **19**, pp. 829–835
13 PAIK, B. S. and OH, J. H.: 'Gain fusion algorithm for decentralized parallel Kalman filters', *IEE Proc. on Control Theory Applications*, 2000, **17**, (1), pp. 97–103
14 RAOL, J. R., and GIRIJA, G.: 'Square-root information filter based sensor data fusion algorithm'. In Proceedings of IEEE conference on *Industrial technology*, Goa, India, January 19–22, 2000
15 VERHAEGEN, M., and VAN DOOREN, P.: 'Numerical aspects of different Kalman filter implementations', *IEEE Trans. on Automatic Control*, 1986, **AC-31**, (10), pp. 107–117
16 KAILATH, T.: 'An innovations approach to least-squares estimation, Part I: Linear filtering in additive white noise', *IEEE Trans. on Automatic Control*, 1968, **AC-13**, (6), pp. 646–655
17 FROST, P. A., and KAILATH, T.: 'An innovations approach to least-squares estimation, Part III: Nonlinear estimation in white gaussian noise', *IEEE Trans. on Automatic Control*, 1971, **AC-16**(3), pp. 214–226
18 SONG, T. L., and SPEYER, J. L.: 'The modified gain EKF and parameter identification in linear systems', *Automatica*, 1986, **22**, (1), pp. 59–75
19 SCHMIDT, G. C.: 'Designing non-linear filters based on Daum's theory', *Journal of Guidance, Control and Dynamics*, 1993, **16**, (2), pp. 371–376
20 GIRIJA, G., and RAOL, J. R.: 'PC based flight path reconstruction using UD factorization filtering algorithms', *Defense Sc. Jl.*, 1993, **43**, pp. 429–447
21 JATEGAONKAR, R. V., and PLAETSCHKE, E.: 'Algorithms for aircraft parameter estimation accounting for process and measurement noise', *Journal of Aircraft*, 1989, **26**, (4), pp. 360–372
22 HASSIBI, B., SAYAD, A.H., and KAILATH, T.: 'Linear estimation in Krein spaces – Part II: Applications', *IEEE Trans. on Automatic Control*, 1996, **41**, (1)

23 JIN, S. H., PARK, J. B., KIM, K. K., and YOON, T. S.: 'Krein space approach to decentralized H_∞ state estimation', *IEE Proc. Control Theory Applications*, 2001, **148**, (6), pp. 502–508
24 BAR-SHALOM, Y., and KIRUBARAJAN, T.: 'Estimation with applications – tracking and navigation theory, algorithms and software' (John Wiley & Sons, Inc., New York, 2001)

4.9 Exercises

Exercise 4.1

Let $z = y + v$. Obtain variance of $z - \hat{z}$. We assume that v is a zero-mean white noise process and z is the vector of measurements.

Exercise 4.2

The transition matrix is defined as $\phi = e^{A\Delta t}$ where A is the state-space system matrix and Δt the sampling interval. Obtain the state transition matrix for $A = \begin{bmatrix} 0 & 1 \\ 0 & -a \end{bmatrix}$, if Δt is small; $a\Delta t$ is small. Use Taylor's series expansion for obtaining ϕ.

Exercise 4.3

Let the scalar discrete-time system be given by

$$x(k+1) = \phi x(k) + bu + gw$$

$$z(k) = cx(k) + v$$

Here, u is the deterministic (control) input to the system and w is the process noise, which is assumed white and Gaussian. Obtain the complete set of Kalman filter equations. What happens to the u term in the covariance update equation?

Exercise 4.4

Let $\dot{x} = Ax + w$ and let the elements of matrix A be unknown. Formulate the state-space model for the joint state and parameter estimation to be used in the EKF.

Exercise 4.5 [3]

Assume that the measurement noise is coloured (non-white) and is given by

$$\dot{v} = A_2 v + w_2$$

Then, append this equation to the state-space model of a linear system and obtain a composite model suitable for the KF. Comment on the structure of the composite system model.

Exercise 4.6

What is the distinction between residual error, prediction error and filtering error in the context of the state/parameter estimation?

Exercise 4.7

What is the purpose/advantage of partitioning the KF algorithm into time propagation and measurement update parts? See eqs (4.17) to (4.22).

Exercise 4.8

We have seen that the covariance matrix of the innovations (i.e., residuals) is $S = HPH^T + R$. We can also compute the residuals empirically from

$$r(k+1) = z(k+1) - H\tilde{x}(k+1)$$

This gives $\text{Cov}(rr^T)$. Explain the significance of both these computations. The matrix S is computed by the Kalman filter algorithm. (Hint: both the computations are for the same random variable r.)

Exercise 4.9

Derive the explicit expression for P, the state covariance matrix of the Kalman filter, taking a scalar problem and comment on the effect of measurement noise variance on $\hat{P}$.

Exercise 4.10

Establish the following relationship:

> Variance of (x) = mean squared value of (x) – square of mean value of (x), for a random variable x.

Exercise 4.11

Under what condition is the RMS value of a signal equal to the standard deviation of the signal?

Exercise 4.12

Why is the UD filtering algorithm square root type without involving the square rooting operation in propagation of the covariance related computations?

Exercise 4.13

Substitute eq. (4.15) for the Kalman filter in eq. (4.13) for the covariance matrix update, and obtain the compact form of $\hat{P}$ as in eq. (4.16), by using only simple algebraic manipulations and no approximations.

Exercise 4.14

Why is the residual process in the KF called the 'innovations' process? (Hint: by innovations, it is meant that some new information is obtained/used.)

Exercise 4.15

Derive recursive expressions for determination of the average value and variance of a variable x.

Chapter 5

Filter error method

5.1 Introduction

The output error method discussed in Chapter 3 is perhaps the most widely used approach for parameter estimation. It has several nice statistical properties and is relatively easy to implement. In particular, it gives good results when the data contain only measurement noise and no process noise. However, when process noise is present in the data, a suitable state estimator is required to obtain the system states from noisy data. For a linear system, the Kalman filter is used, as it happens to be an optimal state estimator. For nonlinear systems, there is no practical optimal state estimator and an approximate filter based on system linearisation is used.

There are two approaches to handle process noise in the data: i) filtering methods, e.g., the extended Kalman filter; and ii) the filter error methods. An optimal nonlinear filter is required for computing the likelihood function exactly. The extended Kalman filter can be used for nonlinear systems and the innovations computed from this approach are likely to be white Gaussian if we can assure that the measurements are frequent.

In Chapter 4, the extended Kalman filter was applied to data with process noise for state as well as parameter estimation. The model parameters in this filtering technique are included as additional state variables (state augmentation). The most attractive feature of this approach is that it is one-pass and therefore computationally less demanding. However, experience with the use of the extended Kalman filter for parameter estimation reveals that the estimated parameter values are very sensitive to the initial values of the measurement noise and the state error covariance matrices. If the filter is not properly tuned, i.e., if the *a priori* values of the noise covariance matrices are not chosen appropriately, an extended Kalman filter can produce unsatisfactory results. Most of the applications with extended Kalman filters reported in the literature relate to state estimation rather than parameter estimation. The filter error method, on the other hand, includes a Kalman filter in the Gauss-Newton method (discussed in Chapter 3) to carry out state estimation. In this approach, the sensitivity of estimated values of the parameters to covariance matrix estimates is not so critical.

The filter error method is the most general approach to parameter estimation that accounts for both the process and the measurement noise. The method was first studied in Reference 1 and since then, various applications of the techniques to estimate parameters from measurements with turbulence (accounting for process noise) have been reported [2, 3]. As mentioned before, the algorithm includes a state estimator (Kalman filter) to obtain filtered data from noisy measurements (see Fig. 5.1).

Three different ways to account for process noise in a linear system have been suggested [4]. All these formulations use the modified Gauss-Newton optimisation to estimate the system parameters and the noise statistics. The major difference among these formulations is the manner in which the noise covariance matrices are estimated. A brief insight into the formulations for linear systems is provided next.

5.2 Process noise algorithms for linear systems

Following the development of the linear model in eq. (3.1), the set of equations for a linear system with stochastic input can be written as:

$$\begin{aligned} \dot{x}(t) &= Ax(t) + Bu(t) + Gw(t) \\ y(t) &= Hx(t) \\ z(k) &= y(k) + \xi v(k) \end{aligned} \tag{5.1}$$

The noise vectors w and v represent the uncorrelated, mutually independent, white Gaussian process and measurement noise sequences with identity spectral density and covariance matrices, respectively. The power spectral density of the process

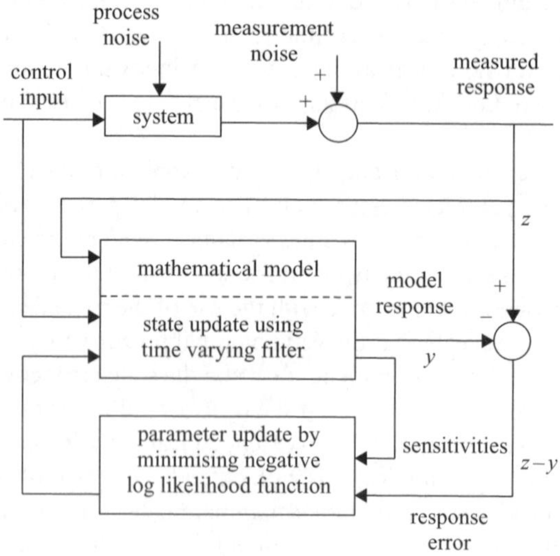

Figure 5.1 Schematic for parameter estimation using filter error method

noise term is given by GG^T and the covariance matrix for the measurement noise term is given by $R = \xi\xi^T$.

Equation (5.1) presents a mixed continuous/discrete form with the state equation expressed as a continuous-time differential equation and the observation equation expressed in the discrete-time form. Such a description of the system is most suitable since the measurements are mostly available at discrete times for analysis on a digital computer. The differential form of the state equation can be solved for x either by numerical integration or by the transition matrix approach (see Section A.43). The continuous-time equation can be regarded as a limiting case of the discrete equation as the sampling interval becomes very small. Working with a purely discrete form of state equation poses no problems. While the discrete form is defined in terms of the transition matrix, the continuous form of state equation is defined in terms of an A matrix. Since the elements of matrix A have more physical meaning attached to them than the elements of the transition matrix, it is easier to work with the mixed form described in eq. (5.1).

A Gauss-Newton optimisation is used to minimise the cost function:

$$J = \frac{1}{2}\sum_{k=1}^{N}[z(k) - y(k)]^T S^{-1}[z(k) - y(k)] + \frac{N}{2}\ln|S| \tag{5.2}$$

where y is the vector of filter predicted observations (see Fig. 5.1) and z is a vector of measured observations sampled at N discrete points. The matrix S denotes the covariance matrix of the residuals (innovations). For the case where the process noise is zero (i.e., $G = 0$ in eq. (5.1)), we have $S = R$ and eq. (5.2) reduces to eq. (3.52). However, if the process noise is not zero, then the Kalman filter is used to obtain the filtered states from the predicted states using the following set of equations [5].

Time propagation

$$\begin{aligned} \tilde{x}(k+1) &= \phi\hat{x}(k) + \psi B u_e(k) \\ y(k+1) &= H\tilde{x}(k+1) \end{aligned} \tag{5.3}$$

Here, $u_e(k) = (u(k) + u(k-1))/2$ denotes the mean value of the control input, ϕ denotes the transition matrix given by $\phi = e^{A\Delta t}$ and ψ is its integral given by $\psi = \int_0^{\Delta t} e^{A\tau}d\tau$. The sampling interval is given by $\Delta t = t_k - t_{k-1}$.

Using Taylor's series expansion, the matrices ϕ and ψ can be written in the following form:

$$\begin{aligned} \phi &\approx I + A\Delta t + A\frac{\Delta t^2}{2!} + \cdots \\ \psi &\approx I\Delta t + A\frac{\Delta t^2}{2!} + A^2\frac{\Delta t^3}{3!} + \cdots \end{aligned} \tag{5.4}$$

Correction

$$\hat{x}(k+1) = \tilde{x}(k+1) + K[z(k+1) - y(k+1)] \tag{5.5}$$

The Kalman gain K and the covariance matrix of residuals S are related to each other by the equation

$$K = PH^T S^{-1} \tag{5.6}$$

The matrix S is a function of the state prediction error covariance P, the measurement noise covariance matrix R and the observation matrix H, and is given by the relation

$$S = HPH^T + R \tag{5.7}$$

Different formulations for process noise handle the computation of matrices K, P, S and R in different ways. For example, a steady state form of the Riccati equation is mostly used to compute the matrix P, while the matrices K and S are computed from eqs (5.6) and (5.7). Another approach is to include the elements of K in the parameter vector Θ to be estimated by minimisation of the cost function using a suitable optimisation technique (e.g., Gauss-Newton optimisation). Some of the main features of the approaches suggested to account for process noise in a linear system [4] are highlighted here.

5.2.1.1 Natural formulation

In this approach, the noise matrices G and ξ in eq. (5.1) are treated as unknowns and estimated along with other system parameters using Gauss-Newton optimisation. The natural formulation has the following features:

- The parameter vector Θ = [elements of A, B, H, G and ξ].
- The covariance matrix of residuals S is computed from eq. (5.7).
- The estimates of ξ from this approach are generally poor, leading to convergence problems. This is in direct contrast to the output error method discussed in Chapter 3 where the estimation of R ($R = \xi\xi^T$) from eq. (3.5) poses no problems.
- This formulation turns out to be time consuming, with the parameter vector Θ having elements of the noise matrices G and ξ in addition to the system parameters. The computation of the gradients with respect to the elements of G and ξ puts further demand on the computer time and memory.

5.2.1.2 Innovation formulation

In this formulation, the matrices S and K are estimated directly rather than from eqs (5.6) and (5.7). This obviates the need to include the elements of the noise matrices G and ξ in the parameter vector Θ. The main features of this formulation are:

- The parameter vector Θ = [elements of A, B, H and K].
- The matrix S is computed from the equation

$$S = \frac{1}{N}\sum_{k=1}^{N}[z(k) - y(k)][z(k) - y(k)]^T \tag{5.8}$$

- The elements of measurement noise matrix ξ can be estimated from the expression: $R = \xi\xi^T = S - HPH^T$. This eliminates the difficulty of estimating ξ directly (as in natural formulation), thereby avoiding convergence problems.

- In this formulation, the inclusion of K in vector Θ can lead to identifiability problems (see Section A.27), particularly for higher order systems. For large systems, the matrix K increases in size and there might not be sufficient information in the data to correctly estimate all the elements of matrix K. Further, since no physical meaning can be attached to the elements of K, it is rather difficult to decide upon the accuracy of its estimated elements.
- Despite the above problem, this approach has better convergence than the natural formulation. This is primarily due to the omission of ξ from the parameter vector Θ.
- The computed value of R from this approach may not always be correct. Therefore a complicated set of constraints has to be followed to ensure a valid solution of R (estimated R should be positive semi-definite).

5.2.1.3 Mixed formulation

This formulation combines the merits of the natural and the innovation formulation and is considered better than the formulations discussed above. In this method, the elements of matrix G are retained in the parameter vector Θ (strong point of the natural formulation) and the matrix S is estimated from eq. (5.8) (strong point of the innovation formulation). Thus, the method takes the best of the natural and the innovation formulation while doing away with the operations that cause problems in convergence or estimation. The main features of this formulation are:

- The parameter vector $\Theta = $ [elements of A, B, H and G].
- The matrix S is estimated as in eq. (5.8).
- After obtaining P by solving the steady-state form of the Riccati equation, K is computed from eq. (5.6). Thus, the problems associated with direct estimation of K in the innovation formulation are avoided in this approach.
- This formulation requires less computer time and has good convergence.
- The inequality constraint of the innovation formulation is retained to ensure a legitimate solution of R. This requires quadratic programming leading to a complex optimisation problem [4].
- Since the update of parameter vector Θ and the covariance matrix S are carried out independently, some convergence problems can arise. A heuristic approach of compensating the G matrix whenever S is revised to take care of this problem is suggested in Reference 4.

Once the filtered states are obtained, the parameter vector update can be computed using the expressions given in eqs (3.54) to (3.56) for the output error method. The only change made in these equations is to replace the measurement noise covariance matrix R by the covariance matrix of residuals S.

The update in the parameter vector is given by

$$\Delta\Theta = \left[\nabla_{\Theta}^{2} J(\Theta)\right]^{-1} \left[\nabla_{\Theta} J(\Theta)\right] \tag{5.9}$$

where the first and the second gradients are defined as

$$\nabla_{\Theta} J(\Theta) = \sum_{k=1}^{N} \left[\frac{\partial y}{\partial \Theta}(k)\right]^T S^{-1}[z(k) - y(k)] \tag{5.10}$$

$$\nabla^2_{\Theta} J(\Theta) = \sum_{k=1}^{N} \left[\frac{\partial y}{\partial \Theta}(k)\right]^T S^{-1} \left[\frac{\partial y}{\partial \Theta}(k)\right] \tag{5.11}$$

The vector $\Theta(i)$ at the ith iteration is updated by $\Delta\Theta$ to obtain $\Theta(i+1)$ at the $(i+1)$th iteration:

$$\Theta(i+1) = \Theta(i) + \Delta\Theta \tag{5.12}$$

As observed from eqs (5.10) and (5.11), the update of the parameter vector would require computation of the sensitivity coefficients $\partial y/\partial\Theta$. The sensitivity coefficients for a linear system can be obtained in a straightforward manner by partial differentiation of the system equations.

Computing $\partial y/\partial\Theta$ from partial differentiation of y w.r.t. Θ in eq. (5.3), we get [5]:

$$\frac{\partial y}{\partial \Theta} = H\frac{\partial \tilde{x}(k)}{\partial \Theta} + \frac{\partial H}{\partial \Theta}\tilde{x}(k) \tag{5.13}$$

The gradient $\partial\tilde{x}/\partial\Theta$ can be obtained from eq. (5.3) as

$$\frac{\partial \tilde{x}(k+1)}{\partial \Theta} = \phi\frac{\partial \hat{x}(k)}{\partial \Theta} + \frac{\partial \phi}{\partial \Theta}\hat{x}(k) + \psi\frac{\partial B}{\partial \Theta}u_e + \frac{\partial \psi}{\partial \Theta}Bu_e \tag{5.14}$$

The gradients $\partial\phi/\partial\Theta$ and $\partial\psi/\partial\Theta$ can be obtained from partial differentiation of eq. (5.4) w.r.t. Θ. The gradient $\partial\hat{x}/\partial\Theta$ is required in eq. (5.14), which can be obtained from partial differentiation of eq. (5.5):

$$\frac{\partial \hat{x}(k)}{\partial \Theta} = \frac{\partial \tilde{x}(k)}{\partial \Theta} + \frac{\partial K}{\partial \Theta}[z(k) - y(k)] - K\frac{\partial y(k)}{\partial \Theta} \tag{5.15}$$

The Kalman gain K is a function of the parameter vector Θ and its gradient w.r.t. Θ can be obtained from eq. (5.6):

$$\frac{\partial K}{\partial \Theta} = \frac{\partial P}{\partial \Theta}H^T S^{-1} + P\left(\frac{\partial H}{\partial \Theta}\right)^T S^{-1} \tag{5.16}$$

While S can be computed from eq. (5.7), the state prediction error covariance matrix P is computed from the continuous-time Riccati equation [5]:

$$AP + PA^T - \frac{PH^T S^{-1} HP}{\Delta t} + GG^T = 0 \tag{5.17}$$

The eigenvector decomposition method [6] can be used to solve for P from the above equation. The gradient $\partial P/\partial\Theta$ required for computing $\partial K/\partial\Theta$ in eq. (5.16) can be obtained by differentiating eq. (5.17) w.r.t. Θ. This leads to a set of Lyapunov equations, which can be solved by a general procedure [4, 5].

To compute the gradient $\partial y/\partial\Theta$, the sensitivity eqs (5.13) to (5.17) are solved for each element of the parameter vector Θ. For a nonlinear system, this scheme of obtaining the gradients from partial differentiation of the system equations will involve a lot of effort on the part of the user, as frequent changes might be required in the model structure. A better approach would be to approximate the sensitivity coefficients by finite differences [7].

Following the development of process noise formulations for linear systems [4, 5], two filtering techniques (the steady state filter and the time varying filter) were proposed [7] to handle process noise for nonlinear systems. In both these techniques, the nonlinear filters for the state estimation were implemented in an iterative Gauss-Newton optimisation method. This makes the application of these techniques to parameter estimation problems simple, particularly for users who are familiar with the output error method. However, the implementation of these techniques, specifically the time varying filter, is quite complex. The computational requirements of the time varying filter are also high, but the advantages it offers in terms of reliable parameter estimation far outweigh the disadvantages associated with the high computational cost of the approach.

The steady state and the time varying filters for state estimation in nonlinear systems are described next.

5.3 Process noise algorithms for nonlinear systems

A nonlinear dynamic system with process noise can be represented by the following set of stochastic equations:

$$\dot{x}(t) = f[x(t), u(t), \Theta] + Gw(t) \qquad \text{with initial } x(0) = x_0 \tag{5.18}$$

$$y(t) = h[x(t), u(t), \Theta] \tag{5.19}$$

$$z(k) = y(k) + \xi v(k) \tag{5.20}$$

In the above equation, f and h are general nonlinear vector-valued functions. The w and v are white Gaussian, additive process and measurement noises, respectively, characterised by zero mean. The parameter vector Θ to be estimated consists of the system parameters β, the initial values x_0 of the states and the elements of the process noise matrix G. Computation of the measurement noise matrix ξ or the measurement noise covariance matrix R (where $R = \xi\xi^T$) is discussed later in Section 5.3.2.

The parameter vector Θ to be estimated is expressed as

$$\Theta^T = [\beta^T, x_0^T, G^T] \tag{5.21}$$

In practice, only the diagonal elements of matrix G are included in Θ for estimation. This reduces the computational burden without affecting the accuracy of the system parameter estimates. Frequently, one also needs to estimate the nuisance parameters like the biases in the measurements and control inputs in order to get improved estimates of the system coefficients.

5.3.1 *Steady state filter*

The cost function to be minimised in the steady state filter algorithm is given by eq. (5.2) and the parameter vector update steps are the same as those described in eqs (5.9) to (5.12). The time propagation and state corrections in eqs (5.3) and (5.5) for linear systems are now replaced by the following set of equations for nonlinear systems.

Time propagation

$$\tilde{x}(k) = \hat{x}(k-1) + \int_{t_{k-1}}^{t_k} f[x(t), u_e(k), \Theta]\, dt \tag{5.22}$$

$$\tilde{y}(k) = h[\tilde{x}(k), u(k), \Theta] \tag{5.23}$$

Correction

$$\hat{x}(k) = \tilde{x}(k) + K[z(k) - \tilde{y}(k)] \tag{5.24}$$

As for the state estimation in linear systems, the steady state filter for nonlinear systems computes the matrices K, S and P from eqs (5.6), (5.8) and (5.17), respectively.

The state estimation of nonlinear systems differs from that of linear systems in the following aspects:

1 Estimation of the initial conditions of the state x_0.
2 Linearisation of eqs (5.18) and (5.19) w.r.t. x to obtain the system matrices A and H. The system equations, in the steady state filter, are linearised at each iteration about x_0. This yields the time-invariant matrices A and H (computed only once in each iteration), to obtain the steady state matrices K and P.

$$A(k) = \left.\frac{\partial f(x(t), u(t), \Theta)}{\partial x}\right|_{x=x_0} \tag{5.25}$$

$$H(k) = \left.\frac{\partial h[x(t), u(t), \Theta]}{\partial x}\right|_{x=x_0} \tag{5.26}$$

3 The response gradients $\partial y/\partial\Theta$ required to update the parameter vector in eqs (5.10) and (5.11), and the gradients in eqs (5.25) and (5.26) required to compute the system matrices are obtained by the finite difference approximation method instead of partial differentiation of the system equations.

Gradient computation

Assuming a small perturbation $\Delta x_j (\approx 10^{-5} x_j)$ in the variable x_j of the state vector x, the following expression for the matrices A and H can be obtained using central differencing:

$$A_{ij} \approx \left. \frac{f_i[x_j + \Delta x_j, u(k), \Theta] - f_i[x_j - \Delta x_j, u(k), \Theta]}{2\Delta x_j} \right|_{x=x_0}; \quad \text{for } i, j = 1, \ldots, n \tag{5.27}$$

$$H_{ij} \approx \left. \frac{h_i[x_j + \Delta x_j, u(k), \Theta] - h_i[x_j - \Delta x_j, u(k), \Theta]}{2\Delta x_j} \right|_{x=x_0}; \quad \text{for } i = 1, \ldots, m \text{ and } j = 1, \ldots, n \tag{5.28}$$

where n is the number of states and m is the number of observations in the nonlinear system.

In a similar fashion, using eqs (5.22) to (5.24), the gradients $\partial y/\partial \Theta$ can be obtained by introducing a small perturbation in each of the system parameters one at a time. The change in the system response due to a small change in the parameters can be obtained from the following equations:

$$\tilde{x}_c(k) = \hat{x}_c(k-1) + \int_{t_{k-1}}^{t_k} f[x_c(t), u_e(k), \Theta + \Delta\Theta]\, dt \tag{5.29}$$

$$y_c(k) = h[\tilde{x}_c(k), u(k), \Theta + \Delta\Theta] \tag{5.30}$$

$$\hat{x}_c(k) = \tilde{x}_c(k) + K_c[z(k) - y_c(k)] \tag{5.31}$$

where subscript c represents the change in the vector or matrix due to a small change in the system parameters. Note that the computation of the change in the state variable in eq. (5.31) requires the perturbed gain matrix K_c, which can be obtained from eq. (5.6) as

$$K_c = P_c H_c^T S^{-1} \tag{5.32}$$

For the perturbed parameters, the changed system matrices (A_c and H_c) can be computed from eqs (5.27) and (5.28). These need to be computed only once in an iteration about the point x_0. The changed state error covariance matrix P_c, required for computing K_c in eq. (5.32), can be obtained from eq. (5.17), which now will make use of the changed system matrices A_c and H_c.

Once the changed system response y_c is obtained using the above set of perturbation equations, the gradient $\partial y/\partial \Theta$ can be easily computed. Assuming that y_{ci} represents the change in the ith component of the measurement vector y corresponding to perturbation in parameter Θ_j, the gradient $\partial y/\partial\Theta$ is given by

$$\left(\frac{\partial y(k)}{\partial \Theta}\right)_{ij} \approx \frac{y_{ci}(k) - y_i(k)}{\Delta\Theta_j} \qquad \text{for } i = 1, \ldots, m \text{ and } j = 1, \ldots, q \tag{5.33}$$

where q represents the dimension of the parameter vector Θ.

Thus, we see that the partial differential equations (eqs (5.13) to (5.16)) for computing the gradients in a linear system are replaced by a set of perturbation equations in the case of a nonlinear system. There is no need to explicitly compute the gradients like $\partial\hat{x}/\partial\Theta$, $\partial K/\partial\Theta$ and $\partial P/\partial\Theta$ for nonlinear systems, as these are implicitly taken care of while solving the perturbed system equations. This also implies that the set of Lyapunov equations for computing the gradient of P (as in case of the linear systems) is no longer required for nonlinear system state estimation.

Having obtained the covariance matrix of innovations S from eq. (5.8), the measurement noise covariance matrix can be obtained as

$$R = S - HPH^T \tag{5.34}$$

We see that this procedure of obtaining the elements of R (and therefore ξ) is similar to the one outlined in the mixed process noise formulation for linear systems. As such, this approach faces the same problems as discussed in the mixed formulation. It means that the estimates of ξ might not be legitimate and a constrained optimisation will have to be carried out to ensure that R turns out to be positive semi-definite. Further, as with the mixed formulation for linear systems, the steady state filter algorithm for a nonlinear system also requires compensation of the G matrix whenever S is updated [7].

The steady state process noise filter is adequate for most of the applications encountered in practice. For large oscillatory motions or when the system response shows a highly nonlinear behaviour, the use of a time varying filter is more likely to produce better parameter estimates than a steady state filter.

5.3.2 *Time varying filter*

Of all the process noise algorithms discussed so far, the time varying filter (TVF) is the most complex to implement, although the formulation runs parallel to that of the steady state filter. Unlike the steady state filter, the matrices S, K and P in the time varying filter are computed at each discrete time point k. Similarly, the matrices A and H obtained from the first order linearisation of the system equations are computed at every data point in an iteration. This puts a lot of burden on the computer time and memory.

Following the equations developed for the steady state filter, the time varying filter is formulated as follows. The cost function to be minimised in the time varying filter is given by

$$J = \frac{1}{2}\sum_{k=1}^{N}[z(k) - y(k)]^T S^{-1}(k)[z(k) - y(k)] + \sum_{k=1}^{N}\frac{1}{2}\ln|S(k)| \tag{5.35}$$

where the covariance matrix of innovations S is revised at discrete time point k.

The Gauss-Newton optimisation equations for parameter vector update also use the revised values of $S(k)$ instead of the constant value of S.

$$\nabla_\Theta J(\Theta) = \sum_{k=1}^{N} \left[\frac{\partial y}{\partial \Theta}(k)\right]^T S^{-1}(k)[z(k) - y(k)] \tag{5.36}$$

$$\nabla_\Theta^2 J(\Theta) = \sum_{k=1}^{N} \left[\frac{\partial y}{\partial \Theta}(k)\right]^T S^{-1}(k) \left[\frac{\partial y}{\partial \Theta}(k)\right] \tag{5.37}$$

$$\Delta\Theta = \left[\nabla_\Theta^2 J(\Theta)\right]^{-1} [\nabla_\Theta J(\Theta)] \tag{5.38}$$

$$\Theta(i+1) = \Theta(i) + \Delta\Theta \tag{5.39}$$

The time propagation (prediction) and the correction steps used to obtain the updated values of the state $\hat{x}$ and the state error covariance matrix $\hat{P}$ are given below.

Time propagation

$$\tilde{x}(k) = \hat{x}(k-1) + \int_{t_{k-1}}^{t_k} f[x(t), u_e(t), \beta]\, dt \tag{5.40}$$

$$\tilde{y}(k) = h[\tilde{x}(k), u(k), \beta] \tag{5.41}$$

Assuming Δt to be small, the predicted matrix P can be approximated as [8]:

$$\tilde{P}(k) \approx \Phi \hat{P}(k-1)\Phi^T + \Delta t GG^T \tag{5.42}$$

Correction

$$K(k) = \tilde{P}(k)H^T(k)[H(k)\tilde{P}(k)H^T(k) + R]^{-1} \tag{5.43}$$

$$\hat{x}(k) = \tilde{x}(k) + K(k)[z(k) - y(k)] \tag{5.44}$$

$$\begin{aligned} \hat{P}(k) &= [I - K(k)H(k)]\tilde{P}(k) \\ &= [I - K(k)H(k)]\tilde{P}(k)[I - K(k)H(k)]^T + K(k)RK^T(k) \end{aligned} \tag{5.45}$$

The expression for $\hat{P}$ in eq. (5.45) with the longer form on the right hand side of the equation is usually preferred because it is numerically stable and gives better convergence.

The state matrix A at the kth data point is obtained by linearising eq. (5.18) about $\hat{x}(k-1)$:

$$A(k) = \left.\frac{\partial f(x(t), u(t), \beta)}{\partial x}\right|_{x=\hat{x}(k-1)} \tag{5.46}$$

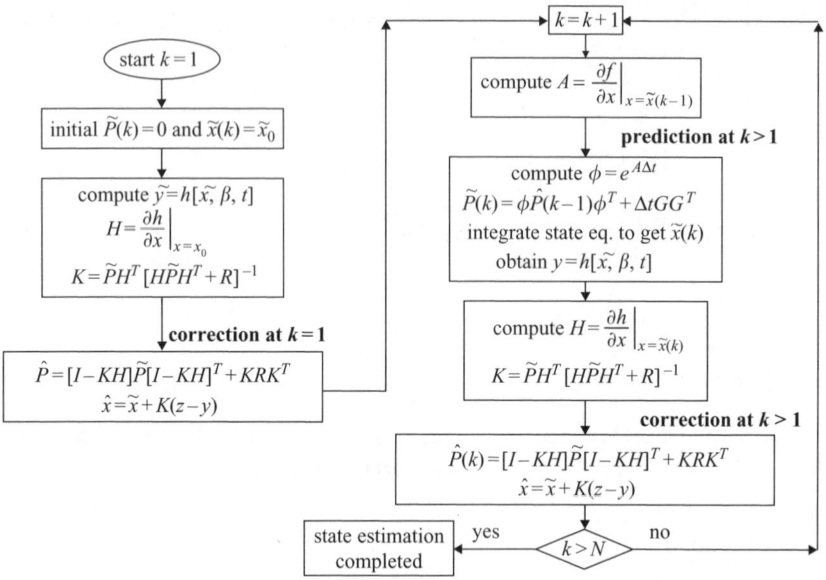

Figure 5.2 Flow diagram showing the prediction and correction steps of TVF

Similarly, the observation matrix H at the discrete time point k can be obtained by linearising eq. (5.19) about $x = \tilde{x}(k)$:

$$H(k) = \frac{\partial h[x(t), u(t), \beta]}{\partial x}\bigg|_{x=\tilde{x}(k)} \tag{5.47}$$

The transition matrix ϕ is the same as defined in eq. (5.4). Starting with suitable guess values of system parameters and state variables, the parameter vector Θ (consisting of the elements of β, the diagonal elements of matrix G and the initial conditions x_0) is updated during each iteration until a certain convergence criterion is satisfied. Further, it is a common practice to start with zero value of state error covariance matrix P, and then use the prediction and correction steps in eqs (5.40) to (5.45), to obtain updates in x and P. The flow diagram in Fig. 5.2 shows the prediction and correction steps of state estimation with TVF.

The gradient computation in TVF is similar to that described in eqs (5.27) to (5.33) for a steady state filter. Using central differencing, the system matrices A and H can be obtained from the expressions

$$A_{ij}(k) \approx \frac{f_i[x_j + \Delta x_j, u(k), \beta] - f_i[x_j - \Delta x_j, u(k), \beta]}{2\Delta x_j}\bigg|_{x=\hat{x}(k-1)};$$

$$\text{for } i, j = 1, \ldots, n \tag{5.48}$$

$$H_{ij}(k) \approx \left. \frac{h_i[x_j + \Delta x_j, u(k), \beta] - h_i[x_j - \Delta x_j, u(k), \beta]}{2\Delta x_j} \right|_{x=\tilde{x}(k)};$$

$$\text{for } i = 1, \ldots, m \text{ and } j = 1, \ldots, n \tag{5.49}$$

Following the procedure outlined in the steady state filter, the response gradient $(\partial y/\partial \Theta)$ can be obtained by introducing a small perturbation in each of the parameters to be estimated, one at a time, and using eqs (5.40) to (5.45) to compute the change in each component y_i of the vector y. Equation (5.33) gives the value for $(\partial y/\partial \Theta)$.

Note that the time varying filter computes the matrix S directly from eq. (5.43) at no extra cost:

$$S = H(k)\tilde{P}(k)H^T(k) + R \tag{5.50}$$

However, to compute S from eq. (5.50) necessarily requires the value of the measurement noise covariance matrix R. The time varying filter formulation offers no solution to obtain R. A simple procedure to compute R can be implemented based on estimation of the noise characteristics using Fourier smoothing [9]. In this approach, Fourier series analysis is used to smooth the measured data and separate out the clean signal from noise based on the spectral content. The approach uses a Wiener filter to obtain a smoothed signal which, when subtracted from the noisy data, yields the noise sequence. If v denotes the noise sequence, the noise characteristics (mean $\bar{v}$ and the measurement noise covariance matrix R) can be obtained as follows:

$$\bar{v} = \frac{1}{N} \sum_{k=1}^{N} v(k) \tag{5.51}$$

$$R = \frac{1}{N-1} \sum_{k=1}^{N} [v(k) - \bar{v}]^2 \tag{5.52}$$

where N is the total number of data points. This procedure to compute R is shown to work well when included in the time varying filter [10]. Since the estimated R from this process is accurate, there is no need to impose any kind of inequality constraints as done in the mixed formulation for linear systems and in the steady state filter for nonlinear systems. The elements of state noise matrix G can either be fixed to some previously obtained estimates or determined by including them in the parameter vector Θ.

5.3.2.1 Example 5.1

From the set of nonlinear equations described in Example 3.3 for a light transport aircraft, simulate the longitudinal short period data of the aircraft using the true values of the parameters listed in Table 3.4. Include process noise in this clean simulated data and apply the time varying filter to estimate the non-dimensional derivatives from the aircraft mathematical model. Also, estimate the model parameters using the output error method and compare the results with those obtained from the time varying filter approach.

5.3.2.2 Solution

Data generation step

A doublet elevator control input (with a pulse width of 2 s) is used in the aircraft model equations (state and measurement model) described in Example 3.3 to generate data for 8 s with a sampling time of 0.03 s. The aircraft data with process noise is simulated for moderate turbulence conditions. In order to have a realistic aircraft response in turbulence, a Dryden model is included in the simulation process (see Section B.14).

State estimation

The parameter vector Θ to be estimated consists of the following unknown elements (see eq. (5.21)):

$$\Theta^T = [\beta^T, x_0^T, G^T]$$

where β is the vector of aircraft longitudinal stability and control derivatives:

$$\beta = [C_{x_0}, C_{x_\alpha}, C_{x_{\alpha^2}}, C_{z_0}, C_{z_\alpha}, C_{z_q}, C_{z\delta_e}, C_{m_0}, C_{m_\alpha}, C_{m_{\alpha^2}}, C_{m_q}, C_{m\delta_e}]$$

x_0 is the vector of initial values of the states u, w, q and θ:

$$x_0 = [u_0, w_0, q_0, \theta_0]$$

G is the process noise matrix whose diagonal elements are included in Θ for estimation:

$$G = [G_{11}, G_{22}, G_{33}, G_{44}]$$

The procedure for parameter estimation with time varying filter involves the following steps:

a As a first step, Fourier smoothing is applied to the simulated noisy measured data to estimate the noise characteristics and compute the value of R [9]. This step is executed only once.

Time propagation step

b Predicted response of aircraft states ($\tilde{x} = [u, w, q, \theta]$) is obtained by solving eq. (5.40). Assuming the initial values of the parameters defined in vector β to be 50 per cent off from the true parameter values and choosing suitable values for u, w, q and θ at $t = t_0$, the state model defined in Example 3.3 is integrated using a fourth order Runge-Kutta method to obtain the time response of the states u, w, q and θ.

c Using the measurement model defined in Example 3.3, eq. (5.41) is solved to obtain $\tilde{y} = [u, w, q, \theta, a_x, a_z]$.

d State matrices A and H are obtained by solving eqs (5.48) and (5.49).

e Next, the transition matrix Φ is obtained from eq. (5.4).

f With the initial value of the state error covariance matrix P assumed to be zero and assigning starting values of 0.02 to all the elements of matrix G (any set of small values can be used for G to initiate the parameter estimation procedure), eq. (5.42) is used to compute $\tilde{P}$.

Correction step

g With R, $\tilde{P}(k)$ and H computed, the Kalman gain $K(k)$ is obtained from eq. (5.43).
h Updated state error covariance matrix $\hat{P}(k)$ is computed from eq. (5.45).
i Updated state vector $\hat{x}(k)$ is computed from eq. (5.44).

Parameter vector update

j Perturbing each element Θ_j of the parameter vector Θ one at a time (perturbation $\approx 10^{-7}\Theta_j$), steps (b) to (i) are repeated to compute $y_{ci}(k)$, where $y_{ci}(k)$ represents the changed time history response in each of the components $u, w, q, \theta, a_x, a_z$ due to perturbation in Θ_j. The gradient $\partial y/\partial\Theta$ can now be computed from eq. (5.33).
k The covariance matrix S is computed from eq. (5.50).
l Equations (5.36) to (5.39) are used to update the parameter vector Θ.

Steps (b) to (l) are repeated in each iteration and the iterations are continued until the change in the cost function computed from eq. (5.35) is only marginal.

For parameter estimation with output error method, the procedure outlined in Chapter 3 was applied. The approach does not include the estimation of matrix G. For the simulated measurements with process noise considered in the present investigation, the algorithm is found to converge in 20 to 25 iterations. However, the estimated values of the parameters are far from satisfactory (column 4 of Table 5.1).

Table 5.1 Estimated parameters from aircraft data in turbulence [10] (Example 5.1)

Parameter	True values	Starting values	Estimated values from OEM	Estimated values from TVF
C_{x0}	−0.0540	−0.1	−0.0049	−0.533
$C_{x\alpha}$	0.2330	0.5	0.2493	0.2260
$C_{x\alpha^2}$	3.6089	1.0	2.6763	3.6262
C_{z0}	−0.1200	−0.25	−0.3794	−0.1124
$C_{z\alpha}$	−5.6800	−2.0	−4.0595	−5.6770
C_{zq}	−4.3200	−8.0	1.8243	−2.7349
$C_{z\delta}$	−0.4070	−1.0	0.7410	−0.3326
C_{m0}	0.0550	0.1	−0.0216	0.0556
$C_{m\alpha}$	−0.7290	−1.5	−0.3133	−0.7296
$C_{m\alpha^2}$	−1.7150	−2.5	−1.5079	−1.7139
C_{mq}	−16.3	−10.0	−10.8531	−16.1744
$C_{m\delta}$	−1.9400	−5.0	−1.6389	−1.9347
G_{11}	–	0.02	–	5.7607
G_{22}	–	0.02	–	−6.4014
G_{33}	–	0.02	–	5.3867
G_{44}	–	0.02	–	2.1719
PEEN (%)	–	–	46.412	9.054

This is in direct contrast to the excellent results obtained with the output error approach (see Table 3.4). This is because the data in Example 3.3 did not have any process noise and as such the output error method gave reliable parameter estimates (see Section B.13) and an excellent match between the measured and model-estimated responses. On the other hand, the response match between the measured and estimated time histories of the flight variables in the present case shows significant differences, also reflected in the high value of $|R|$.

Parameter estimation results with the time varying filter show that the approach converges in about four iterations with adequate agreement between the estimated and measured responses. The estimated parameters from the time varying filter in Table 5.1 compare well with the true parameter values [10]. During the course of investigations with the time varying filter, it was also observed that, for different guesstimates of G, the final estimated values of G were not always the same. However, this had no bearing on the estimated values of the system parameters (vector β), which always converged close to the true parameter values. It is difficult to assign any physical meaning to the estimates of the G matrix, but this is of little significance considering that we are only interested in the estimated values of derivatives that characterise the aircraft motion. Figure 5.3 shows the longitudinal time history match for the aircraft motion in turbulence, and the estimated derivatives are listed in Table 5.1.

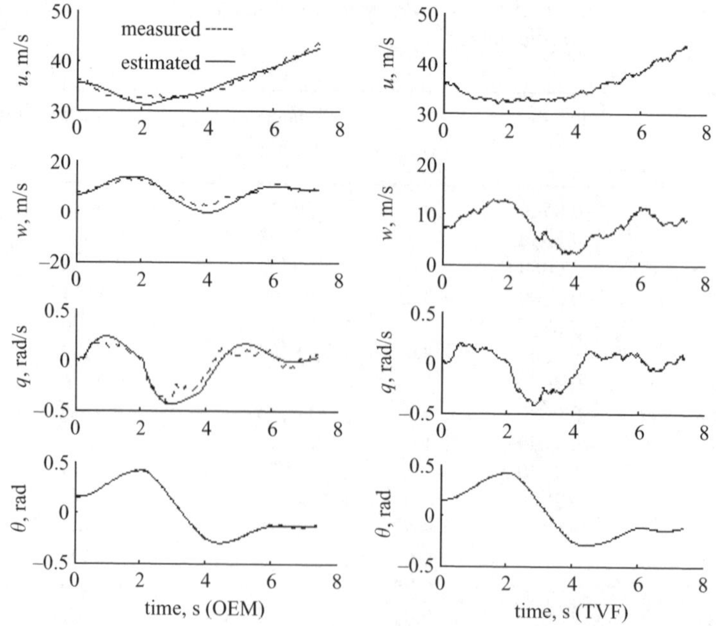

Figure 5.3 Comparison of the measured response in turbulence with the model predicted response from OEM and TVF (Example 5.1)

From the results, it is concluded that the time varying filter is more effective in estimating the parameters from data with turbulence compared with the output error method. Although the time varying filter requires considerably more computational time than the output error method, no convergence problems were encountered during application of this approach to the aircraft data in turbulence.

5.4 Epilogue

The output error method of Chapter 3 accounts for measurement noise only. For parameter estimation from data with appreciable levels of process noise, a filter error method or an extended Kalman filter has to be applied for state estimation. The system parameters and the noise covariances in the filter error method can be estimated by incorporating either a steady state (constant gain) filter or a time varying filter (TVF) in the iterative Gauss-Newton method for optimisation of the cost function. The steady state filter works well for the linear and moderately nonlinear systems, but for a highly nonlinear system, the time varying filter is likely to yield better results. The difficulties arising from complexities in software development and high consumption of CPU time and core (storage/memory) have restricted the use of the time varying filter on a routine basis.

In the field of aircraft parameter estimation, the analysts usually demand the flight manoeuvres to be conducted in calm atmospheric conditions (no process noise). However, in practice, this may not always be possible since some amount of turbulence will be present in a seemingly steady atmosphere. The filter error method has been extensively applied to aircraft parameter estimation problems [11,12]. The extended Kalman filter (EKF) is another approach, which can be used to obtain the filtered states from noisy data. EKF is generally used for checking the kinematic consistency of the measured data [13].

5.5 References

1 BALAKRISHNAN, A. V.: 'Stochastic system identification techniques', in KARREMAN, H. F. (Ed.): 'Stochastic optimisation and control' (Wiley, London, 1968)
2 MEHRA, R. K.: 'Identification of stochastic linear dynamic systems using Kalman filter representation', *AIAA Journal*, 1971, **9**, pp. 28–31
3 YAZAWA, K.: 'Identification of aircraft stability and control derivatives in the presence of turbulence', AIAA Paper 77-1134, August 1977
4 MAINE, R. E., and ILIFF, K. W.: 'Formulation and implementation of a practical algorithm for parameter estimation with process and measurement noise', *SIAM Journal on Applied Mathematics*, 1981, **41**, pp. 558–579
5 JATEGAONKAR, R. V., and PLAETSCHKE, E.: 'Maximum likelihood estimation of parameters in linear systems with process and measurement noise', DFVLR-FB 87-20, June 1987

6 POTTER, J. E.: 'Matrix quadratic solutions', *SIAM Journal Appl. Math.*, 1966, **14**, pp. 496–501
7 JATEGAONKAR, R. V., and PLAETSCHKE, E.: 'Algorithms for aircraft parameter estimation accounting for process and measurement noise', *Journal of Aircraft*, 1989, **26**, (4), pp. 360–372
8 MAINE, R. E., and ILIFF, K. W.: 'Identification of dynamic systems', AGARD AG-300, vol. 2, 1985
9 MORELLI, E. A.: 'Estimating noise characteristics from flight test data using optimal Fourier smoothing', *Journal of Aircraft*, 1995, **32**, (4), pp. 689–695
10 SINGH, J.: 'Application of time varying filter to aircraft data in turbulence', *Journal of Institution of Engineers (India), Aerospace, AS/1*, 1999, **80**, pp. 7–17
11 MAINE, R. E., and ILIFF, K. W.: 'User's manual for MMLE3 – a general FORTRAN program for maximum likelihood parameter estimation', NASA TP-1563, 1980
12 JATEGAONKAR, R. V., and PLAETSCHKE, E.: 'A FORTRAN program for maximum likelihood estimation of parameters in linear systems with process and measurement noise – user's manual', DFVLR-IB, 111-87/21, 1987
13 PARAMESWARAN, V., and PLAETSCHKE, E.: 'Flight path reconstruction using extended Kalman filtering techniques', DLR-FB 90-41, August 1990

5.6 Exercises

Exercise 5.1

Let $P - \phi^{-1}P(\phi^T)^{-1}$ be given. This often occurs in the solution of the continuous-time Riccati equation. Use the definition of the transition matrix $\phi = e^{F\Delta t}$ and its first order approximation to obtain $P - \phi^{-1}P(\phi^T)^{-1} = (FP + PF^T)\Delta t$.

Exercise 5.2

We have seen in the development of the Kalman filter that the *a posteriori* state covariance matrix is given as $\hat{P} = (I - KH)\tilde{P}$ (see eq. (5.45)). Why should the eigenvalues of KH be less than or at least equal to 1? (Hint: study the definition of P; see Appendix for covariance matrix.)

Chapter 6

Determination of model order and structure

6.1 Introduction

The time-series methods have gained considerable acceptance in system identification literature in view of their inherent simplicity and flexibility [1–3]. These techniques provide external descriptions of systems under study and lead to parsimonious, minimum parameterisation representation of the process. The accurate determination of the dynamic order of the time-series models is a necessary first step in system identification.

Many statistical tests are available in the literature which can be used to find the model order for any given process. Selection of a reliable and efficient test criterion has been generally elusive, since most criteria are sensitive to statistical properties of the process. These properties are often unknown. Validation of most of the available criteria has generally been via simulated data. However, these order determination techniques have to be used with practical systems with unknown structures and finite data. It is therefore necessary to validate any model order criterion using a wide variety of data sets from differing dynamic systems.

The aspects of time-series/transfer function modelling are included here from the perspective of them being special cases of specialised representations of the general parameter estimation problems. The coefficients of time-series models are the parameters, which can be estimated by using the basic least squares, and maximum likelihood methods discussed in Chapters 2 and 3. In addition, some of the model selection criteria are used in EBM procedure for parameter estimation discussed in Chapter 7, and hence the emphasis on model selection criteria in the present chapter.

6.2 Time-series models

The time-series modelling is one of the specialised aspects of system identification/parameter estimation study. It addresses the problem of determining coefficients of a differential or difference equations, which can be fitted to the empirical data, or obtaining coefficients of a transfer function model of a system from its input-output data. One of the main aims of time-series modelling is the use of the model for prediction of the future behaviour of the system or phenomena. One of the major applications of this approach is to understand various natural phenomena, e.g., rainfall-runoff prediction. In general, time-series models are a result of stochastic (random) input to some system or some inaccessible random like influence on some phenomenon, e.g., the temperature variation at some point in a room at certain time. Hence, a time-series can be considered as a stochastic phenomenon. The modelling and prediction of the seasonal time-series are equally important and can be handled using extended estimation procedures. Often, assumption of ergodicity (see Section A.13) is made in dealing with time-series modelling aspects.

We will generally deal with discrete-time systems. Although many phenomena occurring in nature are of continuous type and can be described by continuous-time models, the theory of the discrete-time modelling is very handy and the estimation algorithms can be easily implemented using a digital computer. In addition, the discrete-time noise processes can be easily handled and represented by simple models. However, continuous-time phenomena can also be represented by a variety of (similar) time-series models. A general linear stochastic discrete-time system/model is described here with the usual meaning for the variables [2]:

$$\begin{aligned} x(k+1) &= \Phi_k x(k) + Bu(k) + w(k) \\ z(k) &= Hx(k) + Du(k) + v(k) \end{aligned} \tag{6.1}$$

However, for time-series modelling a canonical form (of eq. (6.1)) known as Astrom's model is given as

$$A(q^{-1})z(k) = B(q^{-1})u(k) + C(q^{-1})e(k) \tag{6.2}$$

Here, A, B and C are polynomials in q^{-1} which is a shift operator defined as

$$q^{-n}z(k) = z(k-n) \tag{6.3}$$

For a SISO system, we have the expanded form as

$$\begin{aligned} z(k) &+ a_1 z(k-1) + \cdots + a_n z(k-n) \\ &= b_0 u(k) + b_1 u(k-1) + \cdots + b_m u(k-m) \\ &\quad + e(k) + c_1 e(k-1) + \cdots + c_p e(k-p) \end{aligned} \tag{6.4}$$

where z is the discrete measurement sequence, u is the input sequence and e is the random noise/error sequence.

We have the following equivalence:

$$\begin{aligned} A(q^{-1}) &= 1 + a_1 q^{-1} + \cdots + a_n q^{-n} \\ B(q^{-1}) &= b_0 + b_1 q^{-1} + \cdots + b_n q^{-m} \\ C(q^{-1}) &= 1 + c_1 q^{-1} + \cdots + c_n q^{-p} \end{aligned}$$

Here, a_i, b_i and c_i are the coefficients to be estimated. We also assume here that the noise processes w and v are uncorrelated and white. In addition, we assume that the time-series we deal with are stationary in the sense that first and second order (and higher) statistics are not dependant on time t explicitly. For mildly non-stationary time-series, the appropriate models can be fitted to the segments of such time-series.

Certain special forms are specified next. These models are called time-series models, since the observation process can be considered as a time-series of data that has some dynamic characteristics, affected usually by a random process. We assume here that inputs are such that they excite the modes of the system. This means that the input contains sufficient frequencies to excite the dynamic modes of the system. This will in turn assure that in the output, there is sufficient effect of the modes and hence the information so that from input-output time-series data, one can accurately estimate the characteristics of the process.

Astrom's model

This is the most general linear time-series analysis model, with full form of error/noise model. Given input (u)/output (z) data, the parameters can be estimated by some iterative process, e.g., ML method. The transfer function form is given by:

$$z = \frac{B(q^{-1})}{A(q^{-1})}u + \frac{C(q^{-1})}{A(q^{-1})}e \tag{6.5}$$

This model can be used to fit time-series data, which can be considered to be arising out of some system phenomenon with a controlled input u and a random excitation (see Fig. 6.1).

Autoregressive (AR) model

By assigning $b_i = 0$ and $c_i = 0$ in the Astrom's model, we get:

$$z(k) = -a_1 z(k-1) - \cdots - a_n z(k-n) + e(k) \tag{6.6}$$

The transfer function form can be easily obtained as

$$z = \frac{1}{A(q^{-1})}e \tag{6.7}$$

Here, the output process $z(k)$ depends on its previous values (and hence the name autoregressive) and it is excited by the random signal e. It is assumed that the parameters a_i are constants such that the process z is stationary (see Fig. 6.2).

We can consider that $1/A(q^{-1})$ is an operator, which transforms the process e into the process z. The polynomial A determines the characteristics of the output signal z and the model is called an 'all poles' model. This is because the roots of $A(q^{-1}) = 0$

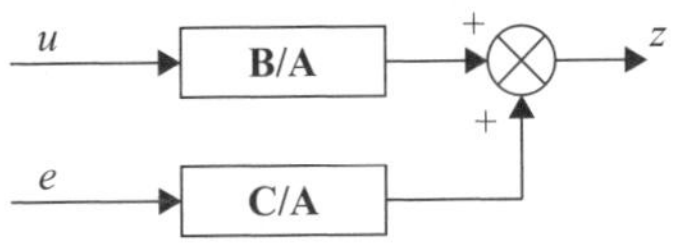

Figure 6.1 Astrom's model

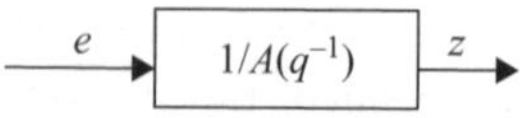

Figure 6.2 *AR model*

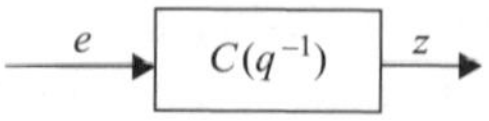

Figure 6.3 *MA model*

are the poles of the transfer function model. The input process e is inaccessible and immeasurable. The parameters of A can be estimated by using the least squares method. In addition, this model is very useful for determining the spectrum of the signal z, if input process e is considered as white process noise, since the parameters of A are estimated and hence known. This method of estimation of spectrum of a signal contrasts with the one using the Fourier transform. However, both the methods are supposed to give similar spectra. It is most likely that the autoregressive spectrum will be smoother compared to the Fourier spectrum.

Moving average (MA) model
If we put $a_i = 0$ and $b_i = 0$ in the Astrom's model, we get:

$$z(k) = e(k) + c_1 e(k-1) + \cdots + c_p e(k-p) \tag{6.8}$$

The process z is now a linear combination of the past and present values of the inaccessible random input process e (see Fig. 6.3).

The roots of $C(q^{-1}) = 0$ are the zeros of the model. The process z is called the MA process and is always stationary since $A(q^{-1}) = 1$. In this form, the output signal does not regress over its past values.

Autoregressive moving average (ARMA) model
Letting $b_i = 0$ in the Astrom's model, we obtain an ARMA model, since it contains both AR and MA parts. We emphasise here that the control input u is absent:

$$z(k) + a_1 z(k-1) + \cdots + a_n z(k-n) = e(k) + c_1 e(k-1) + \cdots + c_p e(k-p) \tag{6.9}$$

$$z = \frac{C(q^{-1})}{A(q^{-1})} e \tag{6.10}$$

So this model is a zero/pole type model and has the structure of the output/input model. More complex time-series can be accurately modelled using this model (see Fig. 6.4).

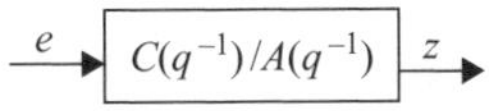

Figure 6.4 ARMA model

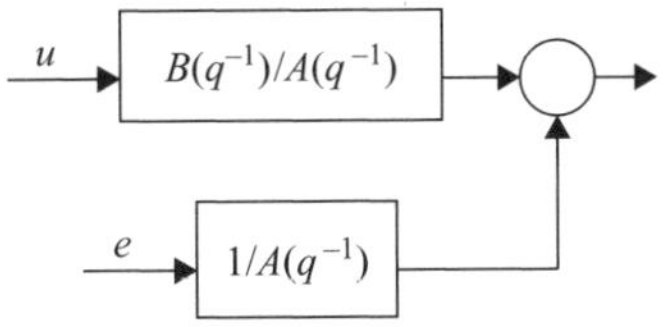

Figure 6.5 LS model

Least squares model
By letting $c_i = 0$ in the Astrom's model, we get

$$\begin{aligned} &z(k) + a_1 z(k-1) + \cdots + a_n z(k-n) \\ &\quad = b_0 u(k) + b_1 u(k-1) + \cdots + b_m u(k-m) + e(k) \end{aligned} \tag{6.11}$$

Here, control input u is present. The model is so called since its parameters can be easily estimated by the LS method. The transfer function form is

$$z = \frac{B(q^{-1})}{A(q^{-1})} u + \frac{1}{A(q^{-1})} e \tag{6.12}$$

It has an AR model for the noise part and the output/input model for the signal part. Determination of $B(q^{-1})/A(q^{-1})$ gives the transfer function model of the system (see Fig. 6.5). One can obtain a discrete Bode diagram of the system from this pulse transfer function and then convert it to the continuous-time domain to interpret the dynamic behaviour of the system. One can use a complex curve fitting technique or bilinear/Padé method [4].

6.2.1 *Time-series model identification*

The estimation of parameters of MA and ARMA can be done using the ML approach, since the unknown parameters appear in the MA part, which represents itself as unknown time-series e. However, parameters of AR and LS models can be estimated using the LS method. Assumption of the identifiability of the coefficients of the postulated models is pre-supposed (see Section A.27).

Let the LS model be given as in eqs (6.11) and (6.12). We define the equation error as shown in Fig. 6.6:

$$\begin{aligned} e(k) &= A(q^{-1}) z(k) - B(q^{-1}) u(k) \\ r(k) &= \hat{A}(q^{-1}) z(k) - \hat{B}(q^{-1}) u(k) \end{aligned} \tag{6.13}$$

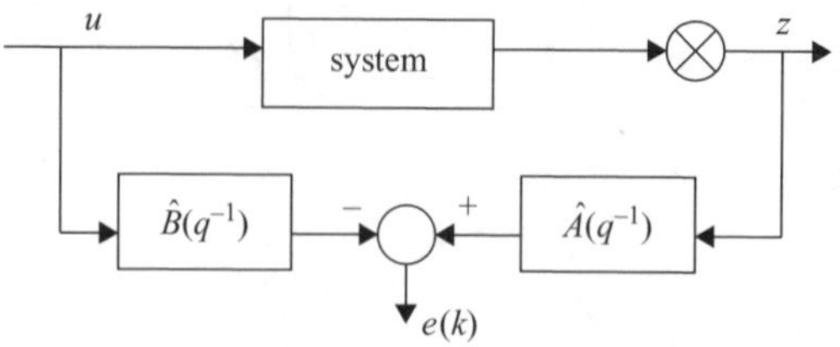

Figure 6.6 Equation error formulation

The above equations can be put in the form: $z = H\beta + e$ where $z = \{z(n+1), z(n+2), \ldots, z(n+N)\}^T$. Also,

$$H = \begin{bmatrix} -z(n) & -z(n-1) & -z(1) & u(n) & u(n-1) & u(1) \\ -z(n+1) & -z(n) & -z(2) & u(n+1) & u(n) & u(2) \\ \vdots & & \vdots & \vdots & & \\ -z(N+n-1) & \cdots & -z(n) & u(N+n-1) & \cdots & u(N) \end{bmatrix} \tag{6.14}$$

N = number of total data used: $m = n$ and $b_0 = 0$.

For example, let $n = 2$ and $m = 1$, then

$$e(k) = z(k) + a_1 z(k-1) + a_2 z(k-2) - b_0 u(k) - b_1 u(k-1) \tag{6.15}$$

$$z(k) = \begin{bmatrix} -z(k-1) & -z(k-2) \end{bmatrix} \begin{bmatrix} a_1 \\ a_2 \end{bmatrix} + [u(k) \quad u(k-1)] \begin{bmatrix} b_0 \\ b_2 \end{bmatrix} + e(k)$$

$$z(k+1) = \begin{bmatrix} -z(k) & -z(k-1) \end{bmatrix} \begin{bmatrix} a_1 \\ a_2 \end{bmatrix} + [u(k+1) \quad u(k)] \begin{bmatrix} b_0 \\ b_2 \end{bmatrix} + e(k+1)$$

The above leads to

$$z = H\beta + e$$

Using the LS method, we get

$$\hat{\beta} = \left\{\hat{a}_1, \hat{a}_2, \ldots, \hat{a}_n \,\vdots\, \hat{b}_1, \hat{b}_2, \ldots, \hat{b}_m\right\} = (H^T H)^{-1} H^T z \tag{6.16}$$

The parameters/coefficients of time-series models can be estimated using the system identification toolbox of MATLAB [2]. The crucial aspect of time-series modelling is that of selection of model structure (AR, MA, ARMA or LS) and the number of coefficients for fitting this model to the time-series data.

6.2.2 Human-operator modelling

Time-series/transfer function modelling has been used in modelling the control activity of the human operator [3] in the manual control experiment of compensatory tracking task in flight research simulators [4]. The empirical time-series based human-operator models (control theoretic models) can be obtained from the input-output

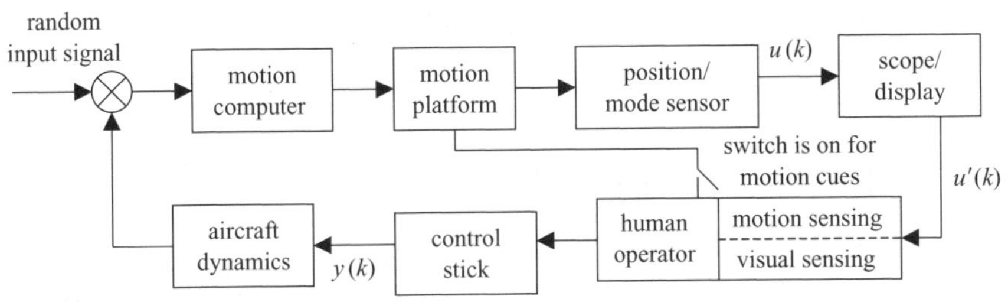

Figure 6.7 Compensatory tracking experiment

data generated while he/she performs a manual control task (either in a fixed based or motion-based flight simulator, see Fig. 6.7). Input to the pilot is in the form of a visual sensory input as derived from the horizon line on an oscilloscope (or some display). This signal is derived from a gyroscope or a pitch attitude sensor (for a motion-based simulator). The actual input is taken from the equivalent electrical input to the display device assuming the dynamics of the display as constant. The output signal is derived from the motion of the stick used by the operator in performing the control task (see Fig. 6.7).

One can define the human-operator model in such a task as the LS model:

$$A(q^{-1})y(k) = B(q^{-1})u(k) + e(k) \tag{6.17}$$

Here, $u(k)$ is the input to the operator, and y is his/her response. An implicit feature of the LS model is that the operator's response naturally separates into the numerator and denominator contributions as shown below [4, 5]:

$$\begin{aligned} H_{sp}(j\omega) &= B(j\omega) \\ H_{EN}(j\omega) &= \frac{1}{A(j\omega)} \end{aligned} \tag{6.18}$$

Thus, H_{sp}, the numerator term can be correlated to the human sensory and prediction part. The denominator H_{EN} term can be correlated to the equalising and the neuromuscular part. In the tracking task, if visual input is viewed as a relatively unpredictable task, then if the motion cue were added (in addition to the visual cues), it will elicit the lead response from the operator. This will show up in the sensory and prediction part of the transfer function H_{sp}. Thus, phase improvement (phase 'lead' in control system jargon) generated by the operator during the congruent motion cues over the visual cues, is attributed to the functioning of the 'predictor operator' in the human pilot. The motion cue is considered as congruent because it is helping or aiding piloting task as the visual cues, and is not contradictory to visual cues.

Thus, it can be seen from the foregoing discussion that simple time-series modelling can be used to isolate the contributions of motion cues, translatory cues and cues from other body sensors to have a better understanding of manual control problems in any environment.

6.3 Model (order) selection criteria

In the absence of *a priori* knowledge, any system that is generating time-series output can be represented by the more popular autoregressive (AR) or a least squares (LS) model structure. Both these structures represent a general nth order discrete linear time invariant system affected by random disturbance. The problem of model order determination is to assign a model dimension so that it adequately represents the unknown system. Model selection procedure involves selecting a model structure and complexity. A model structure can be ascertained based on the knowledge of the physics of the system. For certain processes, if physics is not well understood, then a black-box approach can be used. This will lead to a trial and error iterative procedure. However, in many situations, some knowledge about the system or the process is always available. Then, further refinements can be done using system identification techniques.

Here, we consider the modelling problem in the context of structure and order selection based on well-defined Model Selection Criteria (MSC). We describe several such MSC arising out of various different but related principles of goodness of fit and statistical measures. The criteria are classified based on fit error, number of model parameters, whiteness of residuals and related approaches.

6.3.1 Fit error criteria (FEC)

We describe criteria based on the concept of fit error.

6.3.1.1 Fit error criterion (FEC1)

One of the natural MSC is a measure of the difference between the actual response of the system and estimated response of the postulated/estimated model. Evaluate the FEC as follows [6]:

$$\mathrm{FEC1} = \frac{(1/N)\sum_{k=1}^{N}[z_k - z_k(\hat{\beta}_1)]^2}{(1/N)\sum_{k=1}^{N}[z_k - z_k(\hat{\beta}_2)]^2} \tag{6.19}$$

Apply the decision rule:

If FEC1 < 1 select the model with $\hat{\beta}_1$.
If FEC1 > 1 select the model with $\hat{\beta}_2$.

The ratio FEC can be corrected for the number (n_1, n_2) of unknown parameters in the model by replacing N by $N - n_1$ and $N - n_2$ in the numerator and the denominator of eq. (6.19) respectively. The FEC is considered to be a subjective criterion thereby requiring subjective judgement, i.e., if FEC1 ≈ 1, then both the models would be just as good; one has to prefer a model with fewer coefficients (parameters).

6.3.1.2 Fit error criterion (FEC2)

An alternative FEC, sometimes called prediction fit error (PFE) in the literature, can be used to judge the suitability of the model fit:

$$\mathrm{FEC2} = \frac{(1/N)\sum_{k=1}^{N}[z_k - z_k(\hat{\beta})]^2}{(1/N)\sum_{k=1}^{N} z_k^2} \tag{6.20}$$

Replacing N with $N - n$ can correct the criterion, for the degrees of freedom, in the numerator of eq. (6.20). Essentially, FEC2 compares models based on reduction of residuals to signal power ratio of successive models. Insignificant change in the value of FEC2 determines the order of the model. Essentially, one locates the knee of the curve FEC2 versus model order. Generally, this criterion does not give a sharp knee and hence again requires subjective judgment. In parameter estimation literature (Chapters 2 and 3), this criterion is the usual fit error criterion (often used as percentage fit error: PFE = FEC2 × 100).

6.3.1.3 Residual sum of squares (RSS)

Often, the sum of residuals is used to judge the model adequacy:

$$\mathrm{RSS} = \sum_{k=1}^{N}[z_k - \hat{z}_k(\hat{\beta})]^2 \tag{6.21}$$

If any new parameter enters the model, then there should be significant reduction in RSS, otherwise it is not included in the model.

6.3.1.4 Deterministic fit error (DFE)

For models of input-output type, this is a useful criterion. It accounts for the effects of modelling and computational errors. For the TF type model, the deterministic fit error is given by [7]:

$$\mathrm{DFE} = z - \frac{\hat{B}(q^{-1})}{\hat{A}(q^{-1})}u \tag{6.22}$$

Similar observations as for FEC2 can be made regarding this criterion. The prediction error criteria (PEC) generally provide quantitative means for selecting the models that best support the measured data. The capability of a model to predict the responses of the system for a class of inputs can be judged based on these PECs given next.

6.3.1.5 Prediction error criterion 1 (PEC1)

In this case, the data to be analysed (measured data) are divided into two consecutive segments. The first segment of data is used in identification procedure to estimate the unknown parameters. Then, this model (parameters) is used to predict the response for the second segment and compared with it. The model that predicts this response most accurately is considered an accurate model. Again, subjective judgement is involved since 'most accurately' is not quantified. The PEC1 can be used also as a model validation criterion.

Let the identified model from the first data segment be called $M(\hat{\beta} \mid z_k, k = 1, 2, \ldots, N1)$. Then prediction error time history for the second segment up to $N2$ is generated as:

$$e_z(j) = z_j - \hat{z}_j\{M(\hat{\beta} \mid z_k, k = 1, 2, \ldots, N1)\}; \qquad j = N1 + 1, \ldots, N2 \tag{6.23}$$

Here $N > N1 + N2$.

Further quantification of $e_z(j)$ can be obtained by evaluating its power, i.e., variance as

$$\sigma_{e_z}^2 = \frac{1}{N_2} \sum_{j=1}^{N_2} [e_z(j)]^2 \tag{6.24}$$

Very low value of this variance signifies a good prediction.

6.3.1.6 Prediction error criterion 2 (PEC2)

In this procedure, prediction error is estimated statistically and the criterion is the well-known Akaike's Final Prediction Error (FPE), described next.

6.3.2 Criteria based on fit error and number of model parameters

6.3.2.1 Final prediction error (FPE)

A good estimate of prediction error for a model with n parameters is given by the final prediction error [8]:

$$\text{FPE} = \sigma_r^2(N, \hat{\beta}) \frac{N + n + 1}{N - n - 1}; \qquad \sigma_r^2 = \text{variance of the residuals} \tag{6.25}$$

A minimum is sought with respect to n, the number of parameters. Absolute minimum occurs when σ_r^2 is zero. FPE includes a penalty for large model orders. This means that if n increases, the numerator increases. The penalty is paid in FPE. If n is large, then σ_r^2 will reduce, and hence a compromise is struck. For real data situations local minimum can result. This test is developed for the univariate process corrupted by white noise. The penalty for degrees of freedom is greatly reduced for large N, meaning thereby that FPE is less sensitive to n, if N is large.

6.3.2.2 Akaike's information criterion (AIC/alternatively, it denotes information criterion)

Akaike refined FPE into AIC by extending the maximum likelihood principle and taking into account the parametric dimensionality [9]:

$$\begin{aligned} \text{AIC} = &- 2 \ln(\text{maximum likelihood}) \\ &+ 2(\text{number of independent parameters in the model}) \end{aligned}$$

or

$$\text{AIC} = -2\ln(L) + 2n$$

If the two models are equally likely ($L_1 \approx L_2$), then the one with fewer parameters is chosen. We see from the above expression that if the number of parameters increases, the AIC also increases, and hence the model is less preferable.

For an autoregressive (AR) model of order n we get

$$\text{AIC}(n) = N\ln\sigma_r^2 + 2n \tag{6.26}$$

This is a generalised concept of FPE.

For $n = 0, 1, \ldots$, the value of n, for which the AIC(n) is minimum, is adopted as the true order of the model. However, AIC might not give a consistent model order in a statistical sense. We see from eq. (6.26) that as n increases, the second term increases, but due to fitting with more parameters, the first term decreases, so a compromise is struck.

These criteria, for a given model structure, may not attain unique minimum. Under weak assumptions, they are described by χ^2 distribution. It is well known that FPE and AIC are asymptotically equivalent.

6.3.2.3 Criterion autoregressive transfer function (CAT)

Parzen [10] and Tong [11] advanced these CAT methods for model order determination.

- *Parzen (PCAT1)* This criterion was advanced with a view to obtaining the best finite AR model based on a finite number of measurements used for time-series modelling. The formula for PCAT is given as

$$\text{PCAT1}(n) = 1 - \frac{\hat{\sigma}_\infty^2}{\bar{\sigma}_r^2} + \frac{n}{N}; \qquad n = 0, 1, \ldots \tag{6.27}$$

  where $\hat{\sigma}_\infty^2$ = estimate of the one-step ahead prediction error variance σ^2 and $\bar{\sigma}_r^2$ = unbiased estimate: $(N)/(N-1)\hat{\sigma}_r^2$.

  PCAT1 can be considered asymptotically to obtain the same order estimate as that obtained by AIC [11]. PCAT1 signifies the minimisation of relative mean square error between nth order AR model and theoretical AR model.
- *Parzen (PCAT2)* A modified criterion is given by

$$\text{PCAT2}(n) = \frac{1}{N}\sum_{j=1}^{n}\frac{1}{\bar{\sigma}_j^2} - \frac{1}{\bar{\sigma}_r^2} \tag{6.28}$$

  Here, $\text{PCAT2}(0) = -(1+N)/N$, and minimum is sought.
- A modification of PCAT2 was proposed [11], since for true AR(l) model, PCAT2 may prefer AR(0) model to AR(l) model. Thus, modified criterion which avoids

this ambiguity, is given by

$$\mathrm{MCAT}(n) = \frac{1}{N}\sum_{j=0}^{n}\frac{1}{\bar{\sigma}_j^2} - \frac{1}{\bar{\sigma}_r^2} \tag{6.29}$$

and minimum is sought. It has been shown that MCAT and AIC have identical local behaviour. However, global maxima of MCAT(n) and AIC(n) do not necessarily occur at the same n.

6.3.3 *Tests based on whiteness of residuals*

These tests are used to check whether the residuals of fit are a white noise sequence, thereby asserting independence at different time instants. We describe two such tests.

6.3.3.1 Autocorrelation based whiteness of residuals (ACWRT)

The test is performed as follows:

Estimate the autocorrelation function $R_{rr}(\tau)$ of residual sequence $r(k)$, for lag $\tau = 1, 2, \ldots, \tau_{\max}$

$$\hat{R}_{rr}(\tau) = \frac{1}{N}\sum_{k=\tau}^{N} r(k)r(k-\tau) \tag{6.30}$$

Here it is assumed that $r(k)$ is a zero mean sequence. $\hat{R}_{rr}(\tau)$ is considered asymptotically unbiased and a consistent estimate of true autocorrelation [12]. Also, under null hypothesis, $\hat{R}_{rr}(\tau)$ for $\tau = 1, 2, \ldots$ are asymptotically independent and normal with zero mean and covariance of $1/N$. Thus, they must lie in the band $\pm 1.96/\sqrt{N}$ at least for 95 per cent of the times for the null hypothesis. Usually the normalised ratio is used: $\hat{R}_{rr}(\tau)/\hat{R}_{rr}(0)$. The autocorrelations tend to be an impulse function if the residuals are uncorrelated.

6.3.3.2 Whiteness of residuals (SWRT)

Stoica has proposed another test to check the residual of estimation for whiteness [13]. If a discrete time-series is a white sequence, then

$$\sum_{\tau=1}^{\tau_{\max}} R_{rr}^2(\tau) \le \frac{(k_j + 1.65\sqrt{2k_j})R_{rr}^2(0)}{N} \tag{6.31}$$

$$k_j = \tau_{\max} - n_j - 1; \qquad \tau_{\max} = 20$$

This SWRT test is considered more powerful than the previous test of eq. (6.30).

6.3.4 *F-ratio statistics*

The ratio test is based on the assumption of normally distributed random disturbances and requires *a priori* specifications of acceptance-rejection boundaries.

Due to this, such tests should be used in conjunction with other tests (see Sections A.6 and A.7):

$$F_{n_1 n_2} = \frac{V_{n_1} - V_{n_2}}{V_{n_2}} \frac{N - 2n_2}{2(n_2 - n_1)} \tag{6.32}$$

In the above equation V_{n1} and V_{n2} are the minimum values of the loss function for a model with n_1, n_2 parameters, respectively. The random variable F for large N is asymptotically $F(n_2 - n_1, N - n_2)$ distributed (see Sections A.20 and A.21). When the number of parameters is increased by 2, we have:

$$F(2, 100) = 3.09 \Rightarrow \text{Prob}(F > 3.09) = 0.05$$

and

$$F(2, \infty) = 3.00 \Rightarrow \text{Prob}(F > 3.00) = 0.05$$

Thus, at a risk level of 5 per cent and $N > 100$, the quantity F should be at least 3 for the corresponding reduction in loss function to be significant. A slightly different version of this criterion, where R could be any statistic computed using the square of a variable, e.g., covariance of residual, etc., is given as

$$F(j) = \frac{R(0, \hat{\beta}_j) - R(0, \hat{\beta}_{j+1})}{R(0, \hat{\beta}_{j+1})}(N - n_{j+1} - 1); \qquad j = 1, 2, \ldots \tag{6.33}$$

In the above, $R(0)$ can signify the autocorrelation at zero lag, implying the variance of the residuals.

6.3.5 *Tests based on process/parameter information*

1 Entropy

Entropy signifies disorder in the system (see Section A.16). This test is based on the amount of information measure of an AR process (of order n), which is characterised by the entropy. It is possible to judge the order of the given process before estimating the parameters because computation is based on the correlation matrices of different orders for assumed AR models [14]:

$$E_n(j) = \ln \frac{N - n_j}{N - 2n_j - 1} + \ln |S_{j+1}| - \ln |S_j| \tag{6.34}$$

Here, S_j = correlation matrix with its elements as autocorrelations, $\hat{R}_{rr}(\tau)$; $\tau = 1, 2, \ldots \tau_{\max}$ and $|S|$ = determinant of S.

The value of n_j for which $E_n(j)$ is minimum is selected as the adequate order. This test can be regarded as the pre-estimation criterion. It has to do with the minimisation of the difference in the adjacent entropies. Decrease in entropy signifies the increase of 'order' in the system and hence leads to proper model order of the system.

2 From the definition of the information measure it is known that the amount of uncertainty in estimates and hence dispersion are related to the inverse of the information matrix. Thus, near singularity of this matrix means large standard

deviations of the parameter estimates. Near singularity could also signify that the model structure has been overly large, thereby losing the parameter identifiability property.

6.3.6 Bayesian approach

The criteria based on this approach have been advanced in [15].

1 *Posteriori probability (PP)* This test is based on the Bayesian type procedure for discrimination of structure of the models. If C_j is the class of models, then the appropriateness of a class C_j to represent the given data set z is measured by the *a posteriori* probability $P(C_j \mid z)$. A low value of $P(C_j \mid z)$ indicates that C_j is inappropriate for representing z. This test gives a consistent order selection criterion; the simplified version is given as:

$$\mathrm{PP}(n_j) = -N \ln(\sigma_r^2) - n_j \ln\left(\frac{\sigma_z^2}{\sigma_r^2}\right) - (n_j + 1) \ln N \tag{6.35}$$

Here σ_z^2 = variance of the given time-series. One chooses n_j that gives the largest value of PP.

2 *B-statistic* Another consistent order determination statistic is given as

$$B(n_j) = N \ln(\sigma_r^2) + n_j \ln N \tag{6.36}$$

The model with minimum B is chosen, thus giving an adequate (AR or ARMA) model with n_j coefficients.

3 *C-statistic* It is interesting to note that the B-statistic is similar to another statistic:

$$C(n_j) = N \ln(\sigma_r^2) + n_j \mathbf{h}(N) \tag{6.37}$$

where $\mathbf{h}(N)$ is any monotonically increasing function of number of data, and satisfies the following condition:

$$\lim_{N\to\infty} \left[\frac{\mathbf{h}(N)}{N}\right] \Rightarrow 0$$

The decision rules based on C are statistically consistent [15].

6.3.7 Complexity (COMP)

This criterion is based on a compromise between the whiteness of model residuals and the accuracy of estimated parameters. It must be recalled that a good predictor should incorporate all the available information (residuals being white) and one should include accuracy of the parameter estimates in the model discrimination process.

The criterion is given as [16]:

$$\mathrm{COMP}(n_j) = \frac{1}{n_j}\sum_{j=1}^{n_j} p_{jj}^2 - \left\{\frac{trace(P)}{n_j}\right\}^2 + \frac{2}{n_j}\sum_{j=1}^{n_j}\sum_{l=j+1}^{n_j} p_{jl}^2 + \frac{2}{n_j}\sum_{\tau=1}^{\tau_{\max}} (N-\tau)\hat{R}_{rr}(\tau) \tag{6.38}$$

Here P is the covariance matrix of estimated parameters and p_{jl}, the elements of P.

Within a given structure, with a large number of parameters, increased interactions (P) will tend to positively contribute to COMP. The residuals will tend to be white, thereby making the fourth term decrease. Thus, COMP provides a trade-off between the accuracy of the estimates and whiteness of the residuals. However, computational requirement is more than that for AIC, B-statistic and FPE tests. This COMP criterion can be used for model structure as well as model order determination.

6.3.8 Pole-zero cancellation

For input-output (ARMA; see eq. (6.5)) or transfer function (LS) type models (see eq. (6.12)), the process of cancellation of zeros with poles can provide a model with a lesser degree of complexity. A systematic way of cancellation was given in Reference 17. In the conventional method, the numerator and denominator polynomials are factored and cancellation then becomes obvious. However, subjective judgement is involved, since the cancellation might not be perfect.

6.4 Model selection procedures [18]

The subjective tests have been used in many applications and the main difficulty in using these has been the choice of proper levels of statistical significance. The subjective tests tend to ignore the increase of variability of estimated parameters for large model orders. It is often common to assume a 5 per cent risk level as acceptable for the F-test and whiteness tests arbitrarily. However, the whiteness test-SWR does consider the cumulative effects of autocorrelations of residuals. The pole-zero cancellations are often made visually and are again subjective. A systematic exact pole-zero cancellation is possible, but it is computationally more complex [17]. Fit error methods are useful but again subjective and are only necessary but not sufficient conditions.

In the objective-type tests, an extremum of a criterion function is usually sought. The final prediction error (FPE) criterion due to Akaike is based on one-step-ahead prediction and is essentially designed for white noise corrupted processes. The Akaike information criterion AIC is a generalised concept based on a mean log likelihood function. Both the FPE and AIC depend only on residual variance and the number of estimated parameters. At times, these tests yield multiple minima. The criterion autoregressive transfer function (CAT) due to Parzen has been proposed as the best finite AR model derived from finite sample data generated by the AR model of infinite order. The MCAT is a modification of PCAT2 to account for any ambiguity, which may arise for 'true' first order AR processes due to omission of σ_0^2 terms.

Based on the experience gained, the following working rule is considered adequate for selection of the model order to fit typical experimental data [18].

Order determination:
evaluate entropy criterion (AR only)
evaluate FPE
perform F-test
check for pole-zero cancellations (for input-output model).

Model validation:
time history prediction
test residuals for whiteness
cross validation.

Alternatively, readers can arrive at their own rule based on study of other criteria discussed in this chapter.

6.4.1.1 Example 6.1

Generate data using the following polynomial form:

$$z(k) = -z(k-1) + 1.5z(k-2) - 0.7z(k-3) - 0.09z(k-4) + e(k) \tag{6.39}$$

Generate three sets of time-series data by adding random noise $e(k)$ with variance of 1.0, 0.16 and 0.0016 and using the above polynomial form for the AR model. Characterise the noise in this data using the time-series modelling approach by fitting an AR model to the data and estimate the parameters of the model.

6.4.1.2 Solution

Three sets of time-series data are generated using the function IDSIM of the system identification toolbox of PC MATLAB. Given the time-series data, the objective here is to obtain an estimate of the measurement noise covariance in the data. In general, the order of the model to be fitted to the data will not be known exactly and hence various orders of the AR model should be tried before one can arrive at the adequate order based on certain criteria. Hence, using the function AR, AR models with order $n = 1$ to 6, are used to fit the simulated data. For each order, the quality of fit is evaluated using the following steps:

(i) Function COMPARE to evaluate the quality of the model fit.
(ii) Function COV to find the residual covariance and RESID to plot the correlation function of the residuals.
(iii) Akaike's final prediction error criterion FPE.
(iv) Information theoretic criterion-AIC.
(v) PEEN (percentage estimation error norm).

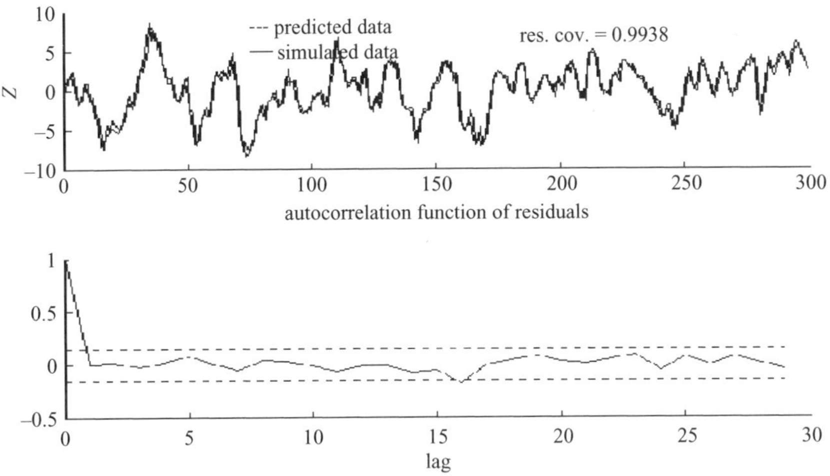

Figure 6.8 *Time-series modelling – 3rd order AR model for data set 1 – noise covariance* = 1 *(Example 6.1)*

The program folder **Ch6ARex1** created using the functions from the system identification toolbox is used for the noise characterisation. Figure 6.8 shows the comparison of model response to the time-series data when the noise variance is 1 and the order of the AR model chosen is 3. It is clear that the residual covariance matches the standard deviation of the noise (1), used in generating the data. The autocorrelation function is also plotted along with bounds. This satisfies the whiteness test for the residuals thereby proving the adequacy of the model to fit the data.

Table 6.1 gives the results of fit error criteria. Since the AR model also gives an estimate of the coefficients of the polynomial and the true values are known (eq. (6.39)), the %PEEN is computed and used as an additional criterion to judge the adequacy of fit in addition to the other fit error criteria. The PEEN indicates a minimum at order 3 and the fit criteria FPE and AIC indicate that even if the order of the model is increased beyond the third, the fit criteria do not show great decrement. Thus, it can be concluded that, for this case of simulated data, the 3rd order AR model gives the best fit and the corresponding RES-COVs give the variance of the noise in the data for all the three cases. It must be emphasised here that this technique of fitting an AR or ARMA model to measurements from sensors and estimating the covariance of the residuals could be used as a tool for characterisation of sensor noise in the measured data.

6.4.1.3 Example 6.2

Simulate data of a target moving with constant acceleration and acted on by an uncorrelated noise, which perturbs the constant acceleration motion. Add measurement noise with standard deviation of 1, 5 and 10 to this data to generate

Table 6.1 Fit criteria – simulated 3rd order AR model data (Example 6.1)

Variance of noise in simulation	Model order	RES-COV (after estimation)	FPE	AIC	%PEEN
1	1	1.4375	1.4568	110.8633	31.8
1	2	1.0021	1.0224	4.6390	8.4
1	**3**	**0.9938**	**1.0206**	**4.1231**	**2.2**
1	4	0.9851	1.0185	3.4971	5.6
1	5	0.9771	1.0170	3.0649	7.8
1	6	0.9719	1.0184	3.4519	8.2
0.16	1	0.2300	0.2331	−438.9112	31.8
0.16	2	0.1603	0.1636	−545.1355	8.4
0.16	**3**	**0.1590**	**0.1633**	**−545.6514**	**2.2**
0.16	4	0.1576	0.1630	−546.2774	5.6
0.16	5	0.1563	0.1628	−546.709	7.8
0.16	6	0.1555	0.1629	−546.222	8.2
0.0016	1	0.0023	0.0023	−1820.4622	31.8
0.0016	2	0.0016	0.0016	−1926.6865	8.4
0.0016	**3**	**0.0016**	**0.0016**	**−1927.2024**	**2.2**
0.0016	4	0.0016	0.0016	−1927.8284	5.6
0.0016	5	0.0016	0.0016	−1928.26	7.8
0.0016	6	0.0016	0.0016	−1927.87	8.2

three sets of data. Fit generalised ARMA models with orders $1, 2, 3, 4, 5, 6$ for each data set to characterise the noise in the data.

6.4.1.4 Solution

The target data is generated using the following state and measurement models:

$$(a) \qquad x(k+1) = \varphi x(k) + Gw(k) \tag{6.40}$$

Here, w is the process noise with $E[w] = 0$ and $Var[w] = Q$ and x is the state vector consisting of target position, velocity and acceleration. ϕ is the state transition matrix given by

$$\phi = \begin{bmatrix} 1 & \Delta t & \dfrac{\Delta t^2}{2} \\ 0 & 1 & \Delta t \\ 0 & 0 & 1 \end{bmatrix}$$

G is a matrix associated with process noise and is given by

$$G = \begin{bmatrix} \dfrac{\Delta t^2}{2} \\ \Delta t \\ 1 \end{bmatrix}$$

$$(b) \qquad z(k) = Hx(k) + v(k) \tag{6.41}$$

Here, H is the observation matrix given by $H = [1 \quad 0]$ so that only the position measurement is available and the noise in the data is to be characterised. v is the measurement noise with $E[v] = 0$ and $Var[v] = R$.

The following initial conditions are used in the simulation: $x_0 = [200 \quad 1 \quad 0.05]$; process noise covariance, $Q = 0.001$ and sampling interval $\Delta t = 1.0$ s.

The data simulation and the estimation programs used for this example are contained in folder **Ch6ARMAex2**. The functions from the system identification toolbox in MATLAB are used for this purpose. Three sets of data are generated by adding Gaussian random noise with standard deviation of 1, 5 and 10 corresponding to the measurement noise variance (R) of 1, 25 and 100. The function ARMAX is used to fit ARMA models of different orders to the data. The results presented in Table 6.2 indicate that the residual covariances match the measurement noise covariances used in the simulation reasonably well. All the three criteria indicate minimum at $n = 6$ for this example. This example amply demonstrates that the technique of using the ARMA models to fit the data can be used for characterising the noise present in any measurement signals, and the estimated covariances can be further used in the Kalman filter, etc.

From the above two examples, it is clear that the RES-COV and FPE have nearly similar values.

6.4.1.5 Example 6.3

Certain criteria for AR/ARMA modelling of time-series data were evaluated with a view to investigating the ability of these tests in assigning a given data set to a particular class of models and to a model within that class.

The results were generated via simulation wherein AR(n) and ARMA(n, m) models were fitted to the AR(2) and ARMA(2,1) process data in a certain specific sequence. These data were generated using Gaussian, zero mean and unit variance random excitation. The model selection criteria were evaluated for ten realisations (using Monte Carlo Simulations; see Section A.31) of each AR/ARMA process. The results are presented in Tables 6.3 to 6.6.

This exercise reveals that the PP and B-statistic criteria perform better than other criteria. Also PP and B-statistic results seem equivalent. The FPE yields over-fitted models. The SWR compares well with PP and B-statistic. The higher order AR model may be adequate to fit the data generated by the ARMA(2,1) process. This seems to agree with the fact that a long AR model can be used to fit an ARMA process data.

Table 6.2 Fit error criteria – simulated data of a moving target (Example 6.2)

Variance of noise in simulation	Model order	RES-COV	FPE	AIC
1	1	3.8019	3.8529	402.6482
1	2	1.5223	1.5531	130.0749
1	3	1.3906	1.4282	104.9189
1	4	1.4397	1.4885	117.3228
1	5	1.3930	1.4499	109.4445
1	6	1.3315	1.3951	97.8960
25	1	40.9705	41.5204	1115
25	2	39.3604	40.1556	1106
25	3	37.5428	38.5575	1094
25	4	32.2598	33.3534	1050
25	5	33.8161	35.1963	1066
25	6	28.3664	29.7218	1015
100	1	137.5646	139.4111	1479
100	2	135.2782	138.0111	1476
100	3	134.8746	138.5198	1477
100	4	122.1087	126.2480	1449
100	5	122.3616	127.3560	1452
100	6	122.0723	127.9051	1435

Table 6.3 Number of realisations in which the criteria have chosen a certain order (of AR model) for AR(2) process data (Example 6.3)

Criterion	AR(1)	AR(2)	AR(3)	AR(4)	Comments
PP	–	10	–	–	PP(i) curve is unimodal
B-statistic	–	10	–	–	Unimodal
SWR	–	10	–	–	–
FPE	–	5	5	–	Local minimum observed
COMP	–	3	2	5	Unexpected results

Table 6.6 indicates that ARMA(3,2) or AR(4) models can adequately fit to the ARMA data but the most suitable model is, of course, ARMA(2,1), as suggested by the first column. This exercise leads to a practical inference that the PP and the B-statistic criteria are very effective not only in selecting a complexity within a given class of

Table 6.4 Number of realisations in which the criteria have chosen a certain order (of ARMA model) for ARMA(2,1) process data (Example 6.3)

Criterion	ARMA(1,0)	ARMA(2,1)	ARMA(3,2)	ARMA(4,3)	Comments
PP	–	9	1	–	Unimodal
B-statistic	–	9	1	–	Unimodal
SWR	1	8	–	1	–
FPE	–	4	5	1	Local minimum in some cases

Table 6.5 Number of realisations in which the criteria have chosen a certain order (of AR model) for ARMA(2,1) process data (Example 6.3)

Criterion	AR(1)	AR(2)	AR(3)	AR(4)	Suggest higher order	Comments
PP	–	3	1	–	6	No sharp maximum
B-statistic	–	3	–	–	7	No sharp minimum
SWR	1	2	2	2	3	–
FPE	–	–	–	–	10	Decreasing

Table 6.6 Number of realisations in which PP and B have preferred the ARMA(n, m) model to the AR(n) model for the ARMA(2,1) process data. Let C1 = ARMA(n, m) and C2 = AR(n), then if PP(C1) > PP(C2), choose C1 and if B(C1) < B(C2), choose C1 (Example 6.3)

Criterion	ARMA(2,1) to AR(2)	ARMA(3,2) to AR(3)	ARMA(4,3) to AR(4)
PP	10	9	3
B-statistic	10	10	4

models but also in assigning a given data set to a certain class of models. Thus, the PP and the B-statistic can be added to the list of suitable working rules of Section 6.4. Interested readers can redo this example using MATLAB toolbox, writing their own modules to code the expressions of various criteria and arrive at their own opinion about the performance of these criteria. Using large number of realisations, say 50 to 100, they can derive inferences on the performance of these criteria based on this study (Monte Carlo simulation; see Section A.31). The present example illustrates one possible evaluation procedure.

6.5 Epilogue

The modelling and estimation aspects for time-series and transfer function analysis have been extensively covered [1, 2]. Three applications of model order estimation have been considered [18]. The data chains for the tests were derived from: i) a simulated second order system; ii) human activity in a fixed base simulator; and iii) forces on a model of aircraft (in a wind tunnel) exposed to mildly turbulent flows. For case i), the AR model identification was carried out using the LS method. Both the objective and subjective order test criteria provided sharp and consistent model order since the simulated response data was statistically well behaved.

For case ii), the time-series data for human response were derived from a compensatory tracking experiment conducted on a fixed base research simulator developed by NAL. Assuming that the human activity could be represented by AR/LS models, the problem of model order determination was addressed. A record length of 500 data points sampled at 50 ms was used for the analysis. The choice of a sixth order AR model for human activity in compensatory tracking task was found suitable. The same data were used to fit LS models with a model order scan from 1 to 8. Based on several criteria, it was confirmed that the second order model was suitable. The discrete Bode diagrams (from discrete-time LS models) were obtained for various models orders. It was found that adequate amplitude ratio (plot versus frequency) was obtained for model order 2. The AR pilot model differs from the LS plot model in model order because the LS model is an input-output model and its degrees of freedom are well taken care of by the numerator part. In the AR model, since there is no numerator part, a longer (large order) model is required. This exercise obtained adequate human pilot models based on time-series analysis. This concept was further expanded to motion-based experiments [4].

Estimation of pitch damping derivatives using random flow fluctuations inherent in the tunnel flow was validated. This experiment used an aircraft's scaled down physical model mounted on a single degree of freedom flexure having a dominant second order response. Since the excitation to the model was inaccessible, and the AR model was the obvious choice, an order test was carried out using a 1000 sample data chain. Since response is known to be dominantly second order, the natural frequency was determined by evaluating the spectra using a frequency transformation of the discrete AR models, obtained by using time-series identification. The estimated natural frequency stabilised for AR(n), $n \geq 10$.

Excellent surveys of system identification can be found [19]. Non-stationary and nonlinear time-series analyses need special treatment and are not considered in the present book. The concept of the 'del' operator is treated in Reference 20. The transfer functions obtained using the 'del' operator are nearer to the continuous-time ones than the pulse transfer functions. The pulse transfer functions show distinctions away from the continuous-time transfer function whereas the 'del' operator shows similarities and brings about the unification of discrete and continuous-time models.

6.6 References

1 BOX, G. E. P., and JENKINS, G. M.: 'Time series: analysis, forecasting and controls' (Holden Day, San Francisco, 1970)
2 LJUNG, L.: 'System identification: theory for the user' (Prentice-Hall, Englewood Cliffs, 1987)
3 SHINNERS, S. M.: 'Modelling of human operator performance utilizing time-series analysis', *IEEE Trans. Systems, Man and Cybernetics*, 1974, **SMC-4**, pp. 446–458
4 BALAKRISHNA, S., RAOL, J. R., and RAJAMURTHY, M. S.: 'Contributions of congruent pitch motion cue to human activity in manual control', *Automatica*, 1983, **19**, (6), pp. 749–754
5 WASHIZU, K., TANAKA, K., ENDO, S., and ITOKE, T.: 'Motion cue effects on human pilot dynamics in manual control'. Proceedings of the 13th Annual conference on *Manual Control*, NASA CR-158107, pp. 403–413, 1977
6 GUPTA, N. K., HULL, W. E., and TRANKLE, T. L.: 'Advanced methods of model structure determination from test data', *Journal of Guidance and Control*, 1978, **1**, pp. 197–204
7 GUSTAVSSON, I.: 'Comparison of different methods for identification of industrial processes', *Automatica*, 1972, **8**, (2), pp. 127–142
8 SODERSTROM, T.: 'On model structure testing in system identification', *Int. Journal of Control*, 1977, **26**, (1), pp. 1–18
9 AKAIKE, H.: 'A new look at the statistical model identification', *IEEE Trans. Automat. Control*, 1974, **AC-19**, pp. 716–722
10 PARZEN, E.: 'Some recent advances in time-series modelling', *IEEE Trans. Automat. Control*, 1974, **AC-19**, pp. 723–730
11 TONG, H.: 'A note on a local equivalence of two recent approaches to autoregressive order determination', *Int. Journal of Control*, 1979, **29**, (3), pp. 441–446
12 MEHRA, R. K., and PESCHON, J.: 'An innovations approach to fault detection in dynamic system', *Automatica*, 1971, **7**, pp. 637–640
13 STOICA, P.: 'A test for whiteness', *IEEE Trans. Automat. Control*, 1977, **AC-22**, pp. 992–993
14 ISHII, N., IWATA, A., and SUZUMURA, N.: 'Evaluation of an autoregressive process by information measure', *Int. Journal of System Sci.*, 1978, **9**, (7), pp. 743–751
15 KASHYAP, R. L.: 'A Bayesian comparison of different classes of dynamic models using the empirical data', *IEEE Trans. Automat. Control*, 1977, **AC-22**, (5), pp. 715–727

16 MAKLAD, M. S., and NICHOLS, S. T.: 'A new approach to model structure determination', *IEEE Trans. Systems, Man and Cybernetics*, 1980, **SMC-10**, (2), pp. 78–84
17 SODERSTROM, T.: 'Test of pole-zero cancellation in estimated models', *Automatica*, 1975, **11**, (5), pp. 537–541
18 JATEGAONKAR, R. V., RAOL, J. R., and BALAKRISHNA, S.: 'Determination of model order for dynamical systems', *IEEE Trans. Systems, Man and Cybernetics*, 1982, **SMC-12**, pp. 56–62
19 ASTROM, K. J., and EYKOFF, P.: 'System identification – a survey', *Automatica*, 1971, **7**, (2), pp. 123–162
20 MIDDLETON, R. H., and GOODWIN, G. C.: 'Digital estimation and control: a unified approach' (Prentice Hall, New Jersey, 1990)

6.7 Exercises

Exercise 6.1

Establish by long division that the LS model of order 1 leads to the AR model of higher order (long AR models).

Exercise 6.2

Obtain transfer function (in frequency domain) for the first order AR time-series model, by replacing q^{-1} by z^{-1}, where $z = \sigma + j\omega$, complex frequency (in z-domain).

Exercise 6.3

Transform the first order LS time-series model to the continuous-time transfer function by using $q^{-1} = e^{-\tau s} \approx 1 - \tau s$, where τ is the sampling interval and $s = \sigma + j\omega$ complex frequency operator (in s-domain, i.e., continuous-time domain).

Exercise 6.4

Repeat Exercise 6.3 with $z^{-1} = e^{-\tau s} \approx (2 - \tau s)/(2 + \tau s)$. What is the name of this transformation?

Exercise 6.5

What is the magnitude and phase of the transformation $z = e^{\tau s} \approx (2 + \tau s)/(2 - \tau s)$? Why would you prefer this transformation compared with the one in Exercise 6.3?

Exercise 6.6

Can you obtain possible operators in the s domain based on i) $q^{-1} \approx 1 - \tau s$, where q^{-1} is a backward shift operator, and ii) $q^{-1} \approx (2 - \tau s)/(2 + \tau s)$?

Exercise 6.7

Establish by simple calculation that the criterion B-statistic, eq. (6.36) puts greater penalty on the number of coefficients in the model than the one in eq. (6.26), the Akaike's information criterion.

Exercise 6.8

Given $z^{-1} = (2 - \tau s)/(2 + \tau s)$, obtain an expression for s.

Exercise 6.9

Given $z = e^{\tau s}$ and $s = \sigma + j\omega$, find expressions for σ and ω. What is the significance of these transformations?

Chapter 7
Estimation before modelling approach

7.1 Introduction

The estimation before modelling (EBM) methodology is essentially a two-step approach [1–3]. In the first step, the extended Kalman filter is used for state estimation. The filtered states or their derivatives/related variables are used in the next step of regression analysis. Thus, the parameter estimation is separated into two independent steps. This is unlike the output error method, where parameter estimation is accomplished in essentially one-step, though in an iterative manner. In the output error method, the model structure has to be defined *a priori* whereas in estimation before modelling, this is taken care of in the second step only. Often smoothing techniques are used in the first step to minimise errors from the extended Kalman filter. The main advantage of the EBM approach is that state estimation is accomplished before any modelling is done. For state estimation, usual system dynamics, which might have only a descriptive mathematical model, is used. In the second step of regression analysis, one can evolve the most suitable detailed mathematical model, the parameters of which are estimated using the least squares method. It is here that model selection criteria play an important role. Another advantage of the estimation before modelling approach is that it can be used to handle data from inherently unstable/augmented systems. In addition, this approach has great utility for aircraft parameter estimation.

In state reconstruction, the nonlinear functions arise due to augmentation of the state vector with unknown sensor bias and scale factors, which also need to be estimated. An extended Kalman filter and a smoother were used to derive smoothed time histories, which in turn were used in the modelling step [2].

7.2 Two-step procedure

In the first step, a combined extended Kalman filter and fixed interval smoother are used. In the second step, the smoothed states along with the measured (control) inputs are used to estimate the parameters of the mathematical model using the stepwise multiple regression method.

The features of this two-step methodology compared to the more often used maximum likelihood-output error method or filter error method are:

1 In the maximum likelihood-output error method, the identified parameters of the mathematical model directly influence the estimated trajectories. If the model structure were good and well known, the method would be very convenient and yield good results. However, often the model structure is not so well known, then alternative models have to be tried leading to a time consuming exercise. This is avoided or greatly reduced in estimation before modelling. Here, many alternative models can be tried in the second step. Model selection criteria can be used to arrive at a most adequate model of the system [4].
2 The maximum likelihood-output error method is a batch-iterative procedure. In estimation before modelling, once the state estimation is accomplished, the second step is a one-shot approach. However, the criteria to select a suitable model (number of coefficients to include in the model) need to be judiciously incorporated in the procedure.
3 Estimation before modelling does not need the starting values of the model parameters unlike the output error method.

7.2.1 Extended Kalman filter/fixed interval smoother

The extended Kalman filter is used for two purposes: i) state estimation; and ii) to estimate parameters that are related to bias, scale factors etc. These parameters are considered as additional states and the combined state vector is estimated. The fixed interval smoother is used for obtaining a smoothed state. The smoother is not treated in this book formally. However, a brief description is given here. The extended Kalman filter equations are the same or almost similar to the ones given in Chapter 4.

In the two-step methodology, the linearisation of the nonlinear functions f_a and h_a is carried out using the finite difference method, thereby generalising the application to any nonlinear problem. This avoids extra coding for evaluation of the partials. There is no need to worry about these partials if any different nonlinear model is to be used.

Often Q and R (see Chapter 4) are assumed diagonal matrices.

7.2.1.1 Smoother

The smoothing process utilises, in principle, more information than the Kalman filter. Smoothing either uses the measurement data and/or it uses the estimated states/covariances from the forward pass of the Kalman filter. The main aim is to obtain better state estimates than the optimal filter. The main process in the smoother is the backward pass starting from the final time to the initial time. Thus, the smoother is a non real-time data processing scheme. Only the noise controllable states are smoothable.

There are three types of smoothing possibilities [5]:

1 The fixed interval is defined as $0 < t < T$ and smoothing is obtained for times t within this interval.
2 Fixed-point smoothing means that a state at a fixed point t is being smoothed as T increases, i.e., more and more data is available.
3 In fixed-lag smoothing, the estimate is being smoothed as time T increases but the lag Δ is fixed between the point at which the smoothing is obtained and T.

Let there be two estimates at time t: one based on forward filtering up to time t and the other being due to backward filtering starting from final time t_f up to the initial time t_0. The idea is to obtain a smoothed/improved estimate by fusion of these two estimates x_f and x_b[5] (see Fig. 7.1):

$$\underline{x} = K_1\hat{x}_f + K_2\hat{x}_b \tag{7.1}$$

$$x_t + \underline{\tilde{x}} = K_1(x_t + \tilde{x}_f) + K_2(x_t + \tilde{x}_b) \tag{7.2}$$

Here, x_t is the true state at time t, and underbar denotes smoothed state/error.

Then, simplifying we get:

$$\underline{\tilde{x}} = (K_1 + K_2 - I)x_t + K_1\tilde{x}_f + K_2\tilde{x}_b \tag{7.3}$$

For unbiased smoothed estimate, we have

$$K_1 + K_2 - I = 0 \quad \Rightarrow \quad K_2 = I - K_1 \tag{7.4}$$

Substituting for K_2 in the above equation for the smoothed estimate, we obtain

$$\underline{x} = K_1\hat{x}_f + (I - K_1)\hat{x}_b$$

or

$$\underline{x} = \hat{x}_b + K_1(\hat{x}_f - \hat{x}_b) \tag{7.5}$$

Thus, we can get an optimal smoothed estimate if we get an optimal gain K_1.

Next, we obtain the covariance matrix of the smoothed estimate error:

$$\underline{\tilde{x}} = K_1\tilde{x}_f + K_2\tilde{x}_b = K_1\tilde{x}_f + (I - K_1)\tilde{x}_b \tag{7.6}$$

$$\text{cov}(\underline{\tilde{x}}\,\underline{\tilde{x}}^T) = (K_1\tilde{x}_f + (I - K_1)\tilde{x}_b)(K_1\tilde{x}_f + (I - K_1)\tilde{x}_b)^T$$

$$P_s = K_1 P_f K_1^T + (I - K_1)P_b(I - K_1)^T \tag{7.7}$$

We have made the assumption that errors $\tilde{x}_f$ and $\tilde{x}_b$ are uncorrelated.

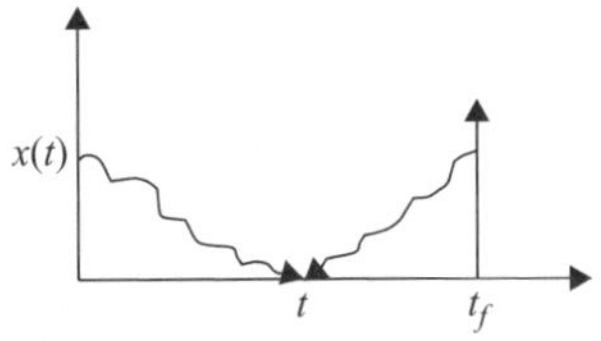

Figure 7.1 Forward and backward filtering

Next, by minimising P_s, we obtain the expression for gain K_1:

$$2K_1 P_f - 2(I - K_1)P_b = 0$$

$$K_1 = P_b(P_f + P_b)^{-1}$$

$$I - K_1 = I - P_b(P_f + P_b)^{-1} = P_f(P_f + P_b)^{-1} \tag{7.8}$$

Thus, we get after simplification [5]:

$$P_s^{-1} = P_f^{-1} + P_b^{-1} \tag{7.9}$$

We take a scalar case to interpret the results:

Let $P_s \rightarrow \sigma_s^2$ and $P_f \rightarrow \sigma_f^2$ and $P_b \rightarrow \sigma_b^2$

Then, we get

$$\left(\sigma_s^2\right)^{-1} = \left(\sigma_f^2\right)^{-1} + \left(\sigma_b^2\right)^{-1}$$

or

$$\sigma_s^2 = \frac{\sigma_f^2 \sigma_b^2}{\sigma_f^2 + \sigma_b^2} \tag{7.10}$$

The above states that the variance of the smoothed estimate state error is less than both the variances σ_f^2 and σ_b^2, thereby suggesting that we have obtained a new estimate with less covariance or uncertainty associated with it.

7.2.1.2 Fixed interval smoother algorithm

The smoother equations are given as in Reference 5:

$$\underline{x}_a(k \mid N) = \hat{x}_a(k) + K_s[\underline{x}_a(k+1 \mid N) - \tilde{x}_a(k+1)] \tag{7.11}$$

Here, K_s is the gain of the smoother algorithm:

$$K_s = \hat{P}(k)\phi^T(k)\tilde{P}^{-1}(k+1) \tag{7.12}$$

The smoother state error covariance matrix is given by:

$$\underline{P}(k \mid N) = \hat{P}(k) + K_s(k)[\underline{P}(k+1 \mid N) - \tilde{P}(k+1)]K_s^T(k) \tag{7.13}$$

Here, a stands for augmented state vector and underbar for smoothed estimates. We note here that this FIS does not use the measurements in the reverse/backward pass. We also note that the smoothed equations use only the state/covariance estimates generated by EKF in the forward pass. So the process is to use EKF starting from initial $\hat{x}_0$ and P_0 and complete one forward pass through all data points sequentially. In the process, all the filtered estimates are stored. The smoother equations are used in the backward pass starting from the final values of the state/covariance estimates and arriving at the initial point. In the process, we obtain smoothed state/covariance estimates. If there are process noise related uncertainties, the smoother is very useful.

7.2.2 *Regression for parameter estimation*

A general form of the model to be identified is given as

$$y(t) = \beta_0 + \beta_1 x_1(t) + \cdots + \beta_{n-1} x_{n-1}(t) + e(t) \tag{7.14}$$

In the above equation, the time history $y(t)$ is available from the first step. Actually, depending upon the problem at hand, the variable $y(t)$ would not be the states directly estimated by EKF. In fact, some intermediate steps would be required to compute y from $\hat{x}$. This will be truer for the aircraft parameter estimation problem as will be discussed subsequently. The intermediate computations will involve all the known constants and variables like x_i and y. What then remains to be done is to determine which parameters should be retained in the model and estimated. The problem is then handled using model order determination criteria and the least squares method for parameter estimation.

Given N observations for $y(t)$ and $x(t)$, the LS estimate of β can be computed by

$$\hat{\beta} = (X^T X)^{-1} X^T Y \tag{7.15}$$

where X and Y are composite data matrices, which have elements from $x(t)$ and $y(t)$, e.g., X is $N \times n$ matrix and Y is $N \times 1$ vector. The covariance matrix of parameter estimation error is given as

$$\text{cov}(\beta - \hat{\beta}) \approx \sigma_r^2 (X^T X)^{-1} \tag{7.16}$$

Here, σ_r^2 is residual variance.

7.2.3 *Model parameter selection procedure*

Several model selection criteria have been discussed in Chapter 6. Although these criteria are presented in the context of time-series identification/model determination, it is possible to use a few of these for the present case: F-statistic, variance of residuals, residual sum of squares and whiteness of residuals, the definitions of which can be found in Chapter 6 or Appendix A.

For selecting an appropriate structure, a stepwise regression method is used. Partial F-statistics are computed to build up the parameter vector by selecting significant parameters in the model one at a time. The process is continued until the model equation is satisfied.

In the first place, it is assumed that the mean of the data is in the model. The estimate of regression is determined. The correction coefficients are computed for each of the independent variables.

$$\rho_{x_j y} = \frac{\sum_{k=1}^{N} x_{kj} y_k}{\sqrt{\sum_{k=1}^{N} x_{kj}^2 \sum_{k=1}^{N} y_k^2}} \tag{7.17}$$

The x_j giving the largest ρ_{xy} is chosen as the first entry into the regression equation. The model is then given as

$$\hat{y} = \hat{\beta}_1 + \hat{\beta}_j x_j + \hat{e} \tag{7.18}$$

Next, the correlation coefficient for each remaining x_i $(i = 2, \ldots, j-1, j+1, \ldots, n)$ is computed on x_j and $\hat{y}$ and is given by

$$\rho_{y_{x_i x_j}} = \frac{\sum_{k=1}^{N} (x_{ki} - x_{kj}\hat{\beta}_j - \hat{\beta}_1)(y_k - \hat{y}_k)}{\sqrt{\sum_{k=1}^{N} (x_{ki} - x_{kj}\hat{\beta}_j - \hat{\beta}_1)^2 \sum_{k=1}^{N} (y_k - \hat{y}_k)^2}} \tag{7.19}$$

The above is the partial correlation of y on x_i, given that x_j is in the regression. The x_i yielding the largest value of $\rho_{y_{x_i x_j}}$ is selected for inclusion in the model:

$$\hat{y} = \hat{\beta}_1 + \hat{\beta}_j x_j + \hat{\beta}_i x_i$$

This process is continued until the remainder of the variables entering in the model do not offer any significant improvement in the model. This is accomplished using the F-statistics:

$$F = \frac{(N-n)\rho_{y_{x_i x_j}}}{(n-1)(1-\rho_{y_{x_i x_j}})} \tag{7.20}$$

This gives a relative statistical significance of each variable in each model, given the fact that other variables are already present in the model. The maximum F value is sought for statistical significance of inclusion of a variable in the regression (it being the correlation coefficient).

In addition, the quantity R^2 can be used:

$$R^2 = \frac{\sum_{k=1}^{N} (\hat{y}_k - \bar{y})^2}{\sum_{k=1}^{N} (y_k - \bar{y})^2} \tag{7.21}$$

the value of which varies from 0 to 1. It is expressed as a percentage of the improvement in R^2 due to the addition of a new parameter in the model and should be of a significant value to justify its inclusion.

The regression method can be implemented using the Householder transformation to obtain an LS solution [6], to avoid matrix ill-conditioning.

Figure 7.2 illustrates the different steps in the EBM procedure for aircraft aerodynamic parameter estimation.

7.2.3.1 Example 7.1

Using the simulated longitudinal short period and lateral-directional data of an aircraft (Appendix B), estimate the aircraft stability and control derivatives using the EBM procedure.

7.2.3.2 Solution

Data generation step

The data for parameter estimation study is generated from a six-degree-of-freedom simulator of an unstable/augmented aircraft. The simulator utilises a nonlinear aerodynamic model consisting of force and moment coefficients defined as functions of α, β, Mach number, thrust and control surface positions. The simulator also uses

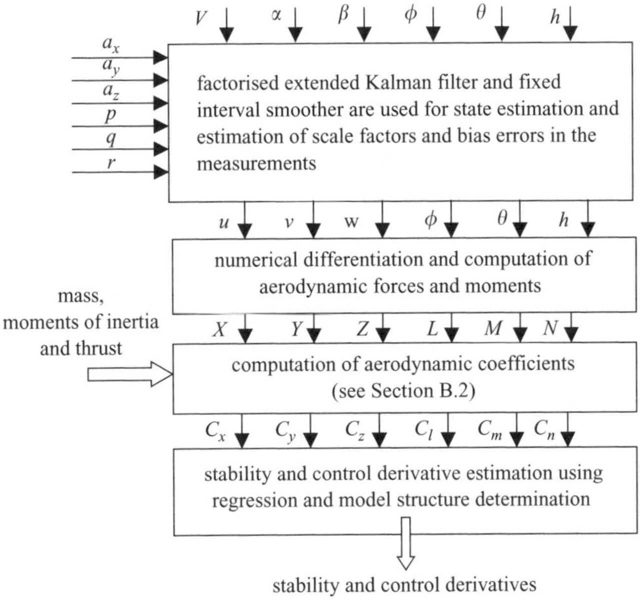

Figure 7.2 Steps in EBM estimation procedure

inputs from sub modules like the actuator dynamics, engine dynamics, weight and inertia module, and atmospheric models, to describe the aircraft closed loop response.

The longitudinal and lateral-directional time histories are generated using the simulator for the flight condition pertaining to Mach $= 0.5$ and altitude $= 4$ km. The longitudinal short period manoeuvre is simulated with a doublet input to the elevator and the Dutch-roll oscillation is simulated with a 10 mm doublet input to the roll stick followed by a 10 mm doublet input to the pilot rudder pedal. The short period manoeuvre is of 8 s duration while the Dutch-roll motion is of 17 s duration. The short period and Dutch-roll motion data are concatenated for the purpose of data compatibility checking which is the first step of the EBM procedure. The data is generated at the rate of 40 samples/s. Additive process noise with $\sigma = 0.001$ is used during the data generation. Measurement noise (SNR $= 10$) is added to $V, \alpha, \beta, \phi, \theta$ and h measurements from the simulator.

Mathematical model formulation for the extended Kalman filter

The first step of estimation of aircraft states is achieved using kinematic consistency check or data compatibility check. This step essentially makes use of the redundancy present in the measured inertial and air data variables to obtain the best state estimates from the dynamic manoeuvre data. Scale factors and bias errors in the sensors (which are used for the measurements) are estimated by expanding the state vector to include these parameters. This process ensures that the data are consistent with the basic underlying kinematic models, which are given below (see Section B.7):

State equations

$$\begin{aligned}
\dot{u} &= -(q - \Delta q)w + (r - \Delta r)v - g\sin\theta + (a_x - \Delta a_x),\\
\dot{v} &= -(r - \Delta r)u + (p - \Delta p)w + g\cos\theta\sin\phi + (a_y - \Delta a_y),\\
\dot{w} &= -(p - \Delta p)v + (q - \Delta q)u + g\cos\theta\cos\phi + (a_z - \Delta a_z)\\
\dot{\phi} &= (p - \Delta p) + (q - \Delta q)\sin\phi\tan\theta + (r - \Delta r)\cos\phi\tan\theta\\
\dot{\theta} &= (q - \Delta q)\cos\phi - (r - \Delta r)\sin\phi\\
\dot{h} &= u\sin\theta - v\cos\theta\sin\phi - w\cos\theta\cos\phi
\end{aligned} \tag{7.22}$$

Observation equations

$$\begin{aligned}
V_m &= \sqrt{u_n^2 + v_n^2 + w_n^2}\\
\alpha_m &= K_\alpha \tan^{-1}\left[\frac{w_n}{u_n}\right]\\
\beta_m &= \sin^{-1}\left[\frac{v_n}{\sqrt{u_n^2 + v_n^2 + w_n^2}}\right]\\
\phi_m &= \phi + \Delta\phi\\
\theta_m &= K_\theta \theta\\
h_m &= h
\end{aligned} \tag{7.23}$$

Here, u_n, v_n, w_n are the velocity components along the three axes at the nose boom of the aircraft:

$$\begin{aligned}
u_n &= u - (r - \Delta r)Y_n + (q - \Delta q)Z_n\\
v_n &= v - (p - \Delta p)Z_n + (r - \Delta r)X_n\\
w_n &= w - (q - \Delta q)X_n + (p - \Delta p)Y_n
\end{aligned} \tag{7.24}$$

State estimation using the extended Kalman filter

For the first step of state estimation using the extended Kalman filter, a model with six states $\{u, v, w, \phi, \theta, \lambda\}$ is formulated. The rates and accelerations are used as inputs to the model resulting in a control input vector $CV = \{p, q, r, a_x, a_y, a_z\}$. It should be mentioned here that measurement noise is added only to the observables $V, \alpha, \beta, \phi, \theta, h$ and no measurement noise is added to the rates and accelerations during data generation for this example. The parameter vector Θ contains seven parameters $\Theta = \{\Delta a_x, \Delta a_z, \Delta p, \Delta q, \Delta r, K_\alpha, K_\theta\}$. (This parameter set was arrived at by integrating the state equations without including any of the scale factors and bias errors in the model and observing the time history match. The parameters found necessary to improve the match are included in the model.) These parameters are included as augmented states along with the six states so that we have a state vector with 13 states and six observations. The above models are used in the EKF (program in folder **Ch7EBMex1**) for obtaining estimates of aircraft states. The fixed interval smoother to obtain smoothed aircraft states has not been used in this example. Further steps of computing forces and moments and subsequent parameter estimation are carried out using the estimated states from the extended Kalman filter. Figure 7.3(a) shows the comparison of the time histories of measured and estimated observables $V, \alpha, \beta, \phi, \theta$, and h. Figure 7.3(b) gives the control vector trajectories, $CV = \{p, q, r, a_x, a_y, a_z\}$. Table 7.1 gives the estimated scale factor and bias errors. It is seen that the scale factors are close to one and most of the bias errors are close to zero for this case. The estimated scale factors and bias values are used to correct the measured data before using it for the computation of the forces and moments.

Computation of forces and moments (intermediate step)

For the computation of the dimensional forces X, Y, Z and moments L, M, N, the rates p, q, r corrected for bias errors and the estimated states u, v, w, ϕ, θ from the state estimation step are used. The time derivatives of u, v, w, p, q and r required for the computations are obtained by using a centrally pivoted five-point algorithm (see Section A.5).

The following equations are used for the computations:

$$\begin{aligned}
X &= \dot{u} - rv + qw + g\sin\theta \\
Y &= \dot{v} - pw + ru - g\cos\theta\sin\phi \\
Z &= \dot{w} - qu + pv - g\cos\theta\cos\phi \\
M &= \dot{q} - prC_4 - (r^2 - p^2)C_5 \\
L + C_3N &= \dot{p} - pqC_1 - qrC_2 \\
N + C_8L &= \dot{r} - pqC_6 - qrC_7
\end{aligned} \tag{7.25}$$

The constant coefficients C_1 to C_8 are given by

$$C_1 = \frac{I_{xz}(I_z + I_x - I_y)}{I_xI_z - I_{xz}^2}; \quad C_2 = \frac{[I_z(I_y - I_z) - I_{xz}^2]}{I_xI_z - I_{xz}^2}; \quad C_3 = \frac{I_{xz}}{I_x};$$

$$C_4 = \frac{I_z - I_x}{I_y}; \quad C_5 = \frac{I_{xz}}{I_y}; \quad C_6 = \frac{[I_x(I_x - I_y) + I_{xz}^2]}{I_xI_z - I_{xz}^2};$$

$$C_7 = \frac{I_{xz}(I_y - I_z - I_x)}{I_xI_z - I_{xz}^2}; \quad C_8 = \frac{I_{xz}}{I_z}$$

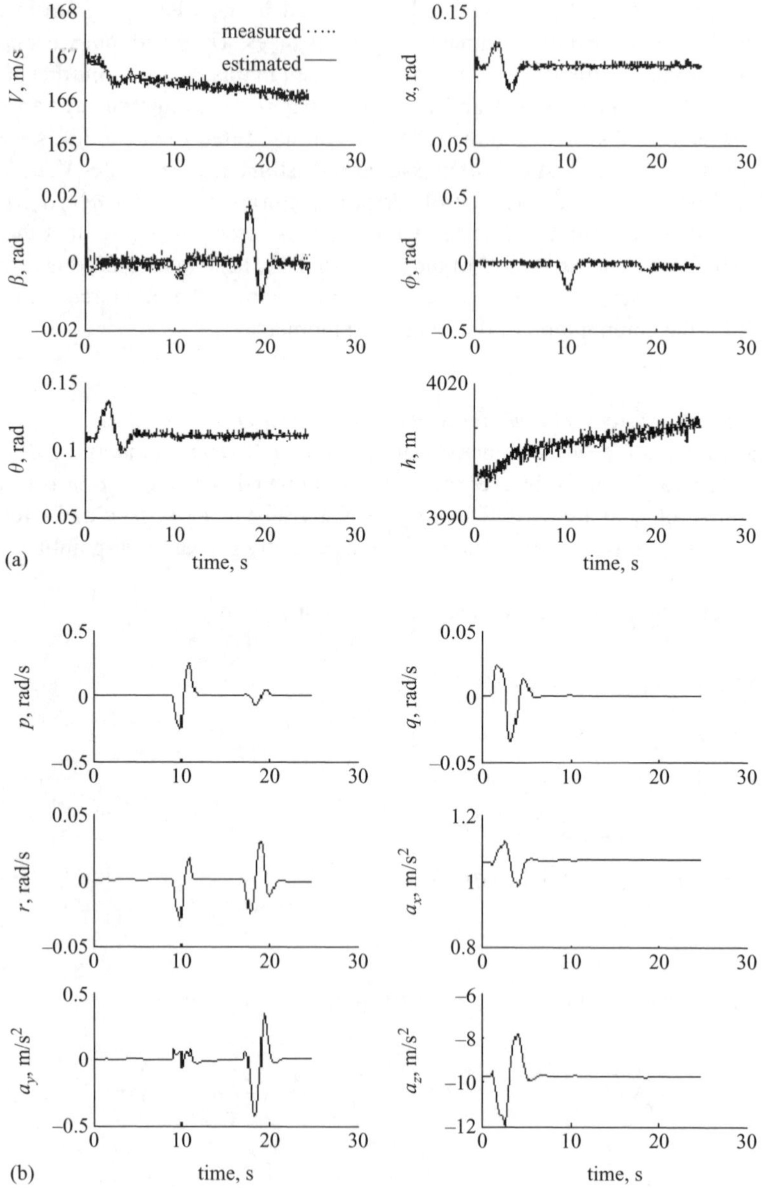

Figure 7.3 *(a) Time history match for the observables (Example 7.1); (b) time histories of control inputs (Example 7.1)*

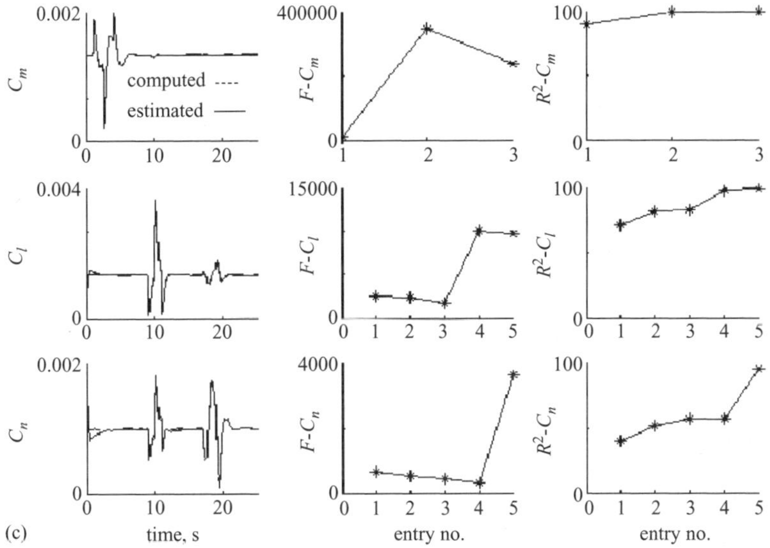

Figure 7.3 Continued. (c) Computed and estimated aerodynamic coefficients, F and R^2 values (Example 7.1)

Table 7.1 Estimates of scale factors and biases (Example 7.1)

Parameter	Data with SNR = 10
Δa_x	0.1137
Δa_z	0.0097
Δp	0.18e−4
Δq	−0.2e−4
Δr	−0.08e−4
K_α	1.1170
K_θ	1.1139

Computation of time histories of aerodynamic coefficients

The following equations are used to generate the time histories of the non-dimensional aerodynamic coefficients $C_x, C_y, C_z, C_l, C_m, C_n$:

$$C_x = \frac{m}{\bar{q}S}\left(X - \frac{T_x}{m}\right)$$

$$C_y = \frac{m}{\bar{q}S}Y$$

$$
\begin{aligned}
C_z &= \frac{m}{\bar{q}S}\left(Z - \frac{T_z}{m}\right); \\
C_l &= \frac{I_x I_z - I_{xz}^2}{I_x I_z}\frac{I_x}{\bar{q}Sb}L \\
C_m &= \left(M - \frac{l_{ze}T_x}{I_y}\right)\frac{I_y}{\bar{q}S\bar{c}} \\
C_n &= \frac{I_x I_z - I_{xz}^2}{I_x I_z}\frac{I_z}{\bar{q}Sb}N
\end{aligned}
\tag{7.26}
$$

Here, T_x, T_z represent the thrust components in the X and Z directions.

Model formulation for stepwise multiple regression method step

Having obtained the time histories of the non-dimensional aerodynamic coefficients as described in the previous section, the stepwise multiple regression method is used to estimate the parameters/coefficients of the aerodynamic model. Since the data pertains to the short period and lateral-directional mode of the aircraft, the forces and moments are not expected to contain any nonlinear terms and hence the following Taylor series expansion of the coefficients has been considered.

$$
\begin{aligned}
C_L &= C_{L0} + C_{L\alpha}\alpha + C_{Lq}\frac{q\bar{c}}{2V} + C_{L\delta_e}\delta_e \\
C_m &= C_{m0} + C_{m\alpha}\alpha + C_{mq}\frac{q\bar{c}}{2V} + C_{m\delta_e}\delta_e \\
C_Y &= C_{Y0} + C_{Y\beta}\beta + C_{Yp}\frac{pb}{2V} + C_{Yr}\frac{rb}{2V} + C_{Y\delta_a}\delta_a + C_{Y\delta_r}\delta_r \\
C_l &= C_{l0} + C_{l\beta}\beta + C_{lp}\frac{pb}{2V} + C_{lr}\frac{rb}{2V} + C_{l\delta_a}\delta_a + C_{l\delta_r}\delta_r \\
C_n &= C_{n0} + C_{n\beta}\beta + C_{np}\frac{pb}{2V} + C_{nr}\frac{rb}{2V} + C_{n\delta_a}\delta_a + C_{n\delta_r}\delta_r
\end{aligned}
\tag{7.27}
$$

This model form was used in the procedure described in Section 7.2.2. Each of the above equations in Taylor's series form is like that of eq. (7.14). The flow angles α, β used in these equations are obtained from the state estimation step and the measured angular rates p, q, r are corrected for bias errors using the values estimated in the same step. The control surface deflections δ_e, δ_a, δ_r are obtained from the simulation data measurements.

Table 7.2 gives the values of the estimated moment derivatives, the standard deviations and the R^2 values. The standard deviations are obtained using the square root of the diagonal elements of the estimation error covariance matrix computed using eq. (2.7). The reference values listed in Table 7.2 are obtained from the simulator aerodynamic database. The pitching moment derivative estimates compare very well with the reference values. For this case the value $R^2 = 99$ also indicates that the model is able to explain the pitching moment coefficient almost completely (99 per cent). However, some of the rolling moment and yawing moment derivative estimates show

Table 7.2 Estimated aerodynamic parameters (Example 7.1)

Parameter	Reference	Estimated
$C_{m_{\delta e}}$	−0.4102	−0.3843 (0.0007)
C_{m_q}	−1.2920	−1.2046 (0.0063)
$C_{m\alpha}$	−0.0012	−0.0012 (0.0002)
$R^2(C_m)$	–	99.86
$C_{l_{\delta a}}$	−0.1895	−0.1640 (0.0008)
C_{l_p}	−0.2181	−0.1863 (0.0023)
C_{l_β}	−0.0867	−0.0679 (0.0009)
$C_{l_{\delta r}}$	0.0222	0.0159 (0.0007)
C_{lr}	0.0912	0.1958 (0.0152)
$R^2(C_l)$	–	97.5
$C_{n_{\delta a}}$	−0.0740	−0.0599 (0.0010)
C_{n_β}	0.1068	0.0911 (0.0011)
$C_{n_{\delta r}}$	−0.0651	−0.0570 (0.0008)
C_{lr}	−0.254	−0.3987 (0.0189)
C_{np}	−0.0154	−0.0148 (0.0028)
$R^2(C_n)$		94.8

some deviations from the reference values. The R^2 also indicates that some more terms may be required to account for the complete variations. The first column of Fig. 7.3(c) shows the comparison of model predicted and computed aerodynamic coefficients C_m, C_l and C_n. It is clear that the estimated aerodynamic coefficients match the computed coefficients fairly accurately. The F and R^2 values versus the entry number into the SMLR algorithm are also plotted in Fig. 7.3(c).

7.3 Computation of dimensional force and moment using the Gauss-Markov process

In Example 7.1, the dimensional force and moment coefficients are computed from eq. (7.25) in the intermediate step. The use of eq. (7.25), however, requires the values of $\dot{u}, \dot{v}, \dot{w}, \dot{p}, \dot{q}$ and $\dot{r}$ which are obtained using a centrally pivoted five-point algorithm (Appendix A). This procedure of computing the dimensional force and moment coefficients can, at times, lead to unsatisfactory results, particularly if the measured data is noisy. In Example 7.1, measurement noise was included only in the observables and not in the rates and accelerations, which act as control inputs in eq. (7.22). In real flight data, all quantities will be corrupted with measurement noise. Numerical differentiation of noisy flight variables might not yield proper values of $\dot{u}, \dot{v}, \dot{w}, \dot{p}, \dot{q}$ and $\dot{r}$, thereby introducing inaccuracies in the computed force and moment coefficients. Filtering the flight measurements before applying numerical differentiation may also fail to yield error free force and moment time histories. The Gauss-Markov process offers a solution to circumvent this problem by doing away with the numerical differentiation scheme. A third order Gauss-Markov model can be described in the following manner [2,7]:

$$\begin{bmatrix} \dot{x} \\ \dot{x}_1 \\ \dot{x}_2 \end{bmatrix} = \begin{bmatrix} 0 & 1 & 0 \\ 0 & 0 & 1 \\ 0 & 0 & 0 \end{bmatrix} \begin{bmatrix} x \\ x_1 \\ x_2 \end{bmatrix}$$

Here, x can be any one of the force or moment coefficients, i.e., X, Y, Z or L, M, N

Consider eq. (7.25) of Example 7.1. The equation can be re-written in the following form:

$$\begin{aligned}
\dot{u} &= rv - qw - g\sin\theta + X \\
\dot{v} &= pw - ru + g\cos\theta\sin\phi + Y \\
\dot{w} &= qu - pv + g\cos\theta\cos\phi + Z \\
\dot{p} &= pqC_1 + qrC_2 + L + C_3N \\
\dot{q} &= prC_4 + (r^2 - p^2)C_5 + M \\
\dot{r} &= pqC_6 + qrC_7 + N + C_8L
\end{aligned} \tag{7.28}$$

Using the third order Gauss-Markov model for the force and moment coefficients gives

$$\begin{aligned}
\dot{X} &= X_1 \\
\dot{X}_1 &= X_2 \\
\dot{X}_2 &= 0 \\
\\
\dot{Y} &= Y_1 \\
\dot{Y}_1 &= Y_2 \\
\dot{Y}_2 &= 0 \\
\\
\dot{Z} &= Z_1 \\
\dot{Z}_1 &= Z_2 \\
\dot{Z}_2 &= 0 \\
\\
\dot{L} &= L_1 \\
\dot{L}_1 &= L_2 \\
\dot{L}_2 &= 0
\end{aligned} \tag{7.29}$$

$$
\begin{aligned}
\dot{M} &= M_1 \\
\dot{M}_1 &= M_2 \\
\dot{M}_2 &= 0
\end{aligned}
$$

$$
\begin{aligned}
\dot{N} &= N_1 \\
\dot{N}_1 &= N_2 \\
\dot{N}_2 &= 0
\end{aligned}
$$

Appending eq. (7.29) to eq. (7.28), the extended Kalman filter method can be applied to the resulting state model to compute the dimensional force and moment coefficients.

With the use of the above procedure to compute X, Y, Z, L, M and N, eq. (7.25) is no longer required. This eliminates the need for numerical differentiation of the variables u, v, w, p, q and r. However, the computational aspects and accuracy of this approach can be studied further [2].

7.4 Epilogue

The fixed interval smoother has two main difficulties: i) inversion of covariance matrix eq. (7.12); and ii) difference of positive semi-definite matrices eq. (7.13). Since the matrices $\hat{P}$ and $\tilde{P}$ originate from KF, they could be erroneous, if the implementation of KF was on a finite-word length computer. This will lead to ill-conditioning of the smoother. A new UD-information based smoother has been devised [8], which overcomes the limitations of Bierman's smoothing algorithm [9] and is computationally more efficient. The EBM seems to have evolved because of a search for an alternative approach to the output error method. More details and applications can be found in References 1–4 and 10. The approach presented in this chapter can also be used to estimate the stability and control derivatives of an aircraft from large amplitude manoeuvres (see Section B.16).

7.5 References

1 STALFORD, H. L.: 'High-alpha aerodynamic identification of T-2C aircraft using EBM method', *Journal of Aircraft*, 1981, **18**, pp. 801–809

2 SRI JAYANTHA, M., and STENGEL, R. F.: 'Determination of non-linear aerodynamic coefficients using estimation-before-modelling method', *Journal of Aircraft*, 1988, **25**, (9), pp. 796–804

3 HOFF, J. C., and COOK, M. V.: 'Aircraft parameter identification using an estimation-before-modelling technique', *Aeronautical Journal*, 1996, pp. 259–268

4 MULDER, J. A., SRIDHAR, J. K., and BREEMAN, J. H.: 'Identification of dynamic systems – applications to aircraft Part 2: nonlinear analysis and manoeuvre design', AGARD-AG-300, **3**, Part 2, 1994

5 GELB, A. (Ed.): 'Applied optimal estimation' (MIT Press, Massachussetts, 1974)

6 BIERMAN, G. J.: 'Factorisation methods for discrete sequential estimation' (Academic Press, New York, 1977)
7 GERLACH, O. H.: 'Determination of performance and stability parameters from unsteady flight manoeuvres', Society of Automotive Engineers, Inc., National Business Aircraft Meeting, Wichita, Kansas, March 18–20, 1970
8 WATANABE, K.: 'A new forward pass fixed interval smoother using the UD information matrix factorisation', *Automatica*, 1986, **22**, (4), pp. 465–475
9 BIERMAN, G. J.: 'A new computationally efficient, fixed-interval, discrete-time smoother', *Automatica*, 1983, **19**, p. 503
10 GIRIJA, G., and RAOL, J. R.: 'Estimation of aerodynamic parameters from dynamic manoeuvres using estimation before modelling procedure', *Journal of Aeronautical Society of India*, 1996, **48**, (2), pp. 110–127

7.6 Exercises

Exercise 7.1

Consider the linear second order model: $m\ddot{x} + d\dot{x} + Kx = u$. Use the finite difference method and convert this model to make it suitable for use in the Kalman filter.

Exercise 7.2 [5]

Assume $\dot{x} = Ax + Bu$. Compute $\dot{\hat{y}}$ if $y = A_2 x$ by using two methods: i) using differentiation of y and; ii) using differentiation of $\hat{x}$, and comment on the resulting expressions.

Exercise 7.3

Establish that if $\hat{\sigma}_x^2 = \tilde{\sigma}_x^2 = \sigma_x^2$, then $\sigma_s^2 = \sigma_x^2$ by using a scalar formulation of smoother covariance of the fixed interval smoother, see eq. (7.13).

Exercise 7.4

Represent the fixed interval smoother in the form of a block diagram.

Exercise 7.5

Using eq. (7.10) for the variance of the smoothed estimate and the concept of information matrix (factor), establish that there is enhancement of information by the smoother, which combines the two estimates.

Chapter 8

Approach based on the concept of model error

8.1 Introduction

There are many real life situations where accurate identification of nonlinear terms (parameters) in the model of a dynamic system is required. In principle as well as in practice, the parameter estimation methods discussed in Chapters 2 to 5 and 7 can be applied to nonlinear problems. We recall here that the estimation before modelling approach uses two steps in the estimation procedure and the extended Kalman filter can be used for joint state/parameter estimation. As such, the Kalman filter cannot determine the deficiency or discrepancy in the model of the system used in the filter, since it pre-supposes availability of an accurate state-space model. Assume a situation where we are given the measurements from a nonlinear dynamic system and we want to determine the state estimates. In this case, we use the extended Kalman filter and we need to have the knowledge of the nonlinear function f and h. Any discrepancy in the model will cause model errors that will tend to create a mismatch of the estimated states with the true state of the system. In the Kalman filter, this is usually handled or circumvented by including the process noise term Q. This artifice would normally work well, but it still could have some problems [1, 2]: i) deviation from the Gaussian assumption might degrade the performance of the algorithm; and ii) the filtering algorithm is dependent on the covariance matrix P of the state estimation error, since this is used for computation of Kalman gain K. Since the process noise is added to this directly, as GQG^T term, one would have some doubt on the accuracy of this approach. In fact, the inclusion of the 'process noise' term in the filter does not improve the model, since the model could be deficient, although the trick can get a good match of the states. Estimates would be more dependent on the current measurements. This approach will work if the measurements are dense in time, i.e., high frequency of measurements, and are accurate.

The above limitations of the Kalman filter can be overcome largely by using the method based on principle of model error [1–6]. This approach not only estimates the states of the dynamic system from its measurements, but also the model discrepancy as a time history. The point is that we can use the known (deficient or linear) model in the state estimation procedure, and determine the deterministic discrepancy of the model, using the measurements in the model error estimation procedure. Once the discrepancy time history is available, one can fit another model to it and estimate its parameters using the regression method. Then combination of the previously used model in the state estimation procedure and the new additional model would yield the accurate model of the underlying (nonlinear) dynamic system, which has generated the data.

This approach will be very useful in modelling of the large flexible structures, robotics and many aerospace dynamic systems, which usually exhibit nonlinear behaviour [3]. Often these systems are linearised leading to approximate linear models with a useful range of operation but with limited validity at far away points from the local linearisation points. Such linear systems can be easily analysed using the simple tools of linear system theory. System identification work generally restricted to such linear and linearised models can lead to modal analysis of the nonlinear systems. However, the linearised models will have a limited range of validity for nonlinear practical data, because certain terms are neglected, in the process of linearisation and approximation. This will produce inaccurate results, and these linearised models will not be able to predict certain behavioural aspects of the system, like drift. In Kalman filter literature, several alternative approaches are available to handle nonlinear state estimation problems: extended Kalman filter, second order Kalman filter, linearised Kalman filter, statistically linearised filter, and so on [7]. In addition, theory of nonlinear filtering on its own merit is very rich. However, most of these approaches still suffer from the point of view of the model error.

The approach studied in this chapter, produces accurate state trajectory, even in the presence of a deficient/inaccurate model and additionally identifies the unknown model (form) as well as its parameters.

The method of model error essentially results in a batch estimation procedure. However, a real-time solution can be obtained using the method of invariant embedding. All these aspects are highlighted in the present chapter.

8.2 Model error philosophy

The main idea is to determine the model error based on the available noisy measurements and in the process the state estimates of the dynamic system.

Let the mathematical description of the nonlinear system be given as

$$\dot{x} = f(x(t), u(t), t) + d(t) \tag{8.1}$$

The unmodelled disturbance is represented by $d(t)$, which is assumed to be piecewise continuous. This is not the process noise term of the Kalman filter theory. Hence, like the output error method, this approach cannot as such handle the true process noise. However, the aim here is different as outlined in the introduction. In control theory, the term $d(t)$ would represent a control force or input which is determined using an optimisation method by minimising the following function [4]:

$$J = \sum_{k=1}^{N} [z(k) - h(\hat{x}(k), k)]^T R^{-1} [z(k) - h(\hat{x}(k), k)] + \int_{t_0}^{t_f} d^T(t) Q d(t)\, dt \tag{8.2}$$

It is assumed that $E\{v(k)\} = 0$; $E\{v(k)v^T(k)\} = R(k)$ which is known. Here, h is the measurement model. The weighting matrix Q plays an important role and is a tuning device for the estimator. One natural way to arrive at Q is to choose it such that the following equality is satisfied:

$$R(k) = [z(k) - h(\hat{x}(k), k)][z(k) - h(\hat{x}(k), k)]^T \tag{8.3}$$

Here, $R(k)$ is the postulated covariance matrix of the measurement noise and the right hand side is the measurement covariance matrix computed using the difference between the actual measurements and the predicted measurements. This equality is called the covariance constraint.

The main advantage of the present approach is that it obtains state estimates in the presence of unmodelled effects as well as accurate estimates of these effects. Except on R, no statistical assumptions are required. The criteria used for estimation are based on least squares and one can obtain a recursive estimator like the Kalman filter after some transformations.

In the process, the model itself is improved, since this estimate of the unmodelled effects can be further modelled and the new model can be obtained as:

Accurate model (of the original system) = deficient model + model fitted to the discrepancy (i.e., unmodelled effects)

The problem of determination of the model deficiency or discrepancy is via minimisation of the cost functional eq. (8.2) which gives rise to the so-called two-point boundary value problem (TPBVP). This is treated in the next section.

8.2.1 Pontryagin's conditions

Let the dynamic system be given as

$$\dot{x} = f(x(t), u(t), t); \qquad x(t_0) = x_0 \tag{8.4}$$

Define a composite performance index as

$$J = \phi(x(t_f), t_f) + \int_{t_0}^{t_f} \psi(x(\tau), u(\tau), \tau)\, d\tau \tag{8.5}$$

The first term is the cost penalty on the final value of the state $x(t_f)$. The term $\psi(\cdot)$ is the cost penalty governing the deviation of $x(t)$ and $u(t)$ from their desired time-histories. The aim is to determine the input $u(t)$, in the interval $t_0 \leq t \leq t_f$, such that the performance index J is minimised, subject to the constraint of eq. (8.4), which states that the state should follow integration of eq. (8.4) with the input thus determined [1].

We use the concept of the Lagrange multiplier (see Section A.28) to handle the constraint within the functional J:

$$J_a = \phi(x(t_f), t_f) + \int_{t_0}^{t_f} [\psi(x(\tau), u(\tau), \tau) + \lambda^T(-f(x(\tau), u(\tau), \tau) + \dot{x})]\, d\tau \tag{8.6}$$

Here λ is the Lagrange multiplier and it facilitates the inclusion of the condition eq. (8.4), which is the constraint on the state of the dynamical system. That is to say, that in the process of determining $u(t)$ by minimisation of J_a, the condition of eq. (8.4) should not be violated. The Lagrange multipliers are known as adjoint variables or co-states. Since, in the sequel, we will have to solve the equations for the Lagrange multipliers, simultaneously with those of state equations, we prefer to use the 'co-state' terminology. If the condition of eq. (8.4) is strictly satisfied, then essentially eqs (8.5) and (8.6) are identical. Equation (8.6) can be rewritten as

$$J_a = \phi(x(t_f), t_f) + \int_{t_0}^{t_f} [H(x(\tau), u(\tau), \tau) - \dot{\lambda}^T(\tau)x(\tau)]\, d\tau + (\lambda^T x)_{t_f} - (\lambda^T x)_{t_0} \tag{8.7}$$

Here,

$$H = \psi(x(\tau), u(\tau), \tau) - \lambda^T(\tau) f(x(\tau), u(\tau), \tau) \tag{8.8}$$

H is called Hamiltonian. The term $\int_{t_0}^{t_f} \lambda^T \dot{x}\, d\tau$ of eq. (8.6) is 'integrated by parts' (see Section A.18) to obtain other terms in eq. (8.7). From eq. (8.7), we obtain, by using the concept of 'differentials'

$$\delta J_a = 0 = \left(\frac{\partial \phi}{\partial x}\delta x\right)_{t_f} + \lambda^T \delta x\Big|_{t_f} - \lambda^T \delta x\Big|_{t_0} + \int_{t_0}^{t_f} \left[\left(\frac{\partial H}{\partial x} - \dot{\lambda}^T\right)\delta x + \frac{\partial H}{\partial u}\delta u\right] d\tau \tag{8.9}$$

From eq. (8.9), the so-called Pontryagin's necessary conditions are

$$\lambda^T(t_f) = -\frac{\partial \phi}{\partial x}\Big|_{t_f} \tag{8.10}$$

$$\frac{\partial H}{\partial x} = \dot{\lambda}^T \tag{8.11}$$

and

$$\frac{\partial H}{\partial u} = 0 \tag{8.12}$$

Here, $\delta x(t_0) = 0$, assuming that the initial conditions $x(t_0)$ are independent of $u(t)$. Equation (8.10) is called the transversality condition.

The eqs (8.1) and (8.10–8.13) define the TPBV problem: the boundary condition for state is specified at t_0 and for the co-state; λ it is specified at t_f (eq. (8.10)).

From eqs (8.8) and (8.11), we obtain

$$\dot{\lambda} = \left(\frac{\partial H}{\partial x}\right)^T = -\left(\frac{\partial f}{\partial x}\right)^T \lambda + \left(\frac{\partial \psi}{\partial x}\right)^T \tag{8.13}$$

$$\frac{\partial H}{\partial u} = 0 = -\left(\frac{\partial f}{\partial u}\right)^T \lambda + \left(\frac{\partial \psi}{\partial u}\right)^T \tag{8.14}$$

One method to solve the TPBVP is to start with a guesstimate on $\lambda(t_0)$ and use $x(t_0)$ to integrate forward to the final time t_f. Then verify the boundary condition $\lambda(t_f) = -(\partial\phi/\partial x)|_{t_f}^T$. If the condition is not satisfied, then iterate once again with new $\lambda(t_0)$ and so on until the convergence of the algorithm is obtained. In the next section, we discuss the method of invariant embedding for solution of the TPBV problem.

8.3 Invariant embedding

Often it is useful to analyse a general process/solution of which our original problem is one particular case [8, 9]. The method of invariant embedding belongs to this category. What it means is that the particular solution we are seeking is embedded in the general class and after the general solution is obtained, our particular solution can be obtained by using the special conditions, which we have kept invariant, in final analysis.

Let the resultant equations from the two-point boundary value problem be given as (see eqs (8.1) and (8.13)):

$$\dot{x} = \Phi(x(t), \lambda(t), t) \tag{8.15}$$

$$\dot{\lambda} = \Psi(x(t), \lambda(t), t) \tag{8.16}$$

We see that the dependencies for Φ and Ψ on $x(t)$ and $\lambda(t)$ arise from the form of eqs (8.1), (8.13) and (8.14), hence, here we have a general two-point boundary value problem with associated boundary conditions as: $\lambda(0) = a$ and $\lambda(t_f) = b$. Now, though the terminal condition $\lambda(t_f) = b$ and time are fixed, we consider them as free variables. This makes the problem more general, which anyway includes our specific problem. We know from the nature of the two-point boundary value problem that the terminal state $x(t_f)$ depends on t_f and $\lambda(t_f)$. Therefore, this dependency can be represented as

$$x(t_f) = r(c, t_f) = r(\lambda(t_f), t_f) \tag{8.17}$$

with $t_f \rightarrow t_f + \Delta t$, and we obtain by neglecting higher order terms:

$$\lambda(t_f + \Delta t) = \lambda(t_f) + \dot{\lambda}(t_f)\Delta t = c + \Delta c \tag{8.18}$$

We also get, using eq. (8.16) in eq. (8.18):

$$c + \Delta c = c + \Psi(x(t_f), \lambda(t_f), t_f)\Delta t \tag{8.19}$$

and therefore, we get

$$\Delta c = \Psi(r, c, t_f)\Delta t \tag{8.20}$$

In addition, we get, like eq. (8.18):

$$x(t_f + \Delta t) = x(t_f) + \dot{x}(t_f)\Delta t = r(c + \Delta c, t_f + \Delta t) \tag{8.21}$$

and hence, using eq. (8.15) in eq. (8.21), we get

$$\begin{aligned} r(c + \Delta c, t_f + \Delta t) &= r(c, t_f) + \Phi(x(t_f), \lambda(t_f), t_f)\Delta t \\ &= r(c, t_f) + \Phi(r, c, t_f)\Delta t \end{aligned} \tag{8.22}$$

Using Taylor's series, we get

$$r(c + \Delta c, t_f + \Delta t) = r(c, t_f) + \frac{\partial r}{\partial c}\Delta c + \frac{\partial r}{\partial t_f}\Delta t \tag{8.23}$$

Comparing eqs (8.22) and (8.23), we get

$$\frac{\partial r}{\partial t_f}\Delta t + \frac{\partial r}{\partial c}\Delta c = \Phi(r, c, t_f)\Delta t \tag{8.24}$$

or, using eq. (8.20) in eq. (8.24), we obtain

$$\frac{\partial r}{\partial t_f}\Delta t + \frac{\partial r}{\partial c}\Psi(r, c, t_f)\Delta t = \Phi(r, c, t_f)\Delta t \tag{8.25}$$

The above equation simplifies to

$$\frac{\partial r}{\partial t_f} + \frac{\partial r}{\partial c}\Psi(r, c, t_f) = \Phi(r, c, t_f) \tag{8.26}$$

Equation (8.26) links the variation of the terminal condition $x(t_f) = r(c, t_f)$ to the state and co-state differential functions, see eqs (8.15) and (8.16). Now in order to find an optimal estimate $\hat{x}(t_f)$, we need to determine $r(b, t_f)$:

$$\hat{x}(t_f) = r(b, t_f) \tag{8.27}$$

Equation (8.26) can be transformed to an initial value problem by using approximation:

$$r(c, t_f) = S(t_f)c + \hat{x}(t_f) \tag{8.28}$$

Substituting eq. (8.28) in eq. (8.26), we get

$$\frac{dS(t_f)}{dt_f}c + \frac{d\hat{x}(t_f)}{dt_f} + S(t_f)\Psi(r, c, t_f) = \Phi(r, c, t_f) \tag{8.29}$$

Next, expanding Φ and Ψ about $\Phi(\hat{x}, b, t_f)$ and $\Psi(\hat{x}, b, t_f)$, we obtain

$$\Phi(r, c, t_f) = \Phi(\hat{x}, b, t_f) + \Phi_{\hat{x}}(\hat{x}, b, t_f)(r(c, t_f) - \hat{x}(t_f))$$
$$= \Phi(\hat{x}, b, t_f) + \Phi_{\hat{x}}(\hat{x}, b, t_f)S(t_f)c \quad (8.30)$$

and

$$\Psi(r, c, t_f) = \Psi(\hat{x}, b, t_f) + \Psi_{\hat{x}}(\hat{x}, b, t_f)S(t_f)c \quad (8.31)$$

Utilising expressions of eqs (8.30) and (8.31), in eq. (8.29), we obtain

$$\frac{dS(t_f)}{dt_f}c + \frac{d\hat{x}(t_f)}{dt_f} + S(t_f)[\Psi(\hat{x}, b, t_f) + \Psi_{\hat{x}}(\hat{x}, b, t_f)S(t_f)c]$$
$$= \Phi(\hat{x}, b, t_f) + \Phi_{\hat{x}}(\hat{x}, b, t_f)S(t_f)c \quad (8.32)$$

Equation (8.32) is in essence a sequential state estimation algorithm but a composite one involving $\hat{x}$ and $S(t_f)$. The above equation can be separated by substituting the specific expressions for Φ and Ψ in eq. (8.32). We do this in the next section after arriving at a two-point boundary value problem for a specific problem at hand, and then using eq. (8.32).

8.4 Continuous-time algorithm

Let the dynamic system be represented by

$$\dot{x} = f(x(t), t) + d(t) \quad (8.33)$$

$$z(t) = Hx(t) + v(t) \quad (8.34)$$

We form the basic cost functional as

$$J = \int_{t_0}^{t_f} [(z(t) - Hx(t))^T R^{-1}(z(t) - Hx(t)) + (d^T(t)Qd(t))]\,dt \quad (8.35)$$

where $d(t)$ is the model discrepancy to be estimated simultaneously with $x(t)$ and $R(t)$ is the spectral density matrix of noise covariance. We reformulate J by using Lagrange multipliers:

$$J_a = \int_{t_0}^{t_f} [(z(t) - Hx(t))^T R^{-1}(z(t) - Hx(t)) + d^T(t)Q_d(t))$$
$$+ \lambda^T(\dot{x}(t) - f(x(t), t) - d(t))]\,dt \quad (8.36)$$

Comparing with eqs (8.7) and (8.8), we get

$$H = (z(t) - Hx(t))^T R^{-1}(z(t) - Hx(t)) + d^T(t)Qd(t) - \lambda^T(f(x(t), t) + d(t))$$
$$= \psi - \lambda^T f_m(x(t), d(t), t) \quad (8.37)$$

By straightforward development paralleling eq. (8.9), we obtain

$$\dot{\lambda}^T = \frac{\partial H}{\partial x} = \frac{\partial \psi}{\partial x} - \lambda^T \frac{\partial f_m}{\partial x} \tag{8.38}$$

$$\dot{\lambda} = \left(\frac{\partial \psi}{\partial x}\right)^T - \left(\frac{\partial f_m}{\partial x}\right)^T \lambda = -f_{\hat{x}}^T \lambda - 2H^T R^{-1}(z(t) - Hx(t)) \tag{8.39}$$

and

$$0 = \frac{\partial H}{\partial d} = 2dQ - \lambda^T$$

leading to

$$d = \tfrac{1}{2} Q^{-1} \lambda \tag{8.40}$$

Thus our two-point boundary value problem is:

$$\begin{aligned} \dot{x} &= f(x(t), t) + d(t) \\ \dot{\lambda} &= -f_{\hat{x}}^T \lambda - 2H^T R^{-1}(z(t) - Hx(t)) \\ d &= \tfrac{1}{2} Q^{-1} \lambda \end{aligned} \tag{8.41}$$

Now, comparing with eqs (8.15) and (8.16), we obtain

$$\Phi(x(t), \lambda(t), t) = f(x(t), t) + d(t) \tag{8.42}$$

and

$$\Psi(x(t), \lambda(t), t) = -f_{\hat{x}}^T \lambda - 2H^T R^{-1}(z(t) - Hx(t)) \tag{8.43}$$

We also have

$$\Psi_{\hat{x}} = 2H^T R^{-1} H - \left[\frac{\delta}{\delta \hat{x}}(\lambda^T f_{\hat{x}})\right] \tag{8.44}$$

and

$$\Phi_{\hat{x}} = f_{\hat{x}} \tag{8.45}$$

Substituting eqs (8.42) to (8.45) in eq. (8.32) and considering t_f as the running time t, we obtain

$$\begin{aligned} \dot{S}(t)\lambda + \dot{\hat{x}}(t) + S(t)\Big[&- f_{\hat{x}}^T \lambda - 2H^T R^{-1}(z(t) - Hx(t)) \\ &+ 2H^T R^{-1} H S(t)\lambda - \frac{\delta}{\delta \hat{x}}(\lambda^T f_{\hat{x}}) S(t)\lambda \Big] \\ &= f(x(t), t) + \tfrac{1}{2} Q^{-1} \lambda + f_{\hat{x}} S(t)\lambda \end{aligned} \tag{8.46}$$

We separate terms related to λ from eq. (8.46) to get

$$\dot{\hat{x}} = f(x(t), t) + 2S(t)H^T R^{-1}(z(t) - Hx(t)) \tag{8.47}$$

$$\dot{S}(t)\lambda = S(t)f_{\hat{x}}^T\lambda + f_{\hat{x}}S(t)\lambda - 2S(t)H^T R^{-1}HS(t)\lambda$$
$$+ \frac{1}{2}Q^{-1}\lambda + S(t)\frac{\delta}{\delta\hat{x}}\left(\lambda^T f_{\hat{x}}\right)S(t)\lambda \tag{8.48}$$

We divide eq. (8.48) by λ and for $\lambda \to 0$, we get

$$\dot{S}(t) = S(t)f_{\hat{x}}^T + f_{\hat{x}}S(t) - 2S(t)H^T R^{-1}HS(t) + \tfrac{1}{2}Q^{-1} \tag{8.49}$$

We also have explicit expressions for the model error (discrepancy), comparing eq. (8.47) to eq. (8.33):

$$\hat{d}(t) = 2S(t)H^T R^{-1}(z(t) - Hx(t)) \tag{8.50}$$

Equations (8.47), (8.49) and (8.50) give the invariant embedding based model error estimation algorithm for continuous-time system of eqs (8.33) and (8.34), in a recursive form. Equation (8.49) is often called the matrix Riccati equation.

In order to implement the algorithm, we need to solve the matrix differential eq. (8.49). We can use the following transformation [10, 11]:

$$a = Sb \tag{8.51}$$

and using eq. (8.49)

$$\dot{S}b = Sf_{\hat{x}}^T b + f_{\hat{x}}Sb - 2SH^T R^{-1}HSb + \tfrac{1}{2}Q^{-1}b \tag{8.52}$$

or

$$\dot{S}b + 2SH^T R^{-1}HSb - Sf_{\hat{x}}^T b = f_{\hat{x}}a + \tfrac{1}{2}Q^{-1}b \tag{8.53}$$

We also have $\dot{a} = \dot{S}b + S\dot{b}$ and $\dot{S}b = \dot{a} - S\dot{b}$.

Using $\dot{S}b$ in eq. (8.53) and defining $\dot{b}$ as in eq. (8.54), we get

$$\dot{b} = -f_{\hat{x}}^T b + 2H^T R^{-1}Ha \tag{8.54}$$

$$\dot{a} = \tfrac{1}{2}Q^{-1}b + f_{\hat{x}}a \tag{8.55}$$

Equations (8.54) and (8.55) are solved by using the transition matrix method (see Section A.43) [11].

We note here that Q is the weighting matrix for the model error term. It provides normalisation to the second part of the cost function eq. (8.36).

8.5 Discrete-time algorithm

Let the true nonlinear system be given as

$$X(k+1) = g(X(k), k) \tag{8.56}$$

$$Z(k) = h(X(k), k) \tag{8.57}$$

Here g is the vector-valued function and Z is the vector of observables defined in the interval $t_0 < t_j < t_N$. Equations (8.56) and (8.57) are rewritten to express explicitly the model error (discrepancy):

$$x(k+1) = f(x(k), k) + d(k) \tag{8.58}$$

$$z(k) = h(x(k), k) + v(k) \tag{8.59}$$

Here f is the nominal model, which is a deficient model. The vector v is measurement noise with zero mean and covariance matrix R. The variable d is the model discrepancy, which is determined by minimising the criterion [9]:

$$J = \sum_{k=0}^{N} [z(k) - h(x(k), k)]^T R^{-1} [z(k) - h(x(k), k)] + d^T(k) Q d(k) \tag{8.60}$$

Minimisation should obtain two things: $\hat{x} \to X$ and estimate $\hat{d}(k)$ for $k = 0, \ldots, N$. By incorporating the constraint eq. (8.58) in eq. (8.60), we get

$$J_a = \sum_{k=0}^{N} [z(k) - h(x(k), k)]^T R^{-1} [z(k) - h(x(k), k)] + d^T(k) Q d(k)$$
$$+ \lambda^T [x(k+1) - f(x(k), k) - d(k)] \tag{8.61}$$

The Euler-Lagrange conditions yield the following [10]:

$$\hat{x}(k+1) = f(\hat{x}(k), k) + \tfrac{1}{2} Q^{-1} \lambda(k) \tag{8.62}$$

$$\lambda(k-1) = f_{\hat{x}}^T(\hat{x}(k), k)\lambda(k) + 2H^T R^{-1} [z(t) - H\hat{x}(k)] \tag{8.63}$$

with

$$H(k) = \left. \frac{\partial h(x(k), k)}{\partial x(k)} \right|_{x(k) = \hat{x}(k)} \quad \text{and} \quad d(k) = \tfrac{1}{2} Q^{-1} \lambda(k)$$

Equations (8.62) and (8.63) constitute a two-point boundary value problem, which is solved by using the invariant embedding method [10]. The resulting recursive algorithm is given as:

$$\hat{x}(k+1) = f_{\hat{x}}(\hat{x}(k), k) + 2S(k+1)H^T(k+1)R^{-1}[z(k+1) - h(\hat{x}(k+1), k+1)] \tag{8.64}$$

$$S(k+1) = \left[I + 2P(k+1)H^T(k+1)R^{-1}H(k+1) \right]^{-1} P(k+1) \tag{8.65}$$

$$P(k+1) = f_{\hat{x}}(\hat{x}(k), k)S(k) f_{\hat{x}}^T(\hat{x}(k), k) + \tfrac{1}{2} Q^{-1} \tag{8.66}$$

and

$$\hat{d}(k) = 2S(k)H^T(k)R^{-1}[z(k) - h(\hat{x}(k), k)] \tag{8.67}$$

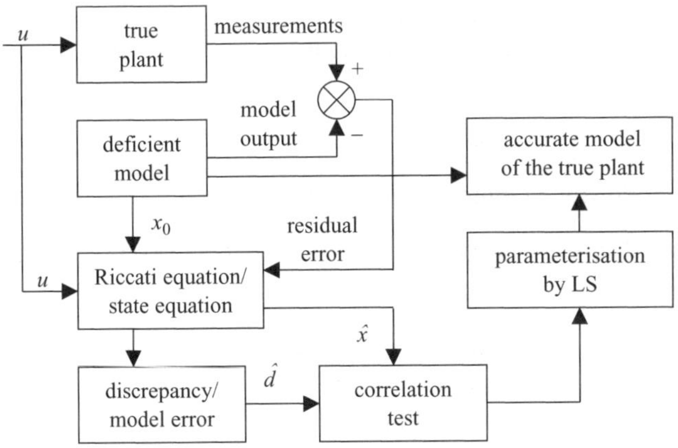

Figure 8.1 Block diagram of the model error estimation algorithm

8.6 Model fitting to the discrepancy or model error

Once we determine the time history of the discrepancy, we need to fit a mathematical model to it in order to estimate the parameters of this model by using a regression method. Figure 8.1 shows the schematic of the invariant embedding based model error estimation.

Assume that the original model of the system is given as

$$z(k) = a_0 + a_1 x_1 + a_2 x_1^2 + a_3 x_2 + a_4 x_2^2$$

Since we would not know the accurate model of the original system, we would use only a deficient model in the system state equations:

$$z(k) = a_0 + a_1 x_1 + a_3 x_2 + a_4 x_2^2 \tag{8.68}$$

The above equation is deficient by the term $a_2 x_1^2$.

When we apply the invariant embedding model error estimation algorithm to determine the discrepancy, we will obtain the time history of d, when we use the deficient model eq. (8.68). Once the d is estimated, a model can be fitted to this d and its parameters estimated (see Chapter 2). In all probability, the estimate of the missing term will be obtained:

$$d(k) = \hat{a}_2 \hat{x}_1^2 \tag{8.69}$$

In the above equation $\hat{x}_1$ is the estimate of state from the model error estimation algorithm. In order to decide which term should be added, a correlation test (Appendix A) can be used. Then the total model can be obtained as:

$$\hat{z}(k) = a_0 + a_1 \hat{x}_1 + \hat{a}_2 \hat{x}_1^2 + a_3 \hat{x}_2 + a_4 \hat{x}_2^2 \tag{8.70}$$

Under the condition that the model error estimation algorithm has converged, we will get $\hat{x} \rightarrow x$ and $\hat{a}_i \rightarrow a_i$, thereby obtaining the correct or adequately accurate model of the system.

8.6.1.1 Example 8.1

Simulate the following nonlinear continuous-time system

$$\dot{X}_1(t) = 2.5\cos(t) - 0.68X_1(t) - X_2(t) - 0.0195X_2^3(t) \tag{8.71}$$

$$\dot{X}_2(t) = X_1(t) \tag{8.72}$$

The above is a modified example of Reference 10.

Estimate the model discrepancy in the above nonlinear equations by eliminating the following terms from eq. (8.71) in turn:

Case (i) $\quad X_2^3$

Case (ii) $\quad X_1, X_2, X_2^3$

Use the invariant embedding model error estimation algorithm to estimate the model discrepancies for each of the cases (i) and (ii).

Fit a model of the form to the discrepancy thus estimated:

$$d(t) = a_1X_1(t) + a_2X_2(t) + a_3X_2^3(t) \tag{8.73}$$

to estimate the parameters of the continuous-time nonlinear system.

8.6.1.2 Solution

Data is generated by integrating eqs (8.71) and (8.72) for a total of 15 s using a sampling time $= 0.05$ s. For case (i), first, a deficient model is formulated by removing the term X_2^3 from eq. (8.71). The deficient model is then used in the invariant embedding model error estimation algorithm as f and the model discrepancy $d(t)$ is estimated. For case (ii), three terms X_1, X_2, X_2^3 are removed from the model to estimate $d(t)$ using the algorithm. Model discrepancies are estimated for each of the cases using the invariant embedding model error estimation files in the folder **Ch8CONTex1**. Values $Q = \text{diag}(0.001, 30)$ and $R = 18$ are used for this example for achieving convergence. The cost function converges to $J = 0.0187$ (for case (ii)).

The parameters are estimated from the model discrepancies using the least squares method. Table 8.1 shows the estimates of the coefficients compared with the true values for the two cases. The estimates compare well with the true values of the parameters. It is to be noted that in all the cases, from the estimated model discrepancy, the parameter that is removed from the model is estimated. Table 8.1 also shows the estimate of a_3 (case (iii)) when only 50 points are used for estimating the model discrepancy by removing the cubic nonlinear term in eq. (8.71). It is clear that the parameter is estimated accurately even when only fewer data points are used in the estimation procedure.

Figure 8.2(a) shows the comparison of the simulated and estimated states for case (ii). Figure 8.2(b) shows the estimated model discrepancies compared with the

Table 8.1 *Nonlinear parameter estimation results – continuous-time (Example 8.1)*

Parameter	a_1 X_1	a_2 X_2	a_3 X_2^3	Terms removed
True values	0.68	1	0.0195	–
Case (i)	(0.68)	(1)	0.0187	X_2^3
Case (ii)	0.5576	0.9647	0.0198	X_1, X_2, X_2^3
Case (iii)*	(0.68)	(1)	0.0220	X_2^3

* estimates with 50 data points, $(\cdot)$ true values retained

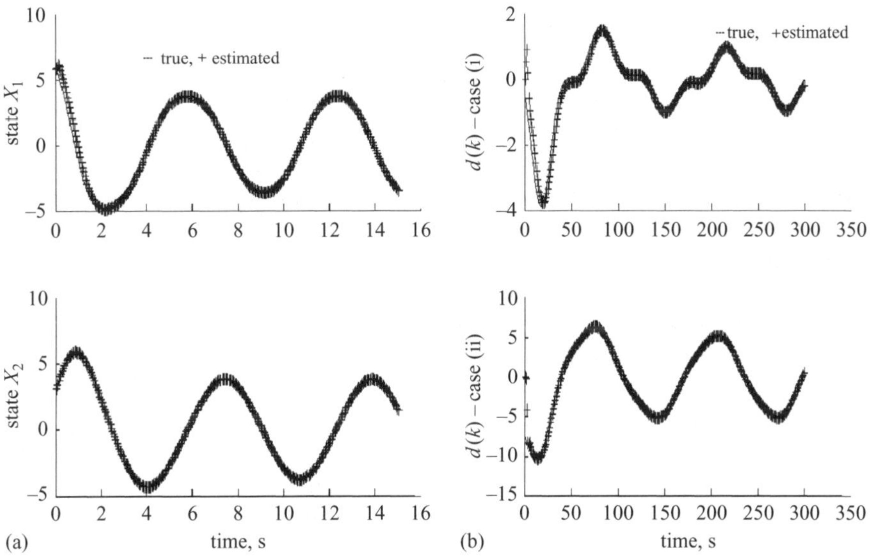

Figure 8.2 *(a) Time history match – states for case (ii) (Example 8.1); (b) time histories of model discrepancies d(k) (Example 8.1)*

true model error for both the cases. The match is very good and it indicates that the model discrepancy is estimated accurately by the algorithm.

8.6.1.3 Example 8.2

Use the simulated short period data of a light transport aircraft to identify and estimate the contribution of nonlinear effects in the aerodynamic model of the aircraft using the model error estimation algorithm. Study the performance of the algorithm when there is measurement noise in the data. Use the geometry and mass parameters given in Example 3.3.

8.6.1.4 Solution

The true data is generated with a sampling interval of 0.03 s by injecting a doublet input to the elevator. The measurements of u, w, q, θ are generated. Random noise with SNR $= 25$ and SNR $= 5$ is added to the measured states to generate two sets of noisy measurements. This example has a similar structure as the one in Reference 10, but the results are re-generated with different SNRs. The estimated model discrepancy does contain noise because the SNRs are low. However, in this case, the discrepancy data was used for parameter estimation using regression and no digital filter was used to filter out the remnant noise as in Reference 10.

For the above exercise, the state and measurement models for estimation of the parameters in the body axis are given in Appendix B.

The aerodynamic model has two nonlinear terms $C_{x_{\alpha 2}}$ and $C_{m_{\alpha 2}}$ in the forward force coefficient and pitching moment coefficient respectively as shown below:

$$C_x = C_{x_0} + C_{x_\alpha}\alpha + C_{x_{\alpha 2}}\alpha^2$$

$$C_m = C_{m_0} + C_{m_\alpha}\alpha + C_{m_{\alpha 2}}\alpha^2 + C_{m_q}\frac{q_m\bar{c}}{2V} + C_{m_{\delta_e}}\delta_e$$

By deleting the two nonlinear terms, the measured data (truth+noise) and the deficient models are used in the model error estimation continuous-time algorithm (folder **Ch8ACONTex2**). $Q = \text{diag}(0.06, 0.06, 0.06, 0.06)$ and $R = \text{diag}(1, 2, 3, 4)$ are used in the program for estimation of model discrepancy. This obtains the discrepancy, which is next modelled using the least squares method.

In order to estimate the parameters responsible for the deficiency, it is necessary to have a functional form relating the estimated states and the model deficiency. The parameters could then be estimated using the least squares method. The functional form is reached by obtaining the correlation coefficients (see Section A.10) between the estimated states and the model deficiency. Several candidate models shown in Table 8.2(a) were tried and correlation coefficients evaluated for each of the models. It is clear from the table that the term involving the state α^2 gives the highest correlation with the estimated deficiency. Table 8.2(b) shows the results of parameter estimation for the nonlinear terms for the case with no noise, SNR $= 25$ and SNR $= 5$. In each case, the true model is obtained using

Estimated true model

$= (\textit{Deficient model}) + (\textit{Estimated model from the model discrepancy})$

It is clear from Table 8.2 that despite the low SNRs, the nonlinear parameters are estimated accurately.

Figure 8.3(a) shows the time histories of the simulated true and deficient states. The continuous-time model error estimation is used to estimate the states recursively. Figure 8.3(b) shows the simulated and estimated states. The good match indicates that the estimated model discrepancy would account for the model deficiency quite accurately.

Table 8.2 *(a) Correlation results; (b) nonlinear parameter estimation results – aircraft data*

(Example 8.2)

(*a*)

ρ for Cm		ρ for Cx	
$Cm_{\alpha 2}$	0.9684	$Cx_{\alpha 2}$	−0.9857
$Cm_{\alpha 3}$	0.9567	$Cx_{\alpha 3}$	−0.9733
$Cm_{\alpha 4}$	0.9326	$Cx_{\alpha 4}$	−0.9486
$Cm_{\alpha 2} + Cm_{\alpha 3}$	0.9678	$Cx_{\alpha 2} + Cx_{\alpha 3}$	−0.9850
$Cm_{\alpha 2} + Cm_{\alpha 4}$	0.9682	$Cx_{\alpha 2} + Cx_{\alpha 4}$	−0.9853
$Cm_{\alpha 3} + Cm_{\alpha 4}$	0.9517	$Cx_{\alpha 3} + Cx_{\alpha 4}$	−0.9839
$Cm_{\alpha 2} + Cm_{\alpha 3} + Cm_{\alpha 4}$	0.9669	$Cx_{\alpha 2} + Cx_{\alpha 3} + Cx_{\alpha 4}$	0.9669

(*b*)

Parameter	$Cx_{\alpha 2}$	$Cm_{\alpha 2}$
True values	3.609	1.715
No noise	3.6370	1.6229
SNR = 25	3.8254	1.7828
SNR = 5	3.9325	1.7562

8.6.1.5 Example 8.3

Simulate the following nonlinear discrete system:

$$X_1(k+1) = 0.8X_1(k) + 0.223X_2(k) + 2.5\cos(0.3k) + 0.8\sin(0.2k) - 0.05X_1^3(k) \tag{8.74}$$

$$X_2(k+1) = 0.5X_2(k) + 0.1\cos(0.4k) \tag{8.75}$$

Estimate the model discrepancy in the above nonlinear equations by eliminating the following terms from eq. (8.74) in turn.

Case (i) X_1^3

Case (ii) X_1, X_1^3

Case (iii) X_1, X_2, X_1^3

Use the invariant embedding model error estimation algorithm to estimate the model discrepancies for each of the cases (i), (ii) and (iii).

To the discrepancy thus estimated, fit a model of the form

$$d(k) = a_1X_1(k) + a_2X_1^2(k) + a_3X_1^3(k) + a_4X_2(k) \tag{8.76}$$

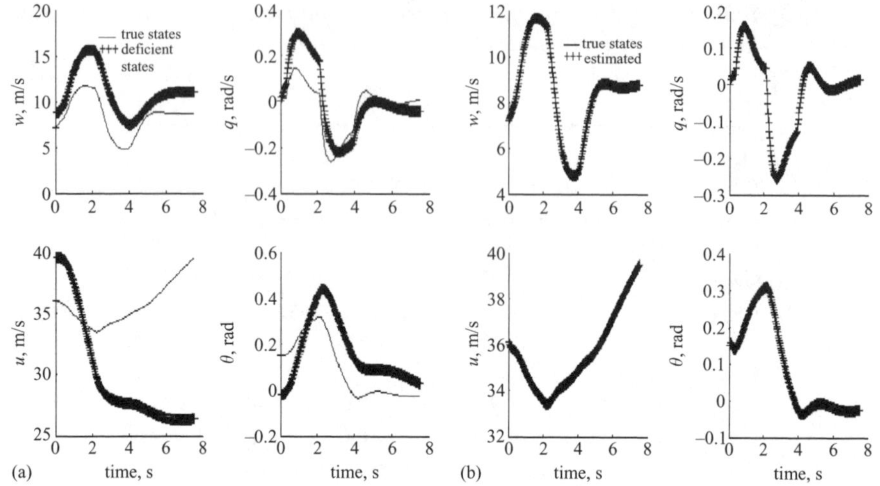

Figure 8.3 (a) True and deficient state time histories (Example 8.2); (b) true and estimated states (after correction for deficiency) (Example 8.2)

to estimate the parameters of the discrete nonlinear system from the estimated model discrepancies $d(k)$.

8.6.1.6 Solution

One hundred samples of data are generated using eqs (8.74) and (8.75). For case (i), a deficient model is formulated by removing the term X_1^3 from the eq. (8.74). The deficient model is used in the invariant embedding model error estimation algorithm as f and the model discrepancy $d(k)$ is estimated. For case (ii), two terms X_1, X_1^3 are removed from the true model eq. (8.74) and for case (iii) three terms X_1, X_2, X_1^3 are removed. Model discrepancies are estimated for each of these cases using the model error estimation files in the folder **Ch8DISCex3**.

Subsequently, a model based on a third order polynomial in X_1 and a first order in X_2 (eq. (8.76)) is fitted to the discrepancy $d(k)$ in each of the cases and the parameters estimated using a least squares method. It is to be noted that although the term containing X_1^2 is not present in the true model of the system, it is included to check the performance of the algorithm. Table 8.3 shows the estimates of the coefficients compared with the true values for the three cases. The estimates compare very well with the true values of the parameters. It is to be noted that in all the cases, from the estimated model discrepancy, the parameter that is removed from the model is estimated. In all the cases, the term a_2 is estimated with a value, which is practically zero since it is anyway not present in the model.

Figure 8.4(a) shows the comparison of the simulated and estimated model states for case (iii). Figure 8.4(b) shows the estimated model discrepancy $d(k)$ compared

Table 8.3 Nonlinear parameter estimation results – discrete-time (Example 8.3)

Parameter	a_1 (X_1)	a_2 (X_1^2)	a_3 (X_1^3)	a_4 (X_2)	Terms removed
True values	0.8	0	−0.05	0.223	–
Case (i)	(0.8)	−1.03e−5	−0.0499	(0.223)	X_1^3
Case (ii)	0.7961	−8.3e−6	−0.0498	(0.223)	X_1, X_1^3
Case (iii)	0.8000	−3.07e−7	−0.0500	0.2224	X_1, X_2, X_1^3

(·) true values used in the model

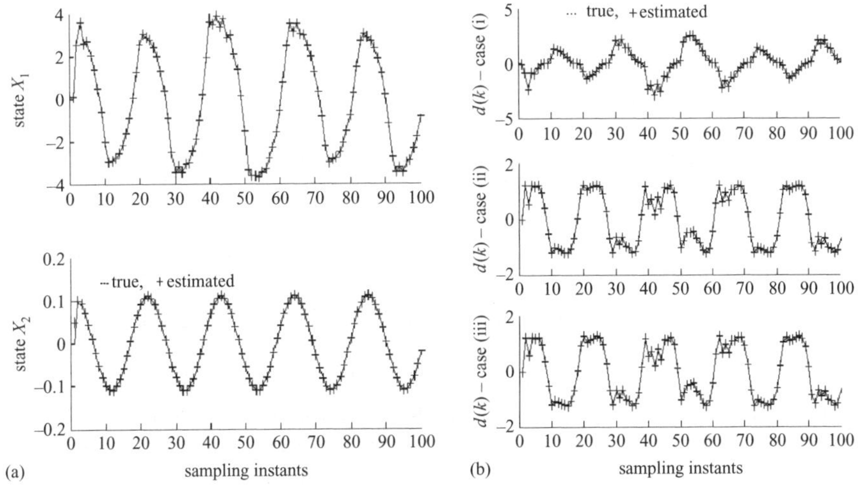

Figure 8.4 (a) Time history match – states for case (iii) (Example 8.3); (b) time histories of model (Example 8.3)

with the true model discrepancies for all the cases. The good match indicates good estimation of the model discrepancy.

8.7 Features of the model error algorithms

First, we emphasise that the matrix $R(t)$ in eq. (8.36) is the spectral density matrix for the covariance of measurement noise. We regard R^{-1} as the weighting matrix in eq. (8.36). We observe here that although the term $d(t)$ or $d(k)$ is called the deterministic discrepancy, the terms related to the residuals appear in it. Two meanings could be attached to the term deterministic:

1 It is not random, since it appears in eq. (8.1) as a model deficiency.
2 It is possible to determine or estimate it from eq. (8.67).

However, the effect of residuals on $d(t)$ or $d(k)$ does not pose any severe problems, because it is further modelled to estimate parameters that fit the model error d.

Some important features of the model error-based solution/algorithm are [1–6]:

1 It does not need initial values of the parameters to fit the model error.
2 It is fairly robust in the presence of noise.
3 It can determine the form of the unknown nonlinearity, and the values of the parameters that will best fit this model. This is made possible by the use of the correlation coefficient, between d and each of the state variable appearing in the model.
4 It requires minimum *a priori* assumptions regarding the model or the system.
5 It gives good results even if few data points are available for the model error time history.

Two important aspects of the algorithm are:

1 Tuning of Q.
2 Proper choice of R.

These can be achieved by using the covariant constraint of eq. (8.3).

8.8 Epilogue

The method of model error estimation has been extensively treated in References 1 to 6, wherein various case studies of deficient models were considered. Very accurate estimates of the parameters from the model error time histories were obtained. The method of invariant embedding has been considered in References 8 and 9. In Reference 6, the authors present a process noise covariance estimator algorithm, which is derived by using the covariance constraint, the unbiased constraint and the Kalman filter. This can be used even if model error is not completely Gaussian. We strongly feel that the model error estimation could emerge as a viable alternative to the output error method and, further, it can give recursive solutions.

8.9 References

1 MOOK, J.: 'Measurement covariance constrained estimation for poorly modelled dynamic system', Ph.D Thesis, Virginia Polytechnic Institute and State University, 1985
2 MOOK, D. J., and JUNKINS, J. L.: 'Minimum model error estimation for poorly modelled dynamic systems', AIAA 25th Aerospace Sciences Meeting, AIAA-87-0173, 1987
3 MOOK, D. J.: 'Estimation and identification of nonlinear dynamic systems', *AIAA Journal*, 1989, **27**, (7), pp. 968–974
4 MAYER, T. J., and MOOK, D. J.: 'Robust identification of nonlinear aerodynamic model structure', AIAA-92-4503-CP, 1992

5 CRASSIDIS, J. L., MARKLEY, F. L., and MOOK, D. J.: 'A real time model error filter and state estimator', Proceedings of AIAA conference on *Guidance, Navigation and Control*, Arizona, USA, Paper no. AIAA-94-3550-CP, August 1–3, 1994

6 MASON, P., and MOOK, D. J.: 'A process noise covariance estimator', Ibid, AIAA-94-3551-CP

7 MAYBECK, P. S.: 'Stochastic modelling, estimation and control', vols 1 and 2 (Academic Press, USA, 1979)

8 DATCHMENDY, D. M., and SRIDHAR, R.: 'Sequential estimation of states and parameters in noisy nonlinear dynamical systems', *Trans. of the ASME, Journal of Basic Engineering*, 1966, pp. 362–368

9 DESAI, R. C., and LALWANI, C. S.: 'Identification techniques' (McGraw-Hill, New Delhi, 1972)

10 PARAMESWARAN, V., and RAOL, J. R.: 'Estimation of model error for nonlinear system identification', *IEE Proc. Control Theory and Applications*, 1994, **141**, (6), pp. 403–408

11 GELB, A. (Ed.): 'Applied optimal estimation' (M.I.T. Press, Cambridge, MA, 1974)

8.10 Exercises

Exercise 8.1

In the expression of J (eq. (8.2)), the weight matrix appears in the second term. Can we call Q as the covariance matrix of some variable? What interpretation can you give to Q?

Exercise 8.2

Consider the second term within the integral sign of eq. (8.6), which apparently shows that the state history seems to be constrained. Explain this in the light of covariance constraint, i.e., eq. (8.3). (Hint: try to establish some logical connection between these two constraints.)

Exercise 8.3

In eq. (8.2), the inverse of R is used as the weighting matrix in the first term. Explain the significance of use of R^{-1} here. (Hint: the terms around R^{-1} signify the covariance of the residuals.)

Exercise 8.4

See eq. (8.3), which states that the theoretical (postulated) covariance matrix is approximately equal to the measurement error covariance matrix and this is called the covariance constraint. Does a similar aspect occur in the context of the Kalman filter theory?

Exercise 8.5

Although d of eq. (8.1) is called the deterministic discrepancy (since the state model does not have process noise), we see from eq. (8.50) that it does contain a residual term, which is a random process. How will this be treated when modelling d?

Exercise 8.6

What simple trick can be used to avoid the errors due to matrix S, eq. (8.49), becoming asymmetrical?

Exercise 8.7

Let $\dot{x} = d(t)$. The measurements are given as $z(k) = x(k) + v(k)$. Formulate the cost function and define Hamiltonian H?

Exercise 8.8

The cost function of eq. (8.6) includes the cost penalty at final time t_f for the state. How will you include the penalty terms for the intermediate points [1] between $t = t_0$ and $t = t_f$.

Exercise 8.9

Obtain $\partial H/\partial x$ from the Hamiltonian equation (see eq. (8.8)) and hence the state space type differential equation for the co-state?

Chapter 9

Parameter estimation approaches for unstable/augmented systems

9.1 Introduction

Parameter estimation of unstable systems is necessary in applications involving adaptive control of processes, satellite launch vehicles or unstable aircraft operating in closed loop. In these applications, under normal conditions, the system operates with the feedback controller and generates controlled responses. The system could become unstable due to sensor failures of critical sensors generating the feedback signals or sudden/unforeseen large dynamic changes in the system. Under these conditions, analysis of the data would give clues to the cause of the failure. This knowledge can be utilised for reconfiguration of the control laws for the systems.

In many applications, it is required to estimate the parameters of the open loop plant from data generated when the system is operating in closed loop. When data for system identification purposes are generated with a dynamic system operating in closed loop, the feedback causes correlations between the input and output variables [1]. This data correlation causes identifiability problems, which result in inaccurate parameter estimates. For estimation of parameters from measured input-output data, it is mandatory that the measured data contain adequate information about the modes of the system being identified. In the case of augmented systems, the measured responses may not display the modes of the system adequately since the feedback is meant to generate controlled responses. It may not be always possible to recover accurately the open loop system dynamics from the identification using closed loop data when conventional approaches of parameter estimation are used. Although some of the conventional parameter estimation techniques are applicable to the augmented systems in principle, a direct application of the techniques might give erroneous results due to correlations among the dynamic variables of the control system.

Thus, the estimation of parameters of open loop plant from the closed loop data is difficult even when the basic plant is stable. The estimation problem complexity is compounded when the basic plant is unstable because the integration of the state model could lead to numerical divergence. In most practical cases, the data could be corrupted by process and measurement noise, which further renders the problem more complex. The problem of parameter estimation of unstable/augmented systems could be handled through the following two approaches:

1 Ignoring the effect of feedback, the open loop data could be used directly. In loosely coupled systems, this approach might work well. However, if the feedback loop is tight, due to data collinearity, this method may give estimates with large uncertainty [2].
2 The models of control system blocks and other nonlinearities could be included to arrive at a complete system model and the closed loop system could be analysed for parameter estimation. In this case, the input-output data of the closed loop system can be used for estimation. However, this approach is complicated since the coupled plant-controller model to be used in the estimation procedure could be of a very high order.

To begin addressing this complex problem, in this chapter, the effect of various feedback types on the parameterisation of the system is reviewed in Section 9.2. In highly unstable systems, the conventional output error parameter estimation procedure (Chapter 3) may not be able to generate useful results because the output response could grow very rapidly. In such cases, for parameter estimation, (i) short data records could be used or (ii) the unstable model could be stabilised by feedback (in the software model) and the open loop characteristics could be obtained from the closed loop data. If limited time records are used, the identification result will be unbiased only when the system is noise free. The equation error method, which does not involve direct integration of the system state equations (Chapter 2), could be used for parameter estimation of unstable systems. However, equation error methods need accurate measurements of state and state derivatives. Alternatively, the Kalman filter could be used for parameter estimation of unstable systems because of its inherent stabilisation properties. The two approaches for parameter estimation of unstable systems (without control augmentation) are discussed in Sections 9.3 and 9.4: i) based on UD factorisation Kalman filtering (applicable to linear as well as nonlinear systems); and ii) an approach based on eigenvalue transformation applicable to linear continuous time systems [3].

Commonly used methods for the detection of collinearity in the data are discussed in Section 9.5. A method of mixed estimation wherein the *a priori* information on some of the parameters is appended in a least squares estimation procedure for parameter estimation from collinear data is discussed in Section 9.6. A recursive solution to the mixed estimation algorithm obtained by incorporating the *a priori* information into the Extended UD Kalman filter structure is given in Section 9.7.

The OEM, which is the most commonly used method for parameter estimation of stable dynamic systems, poses certain difficulties when applied to highly unstable systems since the numerical integration of the unstable state equations leads to diverging solutions. One way to avoid this problem is to provide artificial stabilisation in the mathematical model used for parameter estimation resulting in the feedback-in-model approach. However, practical application of this technique requires some engineering effort. One way to circumvent this problem is to use measured states in the estimation procedure leading to the so-called stabilised output error method (SOEM) [4]. An asymptotic theory of the stabilised output error method [5] is provided in this chapter. The analogy between the Total Least Squares (TLS) [6] approach and the SOEM is also brought out. It is shown that stabilised output error methods emerge as a generalisation of the total least squares method, which in itself is a generalisation of least squares method [7].

Parameter estimation techniques for unstable/augmented systems using the information on dynamics of controllers used for stabilising the unstable plant is discussed in detail. Two approaches are described: i) equivalent model estimation and parameter retrieval approach; and ii) controller augmented modelling approach, and a two-step bootstrap method is presented [8].

Thus, this chapter aims to present a comprehensive study of the problem of parameter estimation of inherently unstable/augmented control systems and provide some further insights and directions. These approaches are also applicable to many aerospace systems: unstable/augmented aircraft, satellite systems etc.

9.2 Problems of unstable/closed loop identification

In Fig. 9.1, the block diagram of a system operating in a closed loop configuration is shown. Measurements of input (at point p1, δ), the error signal input (u at p2) to the plant and the output (z at p3) are generally available. Two approaches to estimate the parameters from the measured data are possible: i) Direct Identification – ignoring the presence of the feedback, a suitable identification method is applied to the data between p2 and p3; and ii) Indirect Identification – the data between p1 and p3 could be analysed to estimate equivalent parameters. In this case, the closed loop system is regarded as a composite system for parameter estimation. The knowledge of the feedback gains and the models of control blocks could then be used to retrieve the parameters of the system from the estimated equivalent model.

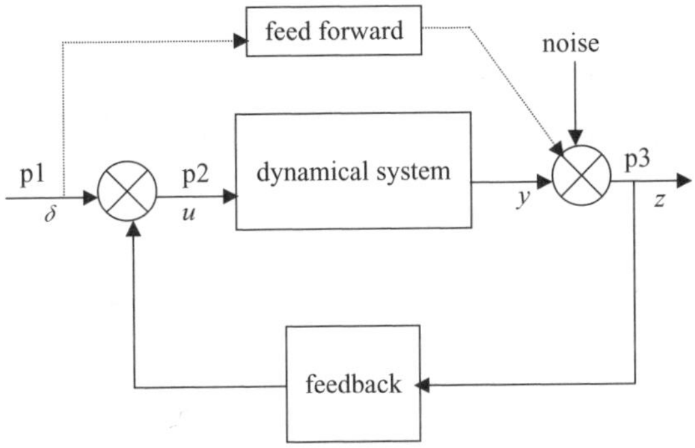

Figure 9.1 Closed loop system

Feedback introduces correlations between the input and output variables. Hence, when the direct identification method is used, the corresponding parameter estimates of the system could be highly correlated. In addition, the noise is correlated with input u due to feedback. As a result, it may not be possible to estimate all the system parameters independently. At best, by fixing some of the parameters at their predicted/analytical values, a degenerate model could be estimated. In addition, due to feedback action constantly trying to generate controlled responses, the measured responses might not properly exhibit modes of the system. Using the conventional methods of analysis, like the output error method and least squares method, it may be possible to obtain accurate estimates of parameters if the control loop system dynamics are only weakly excited during measurement period (if feedback loops are not 'tight'). If feedback were 'tight', data correlations would cause the parameters to be estimated with large uncertainties. Hence, it is necessary to detect the existence and assess the extent of the collinearity in the data. One then uses a suitable method to estimate parameters in the presence of data collinearity.

For unstable plant, the control system blocks augment the plant and this has a direct influence on the structure of the mathematical model [1] of the system as shown in Table 9.1. The basic plant description is given by:

$$\dot{x} = Ax + Bu \tag{9.1}$$

In Table 9.1, δ represents input at point p1 (Fig. 9.1), K is the feedback matrix for constant or proportional feedback systems, L is the matrix associated with differential feedback and F with integrating feedback [1]. From Table 9.1 it is clear that the control system with constant feedback affects only estimates of the elements of system matrix A and does not affect the structure of the system. The state matrix is modified resulting in state equations that represent a system having different dynamics from

Table 9.1 Effect of feedback on the parameters and structure of the mathematical model [1]

Control system type	Input	System states	Changes
Constant feedback	$u = Kx + \delta$	$\dot{x} = (A + BK)x + B\delta$	Coefficients in the column of feedback
Differential feedback	$u = Kx + L\dot{x} + \delta$	$\dot{x} = (I - BL)^{-1} \times[(A + BK)x + B\delta)]$	Almost all coefficients
Integrating feedback	$\dot{u} + Fu = Kx + \delta$	$\begin{bmatrix}\dot{x}\\ \dot{u}\end{bmatrix} = \begin{bmatrix}A & B\\ K & -F\end{bmatrix}\begin{bmatrix}x\\ u\end{bmatrix} + \begin{bmatrix}0\\ 1\end{bmatrix}\delta$	Structure

the original unstable system. With differential feedback, even if only one signal is feedback, all the coefficients are affected, the basic structure remaining the same. The entire structure is changed when the feedback control system has integrators in the feedback loops. The number of poles increases with the number of equations and for a highly augmented system, the overall system order could be very high.

Including the noise w in eq. (9.1), we get

$$\dot{x} = Ax + Bu + w \tag{9.2}$$

If the control system is a constant feedback type, the input u can be represented by

$$u = Kx + \delta \tag{9.3}$$

Here, K is the constant gain associated with the feedback.

Multiplying eq. (9.3) by an arbitrary matrix B_a and adding to eq. (9.2), we get

$$\dot{x} = (A + B_a K)x + (B - B_a)u + w + B_a\delta \tag{9.4}$$

The term $(w + B_a\delta)$ can be regarded as noise and estimates of the parameters are obtained by minimising a quadratic cost function of this noise. If the input δ is large, then the elements of B_a are insignificant and hence they might be neglected. In that case, eqs (9.2) and (9.4) become identical and feedback would have very little influence on estimated results. However, a large δ might excite nonlinear behaviour of the system. If the input δ is small or of short duration, the matrix B_a influences the coefficient matrices of x and u, and the results of identification will be $(A + B_a K)$ and $(B - B_a)$ instead of A and B. This clearly shows that the feedback influences the identifiability of the parameters of the open loop system. This also means that if the input has low intensity, it does not have sufficient power.

When the system responses are correlated due to feedback

$$x = Kx, \qquad K \neq I \tag{9.5}$$

The elements of the K matrix could be the feedback gains. Inserting eq. (9.5) into eq. (9.2) we get

$$\dot{x} = [A + B_a(K - I)]x + Bu + w \tag{9.6}$$

Since B_a is an arbitrary matrix, even here it is difficult to determine elements of A from output responses. Control augmentation is thus found to cause 'near linear' relationships among variables used for parameter estimation which affects the accuracy of the estimates. Hence, it is required to detect this collinearity in the data, assess its extent and accordingly choose an appropriate estimation procedure.

9.3 Extended UD factorisation based Kalman filter for unstable systems

An extended Kalman filter (Chapter 4) could be used for parameter estimation of unstable systems because of the inherent stabilisation present in the filter. As is clear from eq. (4.50), a feedback proportional to the residual error updates the state variables. This feedback numerically stabilises the filter algorithm and improves the convergence of the estimation algorithm. The following example presents the applicability of the extended UD factorisation filter for parameter estimation of an unstable second order dynamical system.

9.3.1.1 Example 9.1

Simulate data of a second order system with the following state and measurement matrices:

$$\begin{bmatrix} \dot{x}_1 \\ \dot{x}_2 \end{bmatrix} = \begin{bmatrix} a_{11} & a_{22} \\ a_{33} & a_{44} \end{bmatrix} \begin{bmatrix} x_1 \\ x_2 \end{bmatrix} + \begin{bmatrix} b_1 \\ b_2 \end{bmatrix} u = \begin{bmatrix} 0.06 & -2.0 \\ 2.8 & 0.08 \end{bmatrix} \begin{bmatrix} x_1 \\ x_2 \end{bmatrix} + \begin{bmatrix} -0.6 \\ 1.5 \end{bmatrix} u \quad (9.7)$$

$$\begin{bmatrix} y_1 \\ y_2 \end{bmatrix} = \begin{bmatrix} 1 & 0 \\ 0 & 1 \end{bmatrix} \begin{bmatrix} x_1 \\ x_2 \end{bmatrix} \quad (9.8)$$

by giving a doublet signal as input to the dynamical system (with sampling interval = 0.05 s). Use UD factorisation based EKF (EUDF) to estimate the parameters of the unstable system. Using $a_{22} = 0.8$ (all other system parameters remaining the same), generate a second data set. Study the effect of measurement noise on the estimation results.

9.3.1.2 Solution

Simulated data for 10 s (with a sampling rate of 20 samples/s), is generated using eqs (9.7) and (9.8) (programs in folder **Ch9SIMex1**). The state model is formulated with the two states x_1, x_2 and the six unknown parameters in eq. (9.7) as augmented states in EUDF (Chapter 4). The measurement model uses the observations y_1 and y_2 generated using eq. (9.8). The parameter estimation programs are contained in the folder **Ch9EUDFex1**.

Table 9.2 gives the eigenvalues of the unstable second order system for the two cases of simulated data obtained by varying the parameter a_{22}. It is clear that for $a_{22} = 0.8$, the instability is higher. Random noise (with SNR = 10) is added to the data to generate two more sets of data for parameter estimation. Table 9.3 shows the results of parameter estimation using EUDF for the four sets of data. The initial guesstimates for the states were chosen to be 20 per cent away from their true values. It is clear that the parameter estimates are very close to the true values in both the cases when there is no noise in the data. However, when there is noise in the data, the

Table 9.2 Eigenvalues of the unstable 2nd order system (Example 9.1)

Case no.	Eigenvalues	Instability
1	$0.0700 \pm j\,2.3664$	Low
2	$0.4300 \pm j\,2.3373$	High

Table 9.3 Parameter estimates (EUDF) – unstable 2nd order system (Example 9.1)

Parameters	Case 1 ($a_{22} = 0.08$)			Case 2 ($a_{22} = 0.8$)		
	True	Estimated (no noise)	Estimated (SNR = 10)	True	Estimated (no noise)	Estimated (SNR = 10)
a_{11}	0.06	0.0602 (0.0011)*	0.0571 (0.0093)	0.06	0.0600 (0.0001)	0.0676 (0.0111)
a_{12}	−2.0	−1.9999 (0.0009)	−1.9047 (0.0568)	−2.0	−2.00 (0.0001)	−1.9193 (0.0624)
a_{21}	2.8	2.8002 (0.0004)	2.9536 (0.0469)	2.8	2.8000 (0.0001)	2.9128 (0.0369)
a_{22}	0.08	0.079 (0.0001)	0.0775 (0.0051)	0.8	0.8 (0.0003)	0.7843 (0.0280)
b_1	−0.6	−0.5923 (0.0004)	−0.5221 (0.0262)	−0.6	−0.5871 (0.0001)	−0.6643 (0.0227)
b_2	1.5	1.5041 (0.0000)	1.5445 (0.0003)	1.5	1.5025 (0.0000)	1.2323 (0.0021)
PEEN %	–	0.2296	5.3078		0.3382	7.9476

* standard deviations of the estimated parameters

estimates show some deviation, which is also reflected in the higher PEEN values for these cases. The estimated parameters are noted down at the last data point (200th point for this case).

Figure 9.2 shows the comparison of the predicted measurements y_1 and y_2 for the case 2 data without noise ($a_{22} = 0.8$) and the estimated parameters using EUDF. From the figure, it is clear that all the estimated parameters converge to the true values. This example clearly illustrates that the EUDF technique is applicable to parameter estimation of unstable systems. It should be noted that when the method is used for parameter estimation from real data, considerable effort would be required to make an appropriate choice of the covariance matrices P, Q and R in addition to reasonably close start up values for the initial values of the states.

9.4 Eigenvalue transformation method for unstable systems

In order that the conventional parameter estimation methods like the output error method could be utilised for parameter estimation of unstable systems when they are operating in open loop, this section presents a technique of transformation of input-output data of a continuous time unstable system. The technique described is applicable to linear continuous time systems. A similar method for transfer function identification of discrete systems is given in Reference 3.

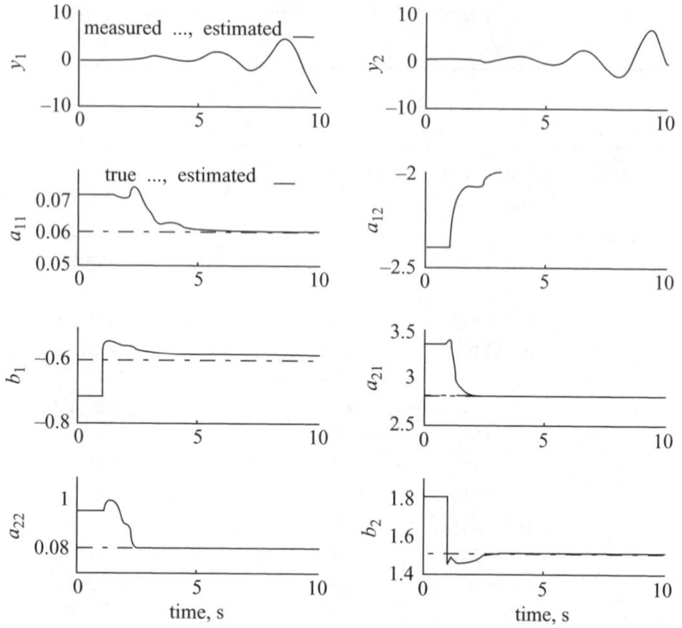

Figure 9.2 *Measurements (y_1, y_2 w/o noise) and estimated parameters (Example 9.1)*

The philosophy involves transformation of the unstable system data into stable time histories by following an appropriate procedure. A transformation parameter, which is based on the real part of the largest unstable eigenvalue of the system, is chosen and is used to transform the system mathematical model as well. By this method, the numerical divergence problem associated with the identification of the unstable system is greatly reduced [9].

A general continuous time linear system is described by

$$\dot{x} = Ax + Bu \qquad \text{with } x(0) = x_0 \tag{9.9}$$

$$y = Hx + v \tag{9.10}$$

Assuming that a suitable parameter δ is available, the states, input and output are transformed to generate transformed variables $\bar{x}$, $\bar{y}$ and $\bar{u}$ using

$$\begin{aligned} &\bar{x}(t) = e^{-\delta t} x(t); \\ &\bar{y}(t) = e^{-\delta t} y(t); \qquad \bar{u}(t) = e^{-\delta t} u(t) \end{aligned} \tag{9.11}$$

This could also be written as

$$x(t) = \bar{x}(t) e^{\delta t}; \tag{9.12}$$

$$y(t) = \bar{y}(t) e^{\delta t}; \qquad u(t) = \bar{u}(t) e^{\delta t} \tag{9.13}$$

Here, overbar represents the transformed variables.

From eq. (9.12), we have

$$\dot{x}(t) = \dot{\bar{x}}(t)e^{\delta t} + \delta e^{\delta t}\bar{x}(t) \tag{9.14}$$

Equations (9.12)–(9.14) are used in eqs (9.9)–(9.10) to get

$$\begin{aligned} \dot{\bar{x}}(t)e^{\delta t} + \delta e^{\delta t}\bar{x}(t) &= A\bar{x}(t)e^{\delta t} + B\bar{u}(t)e^{\delta t} \\ \bar{y}(t)e^{\delta t} &= H\bar{x}e^{\delta t} + v \end{aligned} \tag{9.15}$$

Eliminating $e^{\delta t}$, we get

$$\dot{\bar{x}}(t) + \delta\bar{x}(t) = A\bar{x}(t) + B\bar{u}(t) \tag{9.16}$$

$$\dot{\bar{x}}(t) = (A - I\delta)\bar{x}(t) + B\bar{u}(t) = \bar{A}\bar{x}(t) + B\bar{u}(t) \tag{9.17}$$

$$\bar{y} = H\bar{x} + ve^{-\delta t} \tag{9.18}$$

The new system equations are in terms of the transformed data. It is clear that the eigenvalues of the new system are altered because of δ. The transformed matrix $(A - I\delta)$ will have stable eigenvalues if the transformation parameter is chosen appropriately.

To start the parameter estimation procedure, a set of transformed data is obtained from the measurements $z(k)$ (outputs of the unstable dynamical system) using eq. (9.11), which can be represented by

$$\bar{z}(k) = \bar{y}(k) + \bar{v}(k), \qquad k = 1, 2, \ldots, N \tag{9.19}$$

Here, $\bar{v}$ is the measurement noise, with covariance matrix R_m.

The parameter vector to be estimated is given by $\Theta = \{\bar{A}, B, H\}$. The estimates of the parameters Θ are obtained by minimising the cost function defined as

$$E(\Theta) = \frac{1}{2}\sum_{k=1}^{N}[\bar{z}(k) - \bar{y}(k)]^T R_m^{-1}[\bar{z}(k) - \bar{y}(k)] + \frac{N}{2}\ln|R_m| \tag{9.20}$$

Here we note that

$$R_m = \mathrm{cov}(\bar{v}(k)\bar{v}^T(k)) = E[e^{-\delta t}v(k)v^T(k)e^{-\delta t}] = e^{-2\delta t}R \tag{9.21}$$

Hence, in the OEM cost function, R has to be replaced by R_m.

Minimisation of the above cost function w.r.t. Θ yields:

$$\hat{\Theta}_{l+1} = \hat{\Theta}_l + \mu\Delta\Theta_l \tag{9.22}$$

Here,

$$\Delta\Theta_l = \left[\sum_k \left(\frac{\partial\bar{y}(k)}{\partial\Theta}\right)^T R_m^{-1}\left(\frac{\partial\bar{y}(k)}{\partial\Theta}\right)\right]^{-1}\left[\left(\frac{\partial\bar{y}(k)}{\partial\Theta}\right) R_m^{-1}(\bar{z}(k) - \bar{y}(k))\right] \tag{9.23}$$

From the estimated parameters of the transformed system, the estimates of the A matrix of the original system can be retrieved using

$$A = \bar{A} + I\delta \tag{9.24}$$

The matrices B and H remain unaffected. The transformation scalar δ may be taken as the real part of the largest unstable eigenvalue of the system. This information is available from the design considerations of the control system or some *a priori* information. In practice, while handling real data, the value of δ can be obtained from *a priori* information on the system. Alternatively, an approximate value of δ could be obtained by determining the slope from successive values of the peaks of the oscillatory data. This information gives the positive trend of the data, which grows numerically as time elapses. The transformation then effectively tries to remove the trend from the data, which become suitable for use in the output error method.

9.4.1.1 Example 9.2

Use the simulated data of the unstable second order system (eqs (9.7) and (9.8)) of Example 9.1. Demonstrate the use of the eigenvalue transformation technique to estimate the parameters of the unstable system using OEM.

9.4.1.2 Solution

Simulated data of 10 s duration pertaining to the two cases is generated (folder **Ch9SIMex1**). Random noise with SNR $= 10$ is added to generate noisy data for both cases. Using the measurements of y_1 and y_2, the parameters of A and B in eq. (9.7) are estimated using the OEM method (see folder **Ch9OEMex2**).

Next, selecting $\delta =$ real part of the unstable eigenvalue, the measurements y_1 and y_2 are used to generate detrended measurements $\bar{y}_1, \bar{y}_2$. Using $\bar{y}_1, \bar{y}_2$, the parameters of the unstable system are also estimated using the OEM method.

Table 9.4(a) gives the results of parameter estimation using measurements y_1, y_2. It can be clearly seen that the OEM can be used for parameter estimation when there is no noise in the data even when the instability is high. However, it must be noted that as the instability increases, OEM requires closer start up values to ensure convergence. When noisy data is used, despite using very close start up values, the parameter estimates deviate considerably from the true values, which is also clear from the high value of PEEN.

Table 9.4(b) gives results generated using the detrended measurements $\bar{y}_1, \bar{y}_2$. It is clear from the table that the parameter estimates are fairly close to the true values even in the presence of noise in the data. Figure 9.3(a) gives the comparison of the noisy and estimated measurements for case 2 using y_1, y_2 measurements and Fig. 9.3(b) shows the comparison when $\bar{y}_1, \bar{y}_2$ are used as measurements for the same case 2.

Table 9.4 Parameter estimates (OEM) – (a) using measurements y_1, y_2 (Example 9.2); (b) using measurements $\bar{y}_1$, $\bar{y}_2$ (Example 9.2)

Parameters	Case 1 ($a_{22} = 0.08$)			Case 2 ($a_{22} = 0.8$)		
	True	Estimated (no noise)	Estimated (SNR = 10)	True	Estimated (no noise)	Estimated (SNR = 10)
(*a*)						
a_{11}	0.06	0.0558	−0.1056	0.06	0.0599	−0.0684
		(0.0011)	(0.0766)		(0.0001)	(0.0843)
a_{12}	−2.0	−1.9980	−1.9084	−2.0	−2.0000	−1.9556
		(0.0009)	(0.0610)		(0.0001)	(0.0638)
a_{21}	2.8	2.8024	2.9767	2.8	2.8000	2.9510
		(0.0004)	(0.0983)		(0.0002)	(0.0911)
a_{22}	0.08	0.0832	0.2237	0.8	0.8000	0.9220
		(0.0013)	(0.0768)		(0.0002)	(0.0822)
b_1	−0.6	−0.6699	−0.5949	−0.6	−0.6589	−0.3963
		(0.0012)	(0.0610)		(0.0015)	(0.8811)
b_2	1.5	1.4604	1.5974	1.5	1.4725	1.9897
		(0.0015)	(0.1219)		(0.0018)	(1.1294)
PEEN %	–	2.1188	8.1987		1.6732	14.9521
(*b*)						
a_{11}	0.06	0.0526	0.0640	0.06	0.0529	0.1603
		(0.0015)	(0.0746)		(0.0020)	(0.0764)
a_{12}	−2.0	−1.9961	−2.0275	−2.0	−1.9967	−1.9868
		(0.0013)	(0.0639)		(0.0017)	(0.0642)
a_{21}	2.8	2.8047	2.7708	2.8	2.8066	2.7695
		(0.0018)	(0.0870)		(0.0023)	(0.0897)
a_{22}	0.08	0.0860	0.0470	0.8	0.8253	0.7196
		(0.0015)	(0.0749)		(0.0020)	(0.0762)
b_1	−0.6	−0.6714	−0.5826	−0.6	−0.6648	−0.6368
		(0.0013)	(0.0790)		(0.0019)	(0.0761)
b_2	1.5	1.4611	1.4254	1.5	1.4723	1.2827
		(0.0017)	(0.0922)		(0.0023)	(0.0897)
PEEN %	–	2.1588	2.4362		1.8381	6.6228

9.5 Methods for detection of data collinearity

The general mathematical model for parameter estimation (for use in the least squares method or regression) can be written as

$$y = \beta_0 + \beta_1 x_1 + \cdots + \beta_n x_n \tag{9.25}$$

Here, the regressors x_j, $j = 1, 2, \ldots, n$ are the state and input variables or their combinations, y is the dependent variable and $\beta_0, \ldots, \beta_n$ are unknown parameters.

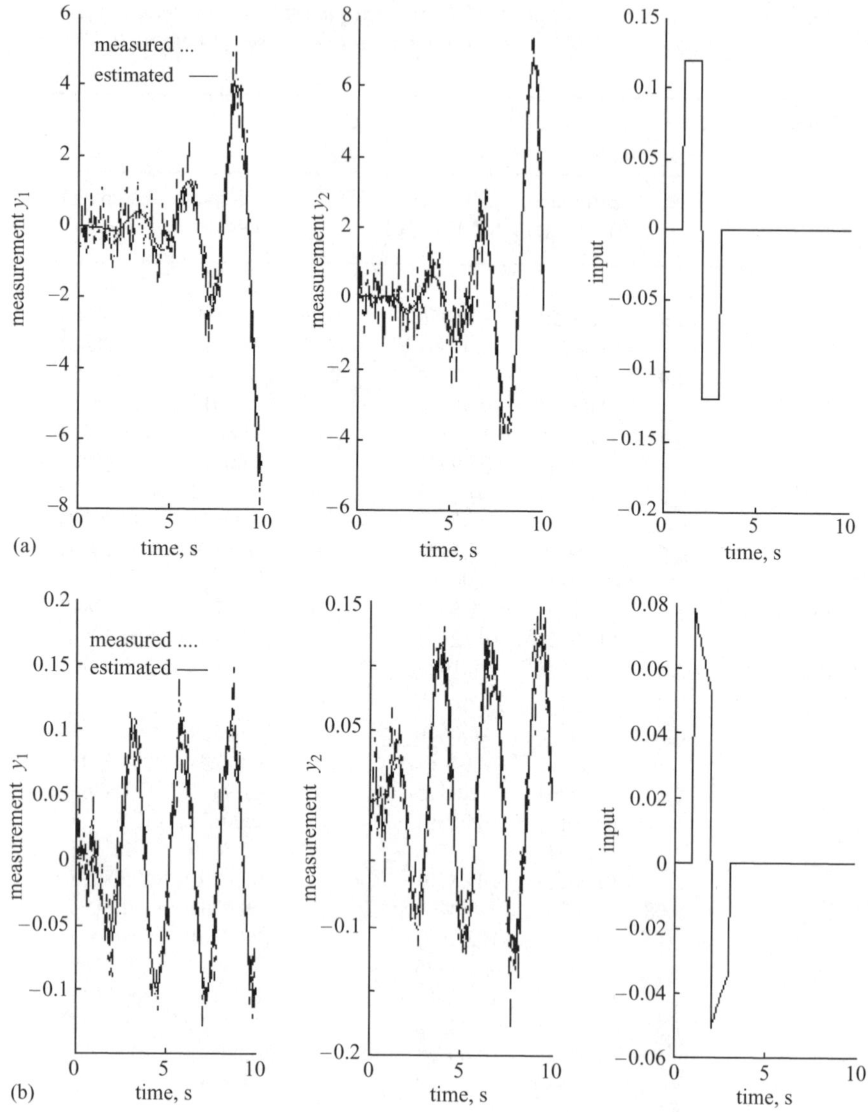

Figure 9.3 Simulated and estimated measurement – (a) unstable data (Example 9.2); (b) data with trend removed (Example 9.2)

Using measured data for y and x, eq. (9.25) can be written as

$$Y = X\beta + v \tag{9.26}$$

Here, Y is the measurement vector, X the matrix of regressors and 1s (1s are to account for the constant term in any regression equation), and β, the unknown parameter vector. The least squares estimates of the parameters β can be obtained using

$$\hat{\beta}_{LS} = (X^T X)^{-1} X^T Y \tag{9.27}$$

Generally, the regressors X are centred and scaled to unit length. If $Xj^{\#}$ denotes the columns of the normalised matrix, collinearity means that for a set of constants k_j not all equal to zero

$$\sum_{j=1}^{n} k_j X_j^{\#} = 0 \tag{9.28}$$

Collinearity could cause computational problems due to ill-conditioning of the matrix in eq. (9.27) and this would result in inaccurate estimates of the parameters. Three commonly used methods for assessing the collinearity among regressors are discussed next [2].

9.5.1.1 Correlation matrix of regressors

The presence of the collinearity can be ascertained by computing the correlation matrix of the regressors. If the correlation coefficients are greater than 0.5, then it indicates the presence of collinearity. However, if there are several co-existing near dependencies among regressors, the correlation matrix may not be able to indicate the same. Hence, its use as a diagnostic should be coupled with other diagnostic measures to be discussed next.

9.5.1.2 Eigen system analysis and singular value decomposition [2]

For assessing the collinearity, the eigensystem analysis and singular value decomposition (SVD; see Sections A.40 and A.41) methods could be used. In this case, the matrix $X^T X$ is decomposed into a product of two matrices: i) a diagonal matrix D with its elements as the eigenvalues λ_j of $X^T X$ and ii) an orthogonal matrix V with the eigenvectors of $X^T X$ as its columns.

$$X^T X = VDV^T \tag{9.29}$$

Near linear dependency in the data is indicated by eigenvalues close to zero or small eigenvalues. Instead of using eigenvalues where it is difficult to define exactly how small the eigenvalue should be, condition number could be used as an indicator of collinearity. Condition number is defined as the ratio of the largest eigenvalue of the system to the eigenvalue pertaining to the regressor j:

$$C_j = \frac{|\lambda_{\max}|}{|\lambda_j|} \tag{9.30}$$

Values of $C_j > 1000$ are indicative of severe collinearity in the data.

When singular value decomposition of matrix X is used to detect collinearity, the matrix X is decomposed as

$$X = USV^T \tag{9.31}$$

Here, U is a $(N \times n)$ matrix and $U^T U = V^T V = I$; S is a $(n \times n)$ diagonal semi-positive definite matrix with elements as the singular values ρ_j of X.

The condition index is defined as the ratio of the largest singular value to the singular value pertaining to the regressor j:

$$CI_j = \frac{\rho_{\max}}{\rho_j} \tag{9.32}$$

It can be used as a measure of collinearity. $CI_j = 5$ to 10 indicates mild collinearity and $CI_j = 30$ to 100 indicates strong collinearity between regressors [2]. SVD is preferred for detection of data collinearity, especially in applications when matrix $X^T X$ is ill-conditioned, because of its better numerical stability.

9.5.1.3 Parameter variance decomposition

An indication of collinearity can be obtained by decomposing the variance of each parameter into a sum of components, each corresponding to only one of the n singular values. The covariance matrix of the parameter estimates θ is given by

$$\operatorname{Cov}(\hat{\theta}) = \sigma_r^2 (X^T X)^{-1} = \sigma_r^2 V D^{-1} V^T \tag{9.33}$$

Here, σ_r^2 is the residual variance.

The variance of each parameter is decomposed into a sum of components, each corresponding to one of the n singular values using the following relation [2]:

$$\sigma_{\theta_j}^2 = \sigma_r^2 \sum_{i=1}^{n} \frac{t_{ji}^2}{\lambda_j} = \sigma_r^2 \sum_{i=1}^{n} \frac{t_{ji}^2}{\rho_j^2} \tag{9.34}$$

Here, t_{ji} are the elements of eigenvector t_j associated with λ_j. It is clear from eq. (9.34) that one or more small singular values can increase the variance of θ_j since ρ_j appears in the denominator. If there is near dependency among variables, the variance of two or more coefficients for the same singular value will indicate unusually high proportions. Define

$$\phi_{ji} = \frac{t_{ji}^2}{\rho_j^2}; \quad \phi_j = \sum_{i=1}^{n} \phi_{ji} \tag{9.35}$$

The j, i variance–decomposition proportion is the proportion of the variance of the jth regression coefficient associated with the ith component of its decomposition in eq. (9.35), and is expressed by

$$\Pi_{ij} = \frac{\phi_{ji}}{\phi_j}; \qquad j, i = 1, 2, \ldots, n \tag{9.36}$$

To create near dependency, two or more regressors are required. Hence, they will reflect high variance–decomposition proportions associated with a singular value. If the variance proportions are greater than 0.5, then the possibility of the collinearity problem is indicated.

9.6 Methods for parameter estimation of unstable/augmented systems

The output error method has been very successfully used for estimation of parameters of linear/nonlinear dynamical systems. However, the method poses difficulties when applied to inherently unstable systems [10]. Even if the basic unstable plant is operating with a stabilising feedback loop, application of the output error method to estimate directly parameters of the state space models of the system from its input-output data is difficult because of the numerical divergence resulting from integration of state equations. Hence, special care has to be taken to avoid this problem. Two approaches are feasible: i) an artificial stabilisation in the mathematical model (called feedback-in-model) used in output error method; and ii) the filter error method (described in Chapter 5).

9.6.1 Feedback-in-model method

This method is based on the fact that the system model used in the parameter estimation (software) can be stabilised by a local feedback in the model [10]. We note that the feedback achieved in this approach is not related to the control system feedback to stabilise the plant (see Fig. 9.1). This observation is also true for the filter error method. The feedback in the feedback-in-model method prevents the numerical divergence and achieves the stabilisation. The method achieves stabilisation of the parameter estimation process, somewhat in a similar fashion as the filter error method. It is applicable to many practical situations if proper care is taken to choose the feedback gain (in the mathematical model of the open-loop unstable plant).

Let the linear system be given by eq. (9.1). Then the predicted state is given by

$$\dot{\hat{x}} = A\hat{x} + Bu \tag{9.37}$$

$$\hat{z} = H\hat{x} \tag{9.38}$$

We see that z is the predicted measurement used in the cost function of the output error method. Now, we suppose that (software) feedback of a state is used in the mathematical model:

$$u = u + K_{sw}\hat{x} \tag{9.39}$$

$$\dot{\hat{x}} = A\hat{x} + Bu + BK_{sw}\hat{x} \tag{9.40}$$

$$\dot{\hat{x}} = (A + BK_{sw})\hat{x} + Bu \tag{9.41}$$

We see from the above equation that the system model can be made stable by proper choice of K_{sw}, if the plant A is unstable.

Next, we show how feedback is achieved in the filter error method. In the filter error method, the Kalman filter is used for prediction/filtering the state and hence obtaining the predicted measurement used in the cost function of eq. (5.2).

$$\dot{\hat{x}} = A\hat{x} + Bu + K(z - H\hat{x}) \tag{9.42}$$

$$\dot{\hat{x}} = (A - KH)\hat{x} + Bu + Kz \tag{9.43}$$

It can be noted from the above equation that unstable A is controlled by the KH term in almost a similar way as done by the term BK_{sw} in the feedback-in-model method.

9.6.2 *Mixed estimation method*

The mixed estimation technique is used for parameter estimation of unstable/augmented systems since it deals with the problem of collinearity in the data in an indirect way [2]. In unstable/augmented systems, due to the linear dependence among the regressors, not all parameters can be estimated independently. The mixed estimation method tries to overcome this linear dependence by using known estimates of certain parameters so that other crucial parameters can be estimated independently. In this method, the measured data is augmented by *a priori* information (see Section B.17) on the parameters directly. Assuming that the prior information on q ($q \leq n$, n the number of parameters to be estimated) of the elements of β is available, the *a priori* information equation (PIE) can be written as

$$a = C_{OE}\beta + \zeta \tag{9.44}$$

Here, a is the q-vector of known *a priori* values, and C_{OE} is a matrix with known constants. This matrix is called the observability enhancement matrix. The matrix C_{OE} is so termed to signify the possible enhancement of the observability of the augmented linear system. By the inclusion of information on β through C_{OE}, the observability of the system is expected to improve. ζ is a random vector with $E(\zeta) = 0$, $E(\zeta v^T) = 0$ and $E\{\zeta\zeta^T\} = \sigma^2 W$, where W is a known weighting matrix. Combining eqs (9.26) and (9.44), the mixed regression model is given by

$$\begin{bmatrix} Y \\ a \end{bmatrix} = \begin{bmatrix} X \\ C_{OE} \end{bmatrix}\beta + \begin{bmatrix} v \\ \zeta \end{bmatrix} \tag{9.45}$$

The mixed estimates are obtained using the least squares method

$$\hat{\beta}_{ME} = \left(X^T X + C_{OE}^T W^{-1} C_{OE}\right)^{-1}\left(X^T Y + C_{OE}^T W^{-1} a\right) \tag{9.46}$$

The covariance matrix is obtained using

$$\mathrm{Cov}(\hat{\beta}_{ME}) = \sigma_r^2\left[X^T X + C_{OE}^T W^{-1} C_{OE}\right]^{-1} \tag{9.47}$$

If the PIE is not known exactly, the resulting estimator could give biased estimates. Generally, the W matrix is diagonal with the elements representing uncertainty of *a priori* values. Here, σ_r^2 is the variance of the residuals:

$$r = \begin{bmatrix} Y \\ a \end{bmatrix} - \begin{bmatrix} X \\ C_{OE} \end{bmatrix}\hat{\beta} \tag{9.48}$$

9.6.2.1 Example 9.3

Simulate short period data of a light transport aircraft using eqs (2.44) and (2.45) with the parameter M_w adjusted to give a system with time to double of 1 s. Feedback the vertical velocity with a gain K to stabilise the system (Fig. 2.7, Chapter 2), using

$$\delta_e = \delta_p + Kw \tag{9.49}$$

Use gain values $K = 0.025$ and $K = 0.25$. Estimate the correlation matrix, condition numbers and variance proportions for the two sets of data. Use least squares and least squares mixed estimation methods to estimate the parameters of the system.

9.6.2.2 Solution

The simulated data is generated by using a doublet input signal (as the pilot stick input) to the model. Two sets of data are generated with gains $K = 0.025$ and $K = 0.25$. Random noise (SNR $= 10$) is added to generate noisy data for the two gain conditions. Correlation matrix, condition numbers and variance proportions are evaluated using the program **lslsme2.m** in folder **Ch9LSMEex3**. The correlation matrix and variance proportions for the case where $K = 0.25$ and SNR $= 10$ are given in Table 9.5. The correlation matrix and variance proportions are computed assuming there is a constant term in the regression equation in addition to the two states α, q and the input δ_e. In Table 9.5(b), condition numbers are also indicated. The correlation matrix indicates a correlation value of 0.8726 between q and α and 0.9682 between α and δ_e and 0.7373 between q and δ_e. The variance proportions corresponding to the condition number $= 988$ indicates collinearity between q, α and δ_e. The computed condition indices (eq. (9.32)) are: 1.0000, 3.9932, 31.4349 and 49.3738, which also indicates the presence of severe collinearity in the data. The least squares method was

Table 9.5 (a) Correlation matrix: K = 0.25 (Example 9.3); (b) variance proportions: K = 0.25 (Example 9.3)

	Const term	α	q	δ_e
(a)				
Const term	1.0000	−0.3693	−0.4497	−0.4055
α	−0.3693	1.0000	0.8726	0.9682
q	−0.4497	0.8726	1.0000	0.7373
δ_e	−0.4055	0.9682	0.7373	1.0000
(b)				
Condition number				
1	0.0000	0.0000	0.2206	0.3594
15.9	0.0000	0.0001	0.6036	0.3451
988.2	0.0000	0.9999	0.1758	0.2955
2437.8	1.0000	0.0000	0.0000	0.0000

Table 9.6 (a) Parameter estimates using least squares method (Example 9.3); (b) parameter estimates using least squares mixed estimation method (Example 9.3)

		$K = 0.025$		$K = 0.25$	
Parameters	True	Estimated (no noise)	Estimated (SNR = 10)	Estimated (no noise)	Estimated (SNR = 10)
(a)					
Z_w	−1.4249	−1.4345	−0.2210	−1.4386	−0.8250
$Z_{\delta e}$	−6.2632	−5.9549	−38.7067	−5.2883	−9.5421
M_w	0.2163	0.2167	0.0799	0.1970	0.1357
M_q	−3.7067	−3.7138	−1.7846	−3.4038	−2.8041
$M_{\delta e}$	−12.7840	−12.7980	−9.0736	−12.5301	−12.1554
PEEN %	–	0.7489	81.4264	2.3780	15.9822
(b)					
Z_w	−1.4249	−1.4362	−1.0035	−1.3976	−1.0404
$Z_{\delta e}$	−6.2632	−5.9008	−6.8167	−5.8923	−5.9488
M_w	0.2163	0.2368	0.1776	0.2598	0.2123
M_q	−3.7067	−3.9908	−3.1359	−3.8190	−3.2525
$M_{\delta e}$	−12.7840	−13.4614	−13.0552	−13.4541	−13.4326
PEEN %	–	1.8224	16.0907	1.6864	11.4771

used for parameter estimation and the results are shown in Table 9.6(a). It is clear from the table that the LS estimates are fairly close to the true values for both cases of $K = 0.025$ and $K = 0.25$ when there is no noise in the data. However, when there is noise in the data, the estimates show a very large deviation from the true values. This is indicated by the high values of the parameter estimation error norm.

Since the parameter most affected by feedback is M_w, it was decided to fix the corresponding control effectiveness parameter, $M_{\delta e}$, at a value equal to 1.05 times of its true value and use the least squares mixed estimation method for the same set of data. Table 9.6(b) gives the least squares mixed estimation estimates. The estimation results indicate considerable improvement when there is noise in the data. It should be noted that for the case when there is no noise in the data, the parameter estimation error norms are a little higher than their corresponding least squares estimates. This is due to the inclusion of an uncertainty of 5 per cent in the control effectiveness derivative.

9.6.2.3 Example 9.4

Simulate the fourth order longitudinal dynamics of an unstable aircraft and the associated filters in the feedback loops of Fig. 9.4 using a doublet pulse input. Assess the extent of collinearity in the data and use the least squares mixed estimation method to estimate the parameters of the open loop plant. Use the following state and measurement models for simulation.

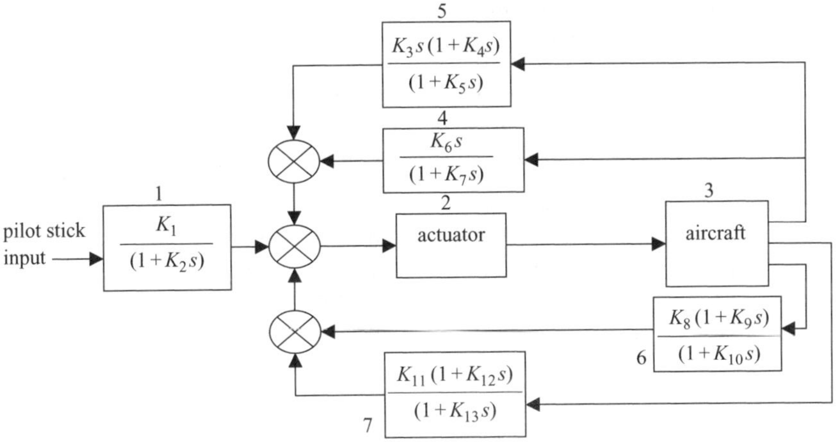

Figure 9.4 Block diagram of an unstable aircraft operating in closed loop

State equations

$$\begin{bmatrix} \dot{\alpha} \\ \dot{q} \\ \dot{\theta} \\ \dot{v}/v_0 \end{bmatrix} = \begin{bmatrix} Z_{\alpha/v_0} & 1 & 0 & Z_{v/v_0} \\ M_\alpha & M_q & 0 & M_{v/v_0} \\ 0 & 1 & 0 & 0 \\ X_\alpha & 0 & X_\theta & X_{v/v_0} \end{bmatrix} \begin{bmatrix} \alpha \\ q \\ \theta \\ v/v_0 \end{bmatrix} + \begin{bmatrix} Z_{\delta_e} \\ M_{\delta_e} \\ 0 \\ X_{\delta_e} \end{bmatrix} \delta_e \tag{9.50}$$

Measurement equations

$$\begin{bmatrix} \alpha \\ q \\ a_x \\ a_z \end{bmatrix} = \begin{bmatrix} 1 & 0 & 0 & 0 \\ 0 & 1 & 0 & 0 \\ C_{31} & 0 & 0 & C_{34} \\ C_{41} & 0 & 0 & C_{44} \end{bmatrix} \begin{bmatrix} \alpha \\ q \\ \theta \\ v/v_0 \end{bmatrix} + \begin{bmatrix} 0 \\ 0 \\ D_{31} \\ D_{41} \end{bmatrix} \delta_e \tag{9.51}$$

Here, $Z_{(\cdot)}, X_{(\cdot)}, M_{(\cdot)}, C_{(\cdot)}, D_{(\cdot)}$ are the aerodynamic parameters to be estimated.

9.6.2.4 Solution

The control blocks and plant given in Fig. 9.4 are realised. The simulated data are generated by using a doublet input signal with sampling interval of 0.1 s. The control system blocks are simulated using the program **Ch9SIMex4**.

Correlation matrix, condition numbers and variance proportions are evaluated using the program **lslsme4.m** in folder **Ch9LSMEex4**. The correlation matrix and variance proportions are given in Table 9.7. The correlation matrix and variance proportions are computed assuming there is a constant term in the regression equation in addition to the three states $\alpha, q, v/v_0$ and the control input δ_e. In Table 9.7(b), condition numbers are also indicated. The correlation matrix indicates a correlation coefficient of -0.76 between the constant term and α, 0.996 between v/v_0 and constant, -0.725 between v/v_0 and α, and -0.697 between δ_e and q. The variance proportions pertaining to the condition number 2331 indicate a value of 0.85

Table 9.7 *(a) Correlation matrix (Example 9.4); (b) variance proportions (Example 9.4)*

	Constant	α	q	v/v_0	δ_e
(a)					
Constant	1.0000	−0.7625	−0.2672	0.9961	−0.2368
α	−0.7625	1.0000	0.5818	−0.7257	0.0548
q	−0.2672	0.5818	1.0000	−0.1819	−0.6972
v/v_0	0.9961	−0.7257	−0.1819	1.0000	−0.3122
δe	−0.2368	0.0548	−0.6972	−0.3122	1.0000
(b)					
Condition number					
1	0.0000	0.0000	0.0000	0.1335	0.0052
14.29	0.0000	0.0463	0.0000	0.0039	0.4497
65.14	0.0000	0.5065	0.01757	0.0131	0.2515
241.8	0.0000	0.3816	0.8306	0.0058	0.2032
2331.1	0.9999	0.0653	0.1517	0.8438	0.0904

for the v/v_0 term and 0.9999 for the constant term, which is an indicator of collinearity in the data. The condition number of 2331 also indicates the presence of high collinearity in this data. The computed condition indices are: 1, 3.78, 8.079, 15.55 and 48.2, which also indicate the presence of severe collinearity in the data.

The LS method was used for parameter estimation and the results are shown in Table 9.8. It was observed that the estimates of M_α, X_α, $X_{v/v0}$ and $X_{\delta e}$ derivatives show deviations from true values. LSME was used for parameter estimation by using *a priori* values on the parameters $Z_{v/v0}$, $Z_{\delta e}$, $M_{v/v0}$, $M_{\delta e}$, $X_{v/v0}$, $X_{\delta e}$ by fixing these derivatives at a value equal to 1.05 times its true value. The LSME estimates are somewhat better than LS estimates as can be seen from Table 9.8. It should be noted that the derivative M_α shows considerable improvement with the LSME method.

9.6.3 Recursive mixed estimation method

In this section, a mixed estimation algorithm that incorporates the *a priori* information of the parameters into the extended Kalman filter (Chapter 4) structure is presented. The *a priori* information equation resembles the conventional measurement model used in the Kalman filter and can be directly appended to the measurement part of the Kalman filter The main advantage of the Kalman filter based mixed estimation algorithm is that it can handle process and measurement noises in addition to giving a recursive solution to the mixed estimation algorithm [11].

Table 9.8 Parameter estimates from least squares (LS) and least squares mixed estimation (LSME) methods (Example 9.4)

Parameter	True values	LS	LSME
$Z_{\alpha/v0}$	−0.771	−0.7820	−0.7735
$Z_{\delta e}$	−0.2989	−0.2837	−0.3000
$Z_{v/v0}$	−0.1905	−0.1734	−0.1800
M_α	0.3794	0.1190	0.3331
M_q	−0.832	−0.7764	−0.8236
$M_{\delta e}$	−9.695	−9.2095	−9.5997
X_α	−0.9371	−0.2309	−0.2120
$X_{v/v0}$	−0.0296	0.1588	−0.0200
$X_{\delta e}$	−0.0422	−0.0142	−0.0400
$M_{v/v0}$	0.0116	0.01189	0.0120
PEEN	–	10.41	7.52

We know that when the Kalman filter is used for parameter estimation, the unknown parameters of the system form part of the augmented state model, (eq. (4.39)). Since the problem now becomes one involving nonlinear terms (products of states), the extended Kalman filter is to be used (Chapter 4). The measurement model has the general form:

$$z(k) = Hx(k) + v(k) \tag{9.52}$$

The *a priori* information equation has the form:

$$a(k) = C_{OE}\beta(k) + \zeta(k) \tag{9.53}$$

Augmenting the measurement equation with *a priori* information equation, we get

$$\begin{bmatrix} z \\ a \end{bmatrix} = \begin{bmatrix} & H \\ 0 & C_{OE} \end{bmatrix} [x_a] + \begin{bmatrix} v(k) \\ \zeta(k) \end{bmatrix} \tag{9.54}$$

Here, x_a represents the augmented state vector, containing states and parameters represented by

$$x_a(k) = \begin{bmatrix} x(k) \\ \beta(k) \end{bmatrix} \tag{9.55}$$

It is assumed that $E\{\zeta v^T\} = 0$ and ζ represents the uncertainty in *a priori* value of the parameters, $\mathrm{cov}(\zeta\zeta^T) = R_a$. The matrix C_{OE} can be such that the *a priori* information on parameters β can be included in a selective way (i.e. a could be of dimension $q < n$). This would render the recursive algorithm conditionally optimal, since Kalman gain will also depend on C_{OE} and R_a.The time propagation equations generally follow eqs (4.48) and (4.49).

The state estimate (augmented state and parameters) related equations are given as follows. The Kalman gain is given by:

$$K = P\begin{bmatrix} H \\ 0 & C_{OE} \end{bmatrix}^T \left[\begin{bmatrix} H \\ 0 & C_{OE} \end{bmatrix} P \begin{bmatrix} H \\ 0 & C_{OE} \end{bmatrix}^T + \begin{bmatrix} R & 0 \\ 0 & R_a \end{bmatrix}\right]^{-1} \tag{9.56}$$

$$\hat{P} = \left[I - K\begin{bmatrix} H \\ 0 & C_{OE} \end{bmatrix}\right]\tilde{P} \tag{9.57}$$

And finally

$$\hat{x}_a(k) = \tilde{x}_a(k) + K\left[\begin{bmatrix} z(k) \\ a \end{bmatrix} - \begin{bmatrix} H \\ 0 & C_{OE} \end{bmatrix}\tilde{x}_a(k)\right] \tag{9.58}$$

It is to be noted that there is no guideline on choice of C_{OE}. The additional *a priori* information acts as a direct measurement of parameters and perhaps enhances the observability of the system.

9.6.3.1 Example 9.5

Simulate the fourth order longitudinal dynamics of the unstable aircraft and the associated filters in the feedback loops of Fig. 9.4 using eqs (9.50) and (9.51). Use a UD based extended Kalman filter (UD) and a UD based mixed estimation Kalman filter (UDME) to estimate the parameters in the eq. (9.50).

9.6.3.2 Solution

Simulated data from Example 9.4 is used for parameter estimation using UD and UDME programs contained in folder **Ch9UDMEex5**. All the collinearity diagnostics had indicated the presence of severe collinearity in the data (Table 9.7). The results of LSME had shown some improvement in the estimates. However, in the presence of measurement noise, the PEENs were still high as seen from Example 9.3 and Table 9.6(b) even for a second order closed loop system. Table 9.9 shows the results of comparison of parameter estimates using UD and UDME filters. *A priori* information on all the control derivatives and the $X_{v/v0}$ derivative was used in the UDME. The uncertainties in these parameters are appended to the measurement noise covariance of the filter (eq. (9.56)). It is to be noted that there is a significant improvement in the estimate of M_α. The study in this section indicates that based on the collinearity diagnostics, when the values of only the control derivatives and the v/v_0 derivatives were fixed at their true values, the UDME gave improved results for almost all the parameters. This is also clear from the low values of PEENs obtained when UDME is used for parameter estimation. Figure 9.5 shows the convergence of some of the estimated parameters ($Z_{\delta e}$, M_q, M_α, $M_{\delta e}$) for the data with SNR = 10. The estimates of the parameters show some discrepancies from their true values for the UD filter whereas when UDME is used, the estimates tend to follow the true values more closely. Thus, UDME gives consistent estimates.

Table 9.9 Parameter estimates UD, UD mixed estimation (UDME) methods (Example 9.5)

		No noise		SNR = 10	
Parameter	True values	UD	UDME	UD	UDME
$Z_{\alpha/v0}$	−0.7710	−0.8332	−0.8406	−0.8830	−0.8905
$Z_{v/v0}$	−0.1905	−0.2030	−0.2013	−0.2018	−0.2002
$Z_{\delta e}$	−0.2989	−0.3377	−0.3000	−0.3391	−0.3000
M_α	0.3794	0.4242	0.3984	0.4296	0.4070
M_q	−0.8320	−0.8836	−0.8558	−0.8525	−0.8263
$M_{v/v0}$	0.0116	0.0134	0.0137	0.0130	0.0132
$M_{\delta e}$	−9.6950	−10.0316	−9.6007	−9.9767	−9.6007
X_α	−0.0937	−0.1008	−0.1017	−0.1037	−0.1045
X_θ	−0.0961	−0.1034	−0.1041	−0.1043	−0.1048
$X_{v/v0}$	−0.0296	−0.0322	−0.0280	−0.0368	−0.0280
$X_{\delta e}$	−0.0422	−0.0462	−0.0400	−0.0461	−0.0400
PEEN%	–	3.5963	1.2494	3.1831	1.5932

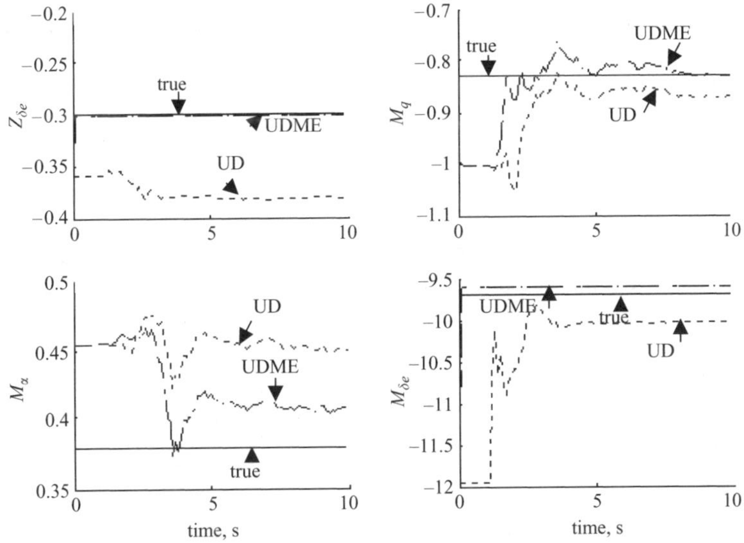

Figure 9.5 Comparison of true parameters, UD and UDME estimates (Example 9.5)

9.7 Stabilised output error methods (SOEMs)

It has been demonstrated in Chapters 2 and 7 that the methods of equation error and regression can be used for estimation of parameters of the system if the measurements of states are available. This principle is extended to the output error method for parameter estimation to arrive at a method called the equation decoupling method, which is directly applicable for parameter estimation of unstable systems [4, 5]. In the equation decoupling method, the system state matrix is decoupled so that one part has only diagonal elements pertaining to each of the integrated states and the off-diagonal elements associated with the states use measured states in the state equations. Due to this, the state equations get decoupled. This decoupling of equations changes the unstable system to a stable one. Thus, it is clear that by incorporating stabilisation into the output error method by means of measured states, the instability caused due to numerical divergence of the integrated states can be overcome. Since the output error algorithm is stabilised by this method, these algorithms are termed stabilised output error methods. The degree of decoupling can be changed depending on the extent of instability in the system. This leads to two types of stabilised output error methods: i) equation decoupling when all the states pertaining to off-diagonal elements are replaced by corresponding measured states; and ii) regression analysis which results when only the states occurring with the parameters, which cause numerical divergence, are replaced by the measured states. It must be noted here that these methods require accurate measurements of states for stabilising the system and estimating the parameters.

Equation decoupling method

The system matrix A is partitioned into two sub-matrices denoted by A_d containing only diagonal elements and A_{od}, containing only off-diagonal elements. When measured states are used, the control input vector u is augmented with the measured states x_m to give

$$\dot{x} = A_d x + [B \quad A_{od}] \begin{bmatrix} \delta \\ x_m \end{bmatrix} \tag{9.59}$$

The integrated variables are present only in the A_d part (supposed to be the stable part) and all off-diagonal variables have measured states. This renders each differential equation to be integrated independently of the others and hence the equations become completely decoupled. The cost function to be minimised would be the same as given in eq. (3.52). The computation of the sensitivity function is carried out using the decoupled matrices A_d and A_{od} and state measurements in addition to the control input variables.

Regression analysis

In this method, measured states are used with those parameters in the state matrix that are responsible for instability in the system and integrated states are used with the remaining parameters. Thus, matrix A is partitioned into two parts, A_s containing the part of matrix A that has parameters not contributing to instability and A_{us} having parameters that do contribute to system instability so that the system equation has the form

$$\dot{x} = A_s x + [B \quad A_{us}] \begin{bmatrix} \delta \\ x_m \end{bmatrix} \tag{9.60}$$

It is clear that integrated states are used for the stable part of the system matrix and measured states for the parameters contributing to the unstable part of the system. Equation (9.60) has a form similar to eq. (9.59) for the equation decoupling method, and the matrix A_d is diagonal whereas matrix A_s will not necessarily be diagonal.

9.7.1 *Asymptotic theory of SOEM*

The equation error method requires measurements of states and derivatives of states for parameter estimation as we have seen in Chapter 2. The output error method uses measurements that are functions of the states of the system and not necessarily the states. The stabilised output error methods require some of the measured states to be used for stabilisation. Thus, the stabilised output error methods seem to fall in between the equation error method and output error method for parameter estimation and can be said to belong to a class of mixed equation error-output error methods. It has been observed that the output error method does not work directly for unstable systems because the numerical integration of the system causes divergence of states. In the case of stabilised output error methods, since the measured states (obtained from the unstable system operating in closed loop) are stable, their use in the estimation process tries to prevent this divergence and at the same time enables parameter estimation of basic unstable systems directly, in a manner similar to that of the output error method for a stable plant [5].

In this section, an analytical basis for the stabilised output error methods is provided by an analysis of the effect of use of measured states on the sensitivity matrix (eq. 3.55) computation and covariance estimation. The analysis is based on the following two assumptions:

1 Analysis for the output error method is valid when applied to a stable system for which the convergence of the algorithm is generally assured.
2 Presented analysis for the stabilised output error method is valid for an unstable system, since the use of measured states stabilises the parameter estimation method.

The analysis is carried out in the discrete-time domain, since it is fairly straightforward to do this. We believe that similar analysis should work well for continuous-time systems, at least for linear estimation problems. In the discrete form, the state and measurement models are given by

$$x(k+1) = \phi x(k) + B_d u(k) \tag{9.61}$$

$$y(k) = Cx(k) + Du(k) \tag{9.62}$$

Here, ϕ denotes the state transition matrix

$$\phi = e^{A\Delta t} = 1 + A\Delta t + A^2 \frac{\Delta t^2}{2!} + \cdots \tag{9.63}$$

Here, B_d denotes the control distribution matrix defined as

$$B_d = \left[I\Delta t + A\frac{\Delta t^2}{2!} + A^2\frac{\Delta t^3}{3!} + \cdots \right] B \tag{9.64}$$

Here, $\Delta t = t(k+1) - t(k)$ is the sampling interval.

It has been shown in Chapter 3 that the parameter improvement $\Delta\Theta$ (for every iteration of the output error algorithm) is obtained by computing the sensitivity matrix. The sensitivity matrix is obtained by partial differentiation of system equations w.r.t. each element of the unknown parameter vector Θ and is given by

$$\left(\frac{\partial y}{\partial \Theta}\right)_{ij} = \frac{\partial y_i}{\partial \Theta_j} \tag{9.65}$$

By differentiating eqs (9.61) and (9.62) with respect to Θ, we get [5]:

$$\frac{\partial x(k+1)}{\partial \Theta} = \phi\frac{\partial x(k)}{\partial \Theta} + \frac{\partial \phi}{\partial \Theta}x(k) + \frac{\partial B_d}{\partial \Theta}u(k) \tag{9.66}$$

$$\frac{\partial y(k)}{\partial \Theta} = C\frac{\partial x(k)}{\partial \Theta} + \frac{\partial C}{\partial \Theta}x(k) + \frac{\partial D}{\partial \Theta}u(k) \tag{9.67}$$

The partial differentiation of u w.r.t. Θ does not figure in these equations, because u is assumed independent of Θ.

Computation of sensitivity matrix in output error method

A simple first order example described by the following state equation is considered to demonstrate the computation of the parameter increments in the output error method and stabilised output error method.

$$\dot{r} = N_r r + N_\delta \delta \tag{9.68}$$

N_r and N_δ are the parameters to be estimated using discrete measurements of the state r and control input δ. With the measurement noise, the measurements are expressed by

$$r_m(k) = r(k) + v(k) \tag{9.69}$$

In eq. (9.69), the system state matrix $A = N_r$; $C = 1$; $B = N_\delta$.

The output error method cost function for this case is given by

$$E(N_r, N_\delta) = \frac{1}{2}\sum_{k=1}^{N}[r_m(k) - r(k)]^2 \tag{9.70}$$

Here, $r(k)$ is the computed response from the algorithm

$$r(k+1) = \phi r(k) + B_d \delta(k) \tag{9.71}$$

Using eqs (9.63) and (9.64), the transition matrix ϕ is given by

$$\phi = 1 + N_r \Delta t \tag{9.72}$$

The control distribution matrix B_d is given by

$$B_d = N_\delta \Delta t \tag{9.73}$$

after neglecting all higher order terms (which is justified for small Δt).

Substituting eqs (9.72) and (9.73) into eq. (9.71), we get

$$r(k+1) = (1 + N_r \Delta t) r(k) + N_\delta \Delta t \delta(k) \tag{9.74}$$

Estimates of N_r and N_δ are obtained by minimising the cost function of eq. (9.70) w.r.t. these parameters. The sensitivity matrix w.r.t. N_r is given by

$$\frac{\partial r(k+1)}{\partial N_r} = \frac{\partial r(k)}{\partial N_r} + N_r \Delta t \frac{\partial r(k)}{\partial N_r} + r(k)\Delta t \tag{9.75}$$

and that with respect to N_δ is given by

$$\frac{\partial r(k+1)}{\partial N_\delta} = \frac{\partial r(k)}{\partial N_\delta} + N_r \Delta t \frac{\partial r(k)}{\partial N_\delta} + \delta(k)\Delta t \tag{9.76}$$

The parameter vector ($\Theta = [N_r, N_\delta]$) and the successive estimates of Θ are obtained by an iterative process (Chapter 3). For the present single state variable case, starting with initial estimates of parameters N_r and N_δ, (Θ_0), the estimates of Θ are obtained by computing first and second gradients of eq. (9.70). The first gradient is given by

$$\nabla E(\Theta) = \begin{bmatrix} \displaystyle\sum_{k=1}^{N} -(r_m(k) - r(k)) \frac{\partial r(k)}{\partial N_r} \\ \displaystyle\sum_{k=1}^{N} -(r_m(k) - r(k)) \frac{\partial r(k)}{\partial N_\delta} \end{bmatrix} \tag{9.77}$$

Substituting eqs (9.75) and (9.76) into eq. (9.77), we get

$$\nabla E(\Theta) = \begin{bmatrix} \displaystyle\sum_{k=1}^{N} -(r_m(k) - r(k)) \left[\frac{\partial r(k-1)}{\partial N_r} + N_r \Delta t \frac{\partial r(k-1)}{\partial N_r} + \Delta t r(k-1) \right] \\ \displaystyle\sum_{k=1}^{N} -(r_m(k) - r(k)) \left[\frac{\partial r(k-1)}{\partial N_\delta} + N_r \Delta t \frac{\partial r(k-1)}{\partial N_\delta} + \Delta t \delta(k-1) \right] \end{bmatrix} \tag{9.78}$$

Computation of sensitivity matrix in stabilised output error method

If the derivative N_r were such that the system becomes unstable, the numerical divergence would be arrested if the measured state were used for the state r in addition to measured control surface deflection δ. In order to analyse the effect of the use of the measured state on sensitivity matrix computations, expressions for the first gradients are evaluated. Using r_m in eq. (9.68), the state equation for r takes the form:

$$\dot{r} = N_r r_m + N_\delta \delta \tag{9.79}$$

Measured r is appended to the measured control surface deflection δ and hence in eq. (9.71), the state matrix $A = 0$ and $B = [N_r, N_\delta]$. Hence, for this case, $\phi = 1$ and $B_d = [N_r N_\delta]\Delta t$.

In the discrete form, eq. (9.79) is represented by

$$r(k+1) = [1]r(k) + \Delta t[N_r \quad N_\delta]\begin{bmatrix} r_m(k) \\ \delta(k) \end{bmatrix} \tag{9.80}$$

The partial differentiation of the control surface deflection with respect to the parameters is not included in the following derivations, since the control surface deflection δ is treated independent of the parameters.

Differentiating eq. (9.80) with respect to Θ, we get the following sensitivity equations:

$$\frac{\partial r(k+1)}{\partial N_r} = \frac{\partial r(k)}{\partial N_r} + N_r \Delta t \frac{\partial r_m(k)}{\partial N_r} + \Delta t r_m(k) \tag{9.81}$$

$$\frac{\partial r(k+1)}{\partial N_\delta} = \frac{\partial r(k)}{\partial N_\delta} + N_r \Delta t \frac{\partial r_m(k)}{\partial N_\delta} + \Delta t \delta(k) \tag{9.82}$$

The measured state can be expressed as a combination of the true state (r_t) and measurement noise (r_n) as

$$r_m = r_t + r_n \tag{9.83}$$

Substituting the above expression into eqs (9.81) and (9.82), we get:

$$\frac{\partial r(k+1)}{\partial N_r} = \frac{\partial r(k)}{\partial N_r} + N_r \Delta t \frac{\partial r_t(k)}{\partial N_r} + N_r \Delta t \frac{\partial r_n(k)}{\partial N_r} + \Delta t r_t(k) + \Delta t r_n(k) \tag{9.84}$$

$$\frac{\partial r(k+1)}{\partial N_\delta} = \frac{\partial r(k)}{\partial N_\delta} + N_r \Delta t \frac{\partial r_t(k)}{\partial N_\delta} + N_r \Delta t \frac{\partial r_n(k)}{\partial N_\delta} + \Delta t \delta(k) \tag{9.85}$$

The first gradient (the subscript s is used to denote the gradient from stabilised output error method), is given by

$$\frac{\nabla E_s(\Theta)}{N-1} = \frac{1}{N-1} \times \begin{bmatrix} -\sum_{k=1}^{N} (r_m(k) - r(k)) \left[\frac{\partial r(k-1)}{\partial Nr} + N_r \Delta t \frac{\partial r_t(k-1)}{\partial N_r} + N_r \Delta t \frac{\partial r_n(k-1)}{\partial N_r} + \Delta t r_t(k-1) + \Delta t r_n(k-1) \right] \\ -\sum_{k=1}^{N} (r_m(k) - r(k)) \left[\frac{\partial r_1(k-1)}{\partial N_\delta} + N_r \Delta t \frac{\partial r_t(k-1)}{\partial N_\delta} + N_r \Delta t \frac{\partial r_n(k-1)}{\partial N_\delta} + \Delta t \delta(k-1) \right] \end{bmatrix} \tag{9.86}$$

The integrated state r figuring in the above equations can also be expressed as the sum of a true state and the error arising due to integration. This in turn could arise due to incorrect initial conditions of the parameters and states:

$$r = r_t + r_i \tag{9.87}$$

Substituting the expression for r_m and r in the first term in the parenthesis of eq. (9.86), we get

$$\frac{\nabla E_s(\Theta)}{N-1} = \frac{1}{N-1} \times \left[\begin{array}{l} -\sum_{k=1}^{N} (r_n(k) - r_i(k)) \left[\frac{\partial r(k-1)}{\partial Nr} + N_r \Delta t \frac{\partial r_t(k-1)}{\partial N_r} + N_r \Delta t \frac{\partial r_n(k-1)}{\partial N_r} + \Delta t r_t(k-1) + \Delta t r_n(k-1) \right] \\ \quad - \sum_{k=1}^{N} (r_n(k) - r_i(k)) \left[\frac{\partial r(k-1)}{\partial N_\delta} + N_r \Delta t \frac{\partial r_t(k-1)}{\partial N_\delta} + N_r \Delta t \frac{\partial r_n(k-1)}{\partial N_\delta} + \Delta t \delta(k-1) \right] \end{array} \right] \tag{9.88}$$

Using eq. (9.87) in eq. (9.78), which is the first gradient of the cost function for output error method, we have,

$$\frac{\nabla E_o(\Theta)}{N-1} = \frac{1}{N-1} \times \left[\begin{array}{l} -\sum_{k=1}^{N} (r_t(k) + r_n(k) - r_t(k) - r_i(k)) \left[\frac{\partial r(k-1)}{\partial N_r} + N_r \Delta t \frac{\partial r_t(k-1)}{\partial N_r} + N_r \Delta t \frac{\partial r_i(k-1)}{\partial Nr} + \Delta t r_t(k-1) + \Delta t r_i(k-1) \right] \\ \quad - \sum_{k=1}^{N} (r_t(k) + r_n(k) - r_t(k) - r_i(k)) \left[\frac{\partial r(k-1)}{\partial N_\delta} + N_r \Delta t \frac{\partial r_t(k-1)}{\partial N_\delta} + N_r \Delta t \frac{\partial r_i(k-1)}{\partial N_\delta} + \Delta t \delta(k-1) \right] \end{array} \right] \tag{9.89}$$

Here, subscript o stands for the output error method.

The integration errors r_i tend to zero as the iterations progress because the initial conditions as well as the parameter estimates improve. Since the noise is independent of parameters, we have from eq. (9.88) (for stabilised output error method):

$$\frac{\nabla E_s(\Theta)}{N-1} = \frac{1}{N-1} \times \left[\begin{array}{l} -\sum_{k=1}^{N} r_n(k) \left[\frac{\partial r(k-1)}{\partial N_r} + N_r \Delta t \frac{\partial r_t(k-1)}{\partial N_r} + \Delta t r_t(k-1) + \Delta t r_n(k-1) \right] \\ \quad - \sum_{k=1}^{N} r_n(k) \left[\frac{\partial r(k-1)}{\partial N_\delta} + N_r \Delta t \frac{\partial r_t(k-1)}{\partial N_\delta} + \Delta t \delta(k-1) \right] \end{array} \right] \tag{9.90}$$

From eq. (9.89) (for output error method), we have

$$\frac{\nabla E_o(\Theta)}{N-1} = \frac{1}{N-1} \times \left[\begin{array}{l} -\sum_{k=1}^{N} r_n(k) \left[\frac{\partial r(k-1)}{\partial N_r} + N_r \Delta t \frac{\partial r_t(k-1)}{\partial N_r} + \Delta t r_t(k-1) \right] \\ -\sum_{k=1}^{N} r_n(k) \left[\frac{\partial r(k-1)}{\partial N_\delta} + N_r \Delta t \frac{\partial r_t(k-1)}{\partial N_\delta} + \Delta t \delta(k-1) \right] \end{array} \right] \tag{9.91}$$

In eq. (9.90), we have the term involving $(1/(N-1))\sum_{k=1}^{N} r_n(k)r_n(k-1)\Delta t$ which tends to zero since the measurement noise r_n is assumed a white process. Hence, in the light of the above observations we get, asymptotically,

$$\frac{\nabla E_s(\Theta)}{N-1} \to \frac{\nabla E_o(\Theta)}{N-1} \tag{9.92}$$

Thus for a good number of iterations, r_i die out quickly and the assumption that r_n is a white process leads to the asymptotic behaviour of the stabilised output error method similar to that of the output error method for this single state case. This is also true for the two-state system [7]. Hence, the result by induction can be considered as valid for n-state systems. Thus, the asymptotic behaviour of the equation decoupling method and regression analysis (stabilised output error methods) is similar to that of the output error method.

It has been established by the asymptotic analysis that stabilised output error methods, when applied to unstable systems, would behave in an almost similar manner to how the output error method would behave when applied to a stable system. This observation puts the stabilised output error methods on a solid foundation and is of fundamental importance.

Intuitive explanation of stabilised output error methods
A second order unstable system described by the following equations is chosen to provide an intuitive explanation of the working of stabilised output error methods:

$$\dot{x}_1 = a_{11}x_1 + a_{12}x_2 + b_1u_1 \tag{9.93}$$

$$\dot{x}_2 = a_{21}x_1 + a_{12}x_2 + b_2u_1 \tag{9.94}$$

Assuming that the parameter a_{21} is responsible for causing instability in the system that causes numerical divergence, if the corresponding state x_1 is replaced by measured x_{1m}, we have the following state equations (with subscript i for integration):

$$\dot{x}_{1_i} = a_{11}x_{1_i} + a_{12}x_{2_i} + b_1u_1 \tag{9.95}$$

$$\dot{x}_{2_i} = a_{21}x_{1m} + a_{12}x_{2_i} + b_2u_1 \tag{9.96}$$

When these equations are integrated, due to use of x_{1m}, divergence of x_2 in eq. (9.96) is arrested and hence that in eq. (9.95) is arrested. Thus, use of the measured state in state equations effectively tries to stabilise the output error cost function. In general, the parameters causing the numerical instability are related to the so-called offending states, which in most of the practical situations are measurable.

9.7.1.1 Example 9.6

Simulate short period (see Appendix B) data of a light transport aircraft using eqs (2.44) and (2.45) with the parameter M_w adjusted to give a system with time to double of 1 s. Feedback the vertical velocity with a gain K to stabilise the system using

$$\delta_e = \delta_p + Kw$$

Use $K = 0.25$. Add noise to generate data with SNR $= 10$. Use the stabilised output error method to estimate the stability and control derivatives (parameters) of the aircraft.

9.7.1.2 Solution

Direct identification between δ_e and output measurements is carried out (see Fig. 2.7). When the output error method is used for parameter estimation, due to the unstable nature of the open loop system, the numerical integration produces divergence in the results. Figure 9.6(a) shows the comparison of the measured and estimated observables. In this case, since the parameter that is causing divergence is M_w, measured state w is used in eq. (2.44) so that the state model for the stabilised output error method becomes

$$\dot{w} = Z_w w + (u_0 + Z_q)q + Z_{\delta_e}\delta_e$$

$$\dot{q} = M_w w_m + M_q q + M_{\delta_e}\delta_e$$

Here, w_m is the measured state.

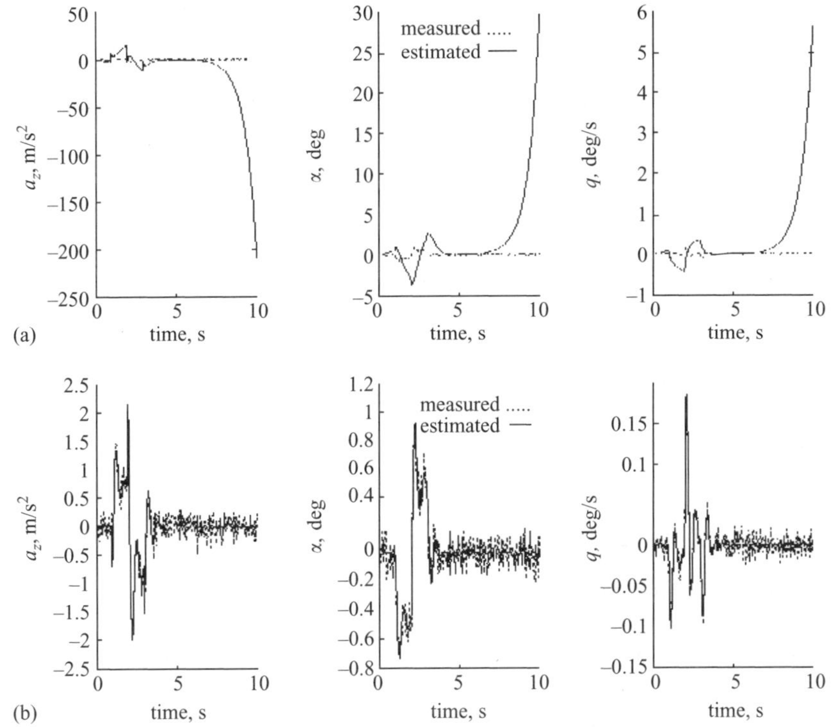

Figure 9.6 (a) Comparison of measured and estimated observables from the output error method (Example 9.6); (b) comparison of measured and estimated observables from the stabilised output error method (Example 9.6)

Table 9.10 Parameter estimates using stabilised output error method ($K = 0.25$, $SNR = 10$) (see also Table 9.6)

Parameters	True	Estimated (SOEM)	Estimated (LS)	Estimated (LSME)
Z_w	−1.4249	−1.3846	−0.8250	−1.0404
$Z_{\delta e}$	−6.2632	−6.1000	−9.5421	−5.9488
M_w	0.2163	0.2222	0.1357	0.2123
M_q	−3.7067	−4.0493	−2.8041	−3.2525
$M_{\delta e}$	−12.7840	−13.3491	−12.1554	−13.4326
PEEN %	–	4.612	15.9822	11.4771

The programs for parameter estimation are contained in folder **Ch9SOEMex6**. Figure 9.6(b) shows the time history match when the stabilised output error method is applied for parameter estimation. Time history match is satisfactory indicating that use of measured states has helped arrest the divergence in the numerical integration procedure. Estimated derivatives are given in Table 9.10. Low parameter estimation error norm indicates the satisfactory performance of the stabilised output error method even when the measurement data is noisy. Results of least squares and least squares mixed estimation methods are also compared in Table 9.10

9.8 Total least squares method and its generalisation

The least squares method gives biased estimates when measurement noise is present in the regressors. The total least squares approach accounts for not only errors in the measurements of output variables but also the errors in state and control variables X appearing in the regression equation [6].

In general, the regression equation is written as

$$Y = X\beta + v \tag{9.97}$$

The least squares methods do not account explicitly for errors in X. The total least squares method addresses this problem.

Next, to arrive at a generalisation theory, in the following discussion, the state and measurement equations of the equation decoupling method are considered. The general form of these equations is given below:

$$\begin{aligned} \dot{x} &= A_d x + [B \quad A_{od}] \begin{bmatrix} u_m \\ x_m \end{bmatrix} \\ y &= Hx + v \end{aligned} \tag{9.98}$$

If $H = I$, the identity matrix, we have

$$y = x + v$$

In discrete form, the above equation can be written as

$$y(k) = \phi_d x(k-1) + [B \quad A_{od}] \begin{bmatrix} u_m(k-1) \\ x_m(k-1) \end{bmatrix} \Delta t + v(k-1) \tag{9.99}$$

The above equation can also be written as

$$y^T_{(k)} = \begin{bmatrix} x(k-1)^T & u^T_m(k-1)\Delta t & x^T_m(k-1)\Delta t \end{bmatrix} \begin{bmatrix} \phi^T_d \\ B^T \\ A^T_{od} \end{bmatrix} + v^T(k-1) \tag{9.100}$$

$$Y = X\Theta + v_m \tag{9.101}$$

Here, X in its expanded form contains state, measured states and control inputs. The Θ is the parameter vector to be estimated. Equation (9.101) has the same general form as the regression eq. (9.97) for the total least squares problem. There are measurement errors in Y of eq. (9.101), and X contains errors due to integration caused by incorrect initial conditions and round off errors. In addition, measurement errors in states x_m and control inputs u_m are present in general. From the above discussions it is clear that equation decoupling formulation of the estimation problem is such that it generalises total least squares problem formulation which itself is known to be a generalisation of the least squares problem. Thus, generalisation of the total least squares problem has been established in terms of the stabilised output error method for which an asymptotic theory has been developed in the previous sections.

9.9 Controller information based methods

As mentioned in the introduction, when information on dynamics of controllers used for stabilising the unstable plant is available, it could be used in the estimation procedure either directly or indirectly. In this section, two approaches to this effect are presented [8].

1 Using the input-output data between p1 and p3, an equivalent parameter set can be estimated. From this set of parameters, the open loop plant parameters can be retrieved from the equivalent parameters by using an appropriate transformation based on the knowledge of the controllers used for stabilisation. If the controller were a complex one, this method would not be feasible as it would be very difficult to retrieve the parameters from the equivalent parameters.
2 Alternatively, a combined mathematical model of the states obtained by combining the system model and the known feedback controllers can be formulated. Keeping the known parameters of the controller fixed in the model, the parameters of the plant can be estimated. This could result in a very high order state-space

model of the combined system when complex controllers are used. In such cases, model reduction techniques could be employed to arrive at a workable solution.

In this section, these two approaches are investigated and the two-step bootstrap method is presented. The two-step bootstrap method utilises the knowledge of the controller and system in an indirect way. It enables smaller order models to be used and has the advantage that it can handle noisy input data. This approach has been earlier used for transfer function estimation of an open loop plant from closed loop data. In this section, it is extended to parameter estimation of state space models.

9.9.1 Equivalent parameter estimation/retrieval approach

Consider a general second order dynamical system given by

$$\begin{bmatrix} \dot{x}_1 \\ \dot{x}_2 \end{bmatrix} = \begin{bmatrix} a_{11} & a_{12} \\ a_{21} & a_{22} \end{bmatrix} \begin{bmatrix} x_1 \\ x_2 \end{bmatrix} + \begin{bmatrix} b_1 \\ b_2 \end{bmatrix} \delta_e \tag{9.102}$$

If the x_2 state is fed back to the input (at p2, Fig. 9.1) through a constant gain K, the proportional controller can be described by

$$\delta_e = Kx_2 + \delta_p \tag{9.103}$$

Here, δ_p is the command input at p1 (Fig. 9.1). Using eq. (9.103) in eq. (9.102), we get

$$\begin{bmatrix} \dot{x}_1 \\ \dot{x}_2 \end{bmatrix} = \begin{bmatrix} a_{11} & b_1 K + \quad a_{12} \\ a_{21} & b_2 K + \quad a_{22} \end{bmatrix} \begin{bmatrix} x_1 \\ x_2 \end{bmatrix} + \begin{bmatrix} b_1 \\ b_2 \end{bmatrix} \delta_p \tag{9.104}$$

It is clear that the coefficients in the second column of the matrix A are affected due to the augmentation. The objective is to estimate the elements of the matrices A and B in eq. (9.102), and an equivalent model for parameter estimation could be formulated as

$$\begin{bmatrix} \dot{x}_1 \\ \dot{x}_2 \end{bmatrix} = \begin{bmatrix} a_{11} & a_{12} \\ a_{21} & a_{22} \end{bmatrix}_{\text{eq}} \begin{bmatrix} x_1 \\ x_2 \end{bmatrix} + \begin{bmatrix} b_1 \\ b_2 \end{bmatrix} \delta_p \tag{9.105}$$

Using the command input δ_p and the measured output y, the equivalent parameters can be estimated. The parameters a_{12} and a_{22} can be computed from the equivalent parameters using the known value of the feedback gain K. For this case, input noise at p1 (in Fig. 9.1) is not considered. Often, equivalent models do not permit accurate determination of the pure aerodynamic effects.

9.9.2 Controller augmented modelling approach

The mathematical model of the plant whose parameters are to be estimated can be augmented to include known models of controller. The model would be easier to augment if the controller is simple. However, it might result in a very high order of system model if the controller is complex. The controller related parameters are kept fixed in the model since they are assumed known, and only the plant parameters are estimated. The controller augmented modelling approach is illustrated by choosing a complex fourth order aircraft longitudinal model augmented by the blocks shown in Fig. 9.4.

The state equations of the basic plant are given by

$$\begin{bmatrix} \dot{\alpha} \\ \dot{q} \\ \dot{\theta} \\ \dot{v/v_0} \end{bmatrix} = \begin{bmatrix} Z_\alpha & 1 & 0 & Z_v/v_0 \\ M_\alpha & M_q & 0 & M_v/v_0 \\ 0 & 1 & 0 & 0 \\ X_\alpha & 0 & X_\theta & X_v/v_0 \end{bmatrix} \begin{bmatrix} \alpha \\ q \\ \theta \\ v/v_0 \end{bmatrix} + \begin{bmatrix} Z_{\delta_e} \\ M_{\delta_e} \\ 0 \\ X_{\delta_e} \end{bmatrix} \tag{9.106}$$

The closed loop model is obtained as

$$\begin{bmatrix} \dot{\alpha} \\ \dot{q} \\ \dot{\theta} \\ \dot{v/v_0} \\ \delta_e \\ \dot{CS_1} \\ \dot{CS_2} \\ \dot{CS_3} \\ \dot{CS_4} \\ \dot{CS_5} \end{bmatrix} = \begin{bmatrix} Z_\alpha & I & 0 & Z_{v/v_0} & Z_{\delta_e}K_{13} & 0 & 0 & 0 & 0 \\ M_\alpha & M_q & 0 & M_{v/v_0} & M_{\delta_e}K_{13} & 0 & 0 & 0 & 0 \\ 0 & 0 & I & 0 & 0 & 0 & 0 & 0 & 0 \\ X_\alpha & 0 & X_\theta & X_{v/v_0} & X_{\delta_e}K_{13} & 0 & 0 & 0 & 0 \\ 0 & 0 & a_{53} & a_{54} & -K_{13} & a_{56} & a_{57} & a_{58} & a_{59} \\ 0 & 0 & 0 & 0 & 0 & a_{66} & 0 & 0 & 0 \\ 0 & 0 & 0 & I & 0 & 0 & a_{77} & 0 & 0 \\ 0 & 0 & 0 & I & 0 & 0 & 0 & a_{88} & 0 \\ 0 & 0 & I & 0 & 0 & 0 & 0 & 0 & a_{99} \\ 0 & 0 & 0 & 0 & 0 & 0 & 0 & 0 & 0 \end{bmatrix}$$

$$\times \begin{bmatrix} \alpha \\ q \\ \theta \\ v/v_0 \\ \delta_e \\ CS_1 \\ CS_2 \\ CS_3 \\ CS_4 \\ CS_5 \end{bmatrix} + \begin{bmatrix} 0 \\ 0 \\ 0 \\ 0 \\ 0 \\ 1 \\ 0 \\ 0 \\ 0 \\ 0 \end{bmatrix} \delta_p \tag{9.107}$$

Here, the variables CS refer to the states pertaining to the blocks 1, 4, 5, 6 and 7. The K_{ij} and a_{ij} are known constants, which implicitly contain the time constants and/or gains of the controller transfer functions. It is seen that the closed loop model for parameter estimation is of a very high order.

In any controller, where signals are fed back, the noise also is fed back and this could result in noise processes, which are not white. In the discussions above, the effect of the feedback of the noise on the mathematical model has not been considered. In the following section, a covariance analysis is carried out to illustrate the effect of the noise feedback on the mathematical models used for parameter estimation.

9.9.3 Covariance analysis of system operating under feedback

When direct identification using measured input and output data (at p2 and p3, Fig. 9.1) is carried out, the correlation between the plant input δ and the output noise v might lead to biased estimates. Also, the signal u could be noisy due to measurement noise of the sensor. This could result in input-output noise correlations in addition to the signal/noise correlation.

To bring about explicitly the modifications in the covariance computations resulting from these correlations, the expressions for the covariance matrix are derived for (i) open loop system with input noise and (ii) closed loop system with input noise.

9.9.3.1 Open loop system with input noise

The analysis is carried out in the discrete domain where the system state and measurements are described by

$$x(k+1) = \varphi x(k) + B_d u(k) + G w(k) \tag{9.108}$$

$$y(k) = H x(k) + v(k) \tag{9.109}$$

Also,

$$\begin{aligned} &E\{x_0\} = \hat{x}_0; \quad P_0 = E\{(\hat{x}_0 - x_0)(\hat{x}_0 - x_0)^T\} \\ &E\{wv^T\} = 0; \quad \hat{x}(0) = \hat{x}_0; \quad \hat{P}(0) = \hat{P}_0 \end{aligned} \tag{9.110}$$

The input signal u can be expressed as a combination of a deterministic part u_d and a non-deterministic part u_n:

$$u(k) = u_d(k) + u_n(k) \tag{9.111}$$

Using eq. (9.111) in eq. (9.108), we get

$$x(k+1) = \varphi x(k) + B_d u_d(k) + B_d u_n(k) + G w(k) \tag{9.112}$$

Combining the last two terms, we get

$$x(k+1) = \varphi x(k) + B_d u_d(k) + [B_d \quad G] \begin{bmatrix} u_n(k) \\ w(k) \end{bmatrix}$$

The above can be written as

$$x(k+1) = \varphi x(k) + B_d u_d(k) + G_a w_a(k) \tag{9.113}$$

Here, the subscript a denotes the augmented effect which is obtained by combining the effects of input noise as part of the process noise.

State estimation error is given by

$$x_e(k) = x(k) - \hat{x}(k) \tag{9.114}$$

Estimation error covariance matrix is given by

$$P(k) = E\{x_e(k) x_e(k)^T\} \tag{9.115}$$

State estimation error at instant $k+1$ is given by

$$x_e(k+1) = x(k+1) - \hat{x}(k+1) \tag{9.116}$$

Substituting for $x(k+1)$ from eq. (9.113) in eq. (9.116), and using the following expression

$$\hat{x}(k+1) = \phi\hat{x}(k) + B_d u_d(k)$$

we get for the state error at $(k+1)$:

$$x_e(k+1) = \varphi x_e(k) + G_a w_a(k) \tag{9.117}$$

Estimation error covariance matrix at $k+1$ is given by

$$\begin{aligned} P(k+1) &= E\{x_e(k+1)x_e(k+1)^T\} \\ &= E\{[\varphi x_e(k) + G_a w_a(k)][\varphi x_e(k) + G_a w_a(k)]^T\} \end{aligned} \tag{9.118}$$

If the estimation error and the (equivalent) process noise $w_a(k)$ are assumed uncorrelated, we get for $P(k+1)$

$$P(k+1) = \phi\hat{P}(k)\phi^T + G_a Q_a G_a^T \tag{9.119}$$

In the above equation, Q_a represents the input noise covariance matrix. From eq. (9.119), it is clear that, when the input is noisy, the process noise covariance matrix will have additional contributions from the input noise.

9.9.3.2 Closed loop system with input noise

When the output y is fed back, the output noise v is correlated with the input signal δ and this process affects the covariance computations. This aspect is illustrated next. Considering the overall closed loop system, the input u (considering the input δ and a feedback resulting from an output y) can be written as

$$u(k) = \delta(k) + Ky(k) + u_n(k) \tag{9.120}$$

Substituting for y from eq. (9.109), we have

$$u(k) = \delta(k) + KHx(k) + Kv(k) + u_n(k) \tag{9.121}$$

Using eq. (9.121) in eq. (9.108), we get

$$\begin{aligned} x(k+1) &= \phi x(k) + B_d\delta(k) + B_d KHx(k) + B_d Kv(k) + B_d u_n(k) + Gw(k) \\ &= (\phi + B_d KH)x(k) + B_d\delta(k) + B_d Kv(k) + G_a w_a(k) \end{aligned} \tag{9.122}$$

Here, the subscript a is used to represent the augmented noise related terms.

The estimate at instant $(k+1)$ is given by

$$\hat{x}(k+1) = \phi\hat{x}(k) + B_d KH\hat{x}(k) + B_d\delta(k) \tag{9.123}$$

Using eqs (9.122) and (9.123), the estimation error can be written as

$$x_e(k+1) = (\phi + B_d KH)x_e(k) + B_d Kv(k) + G_a w_a(k) \tag{9.124}$$

If it is assumed that the estimation state error, the process noise and the measurement noise $v_{(k)}$ are uncorrelated, we get

$$P(k+1) = (\phi + B_d KH)P(k)(\phi + B_d KH)^T + G_a Q_a G_a^T + (B_d K)R(B_d K)^T \tag{9.125}$$

Comparing eqs (9.125) and (9.119), we see that there is an additional term due to the measurement noise covariance when there is feedback and this introduces more uncertainty into the filter computations. In addition, there is a term involving feedback gain implying that the feedback not only causes changes in the elements of the ϕ matrix, but also results in estimation error covariances being higher.

9.9.4 Two-step bootstrap method

If a plant or a system is unstable, it requires stabilisation using a suitable control system. Even otherwise, a control system would be useful to improve the stability or reduce the effect of plant uncertainty on the responses. The identification of such a plant poses the problem that the input signal to the plant is dependent on the output measurement. This poses a problem in parameter estimation as can be seen from the following development [12].

Let the control system be given as in Fig. 9.7. Then,

$$y(s) = Gu(s) + v(s) \tag{9.126}$$

We have

$$\begin{aligned} u(s) &= \delta(s) - Hy(s) \\ &= \delta(s) - H(Gu(s) + v(s)) \\ &= \delta(s) - HGu(s) - Hv(s) \end{aligned} \tag{9.127}$$

From the above, we see that the input u and the measurement noise v are correlated. This circulation of noise in the loop poses identifiability problems. Although, often, H would be a low pass filter, the noise still could prevail at the feedback error point. Thus, before using u for parameter estimation, it may be worthwhile to attempt to reduce the effect of noise further by obtaining the predicted/estimated u.

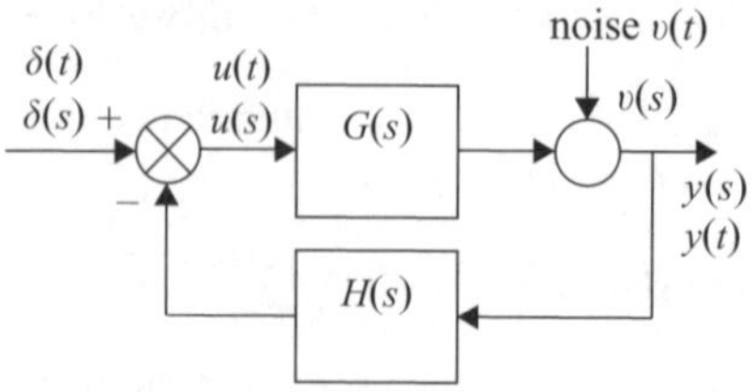

Figure 9.7 Simple control system

We have the sensitivity function of the closed loop system as

$$S = \frac{1}{1 + GH} \tag{9.128}$$

Thus, we have from eq. (9.127):

$$\begin{aligned} u(s) + HGu(s) &= \delta(s) - Hv(s) \\ u(s) &= \frac{1}{1+GH}\delta(s) - \frac{H}{1+GH}v(s) \\ u(s) &= S\delta(s) - HSv(s) \\ y(s) &= Gu(s) + v(s) \end{aligned} \tag{9.129}$$

We see from the above equations that since δ and v are uncorrelated and the measurements of u and δ are available, we can estimate the sensitivity functions. Then, using this form, we can write:

$$\begin{aligned} \hat{u}(s) &= \hat{S}\delta(s) \\ \hat{y}(s) &= G\hat{u}(s) + v(s) \end{aligned} \tag{9.130}$$

Now, since $\hat{u}$ and v are uncorrelated, we can estimate the open loop transfer function G in an open loop way.

The above procedure is next generalised for a continuous-time feedback system.

9.9.4.1 First step

Let the measured input $u(t)$ be treated as the output of the system as shown in Fig. 9.8. The measured output y and the input δ are the inputs to the system. Thus, we have

$$u_m = \delta - \beta y_m \tag{9.131}$$

Here, u_m is the $p \times N$ control input measurement matrix, δ the $p \times N$ reference input matrix and y_m the $n \times N$ measurement data matrix. The unknown parameters are denoted as $\beta(p \times N)$. Since measurements are noisy, we obtain

$$\begin{aligned} u_t + u_n &= \delta - \beta(y_t + y_n) \\ u_t &= \delta - \beta y_t - \beta y_n - u_n = \delta - \beta y_t + v_n \end{aligned} \tag{9.132}$$

Here, v_n denotes a compound noise.

Thus, in the first step, the effect of this noise is minimised and the model that best fits the input is obtained. In case feedback plants are complex, a more generalised

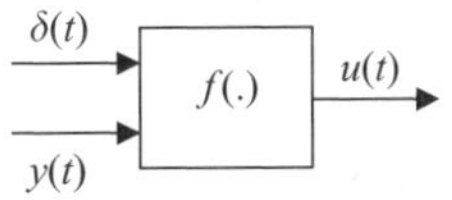

Figure 9.8 Input estimation

model can be used:

$$u = f(y_m, \dot{y}_m, \delta, \dot{\delta}) + noise \tag{9.133}$$

The time-derivatives can be obtained by numerical differentiation of the signals y and r, etc. To the extent possible, a linear or linear-in-parameters model should be fitted in order to keep computations reasonably small. The model is obtained by the LS method to minimise the cost function:

$$J = \frac{1}{2}\sum_{k=1}^{N}[u(k) - f(y(k), \dot{y}(k), \delta, \dot{\delta})]^2 \tag{9.134}$$

Model selection criteria can be used to arrive at an adequate model.

9.9.4.2 Second step

In this step, the system parameters are estimated using the UD filter [8]:

1 Obtain the estimated input trajectories from the first step, say:

$$\hat{u}(k) = \beta_1\hat{y}(k) + \beta_2\dot{\hat{y}}(k) + \beta_3\delta(k) + \beta_4\dot{\delta}(k) \tag{9.135}$$

Here, β_i are estimated from the LS method.

2 Use $\hat{u}(k)$ in the UD filter/extended UD filter algorithms of Chapter 4. Here, the system parameters are considered as unknown and augmented as additional states in the filter. The main advantage of this procedure is that it utilises the estimated feedback error, i.e., $\hat{u}$ as the input to the open loop system and obtains the parameters in recursive manner.

9.10 Filter error method for unstable/augmented aircraft

The filter error method, discussed in Chapter 5, accounts for both process and measurement noise and is, therefore, considered the most general approach to parameter estimation problems. Though primarily used for analysing data in turbulence (process noise), it has also been found to give good results for data without turbulence.

The filter error method has also been used to estimate parameters of unstable systems. In the majority of the parameter estimation applications pertaining to unstable systems, particularly in the field of aircraft flight data analysis, the requirement is to estimate the parameters of the basic unstable plant (open-loop model) rather than obtaining closed loop characteristics of the system. Parameter estimation of open loop unstable models can pose various problems ranging from round off errors to diverging solutions from numerical integration of the unstable system equations. The filter error method is a numerically stable scheme and, as such, easily amenable to unstable systems.

As can be seen from eq. (9.42), the use of the term $[K(k)(z(k) - \tilde{y}(k))]$, which represents a kind of feedback of the fit error $(z(k) - \tilde{y}(k))$ weighted with gain K, renders the filter error algorithm numerically stable. Here, it is interesting to draw a parallel between the stabilised output error method and the filter error method.

In analogy to the filter error method, the stabilised output error method also uses measured states for stabilisation. In fact, filter error method requires the computation of gain K that is quite complex and time consuming. In contrast, the stabilised output error method is easy to implement and can yield good results, particularly if the postulated mathematical model is a good representation of the plant. However, one must remember that measured states will have some noise and the use of such signals for stabilisation in the stabilised output error method will essentially mean that we are introducing an immeasurable stochastic input into the system, which cannot be accounted for in the output error method. The filter error method on the other hand has no such problems.

Next, consider the state equation for the filter error method:

$$\dot{x}(t) = f[x(t), u(t), \Theta] + Gw(t) \tag{9.136}$$

Here, G is the process noise distribution matrix (assumed diagonal) whose elements are unknown and estimated along with other model parameters. Using $G \approx 0$ in parameter estimation with the filter error method will yield results that are similar to those obtained from output error method. On the other hand, estimating G will take care of any modelling errors present in the system equations. It has been argued that the modelling errors arising from the use of linearised or simplified models should be treated as process noise rather than measurement noise. This argument is also supported by the fact that the power spectral densities of the model error and of the response of the system driven by process noise, show similar trends with more power in the lower frequency band.

The model compensation ability of the filter error method through the estimation of distribution matrix G is a useful feature for obtaining parameters of a plant equipped with a controller. The feedback from the controller tends to correlate the input-output variables. The filter error method treats the modelling errors arising from data correlation as process noise, which is suitably accounted for by the algorithm to yield high quality estimates. Parameter estimation of an augmented aircraft equipped with a controller was carried out using output error and filter error methods [13]. It was shown that the feedback signals from the controller and the aileron-rudder interconnect operation cause correlation between the input-output variables that degrade the accuracy of the parameter estimates. The filter error method was found to yield reliable parameter estimates, while the aircraft derivatives estimated from the output error method did not compare well with the reference derivative values.

9.11 Parameter estimation methods for determining drag polars of an unstable/augmented aircraft

The estimation of aircraft lift and drag characteristics (see Section B.19) is an extremely important aspect in any aircraft flight-test program [14, 15]. Using aircraft response measurements, the drag polars are to be obtained throughout the entire mission spectrum. The drag polar data are required to assess the performance capability of the aircraft. A commonly used method for determination of the drag polars

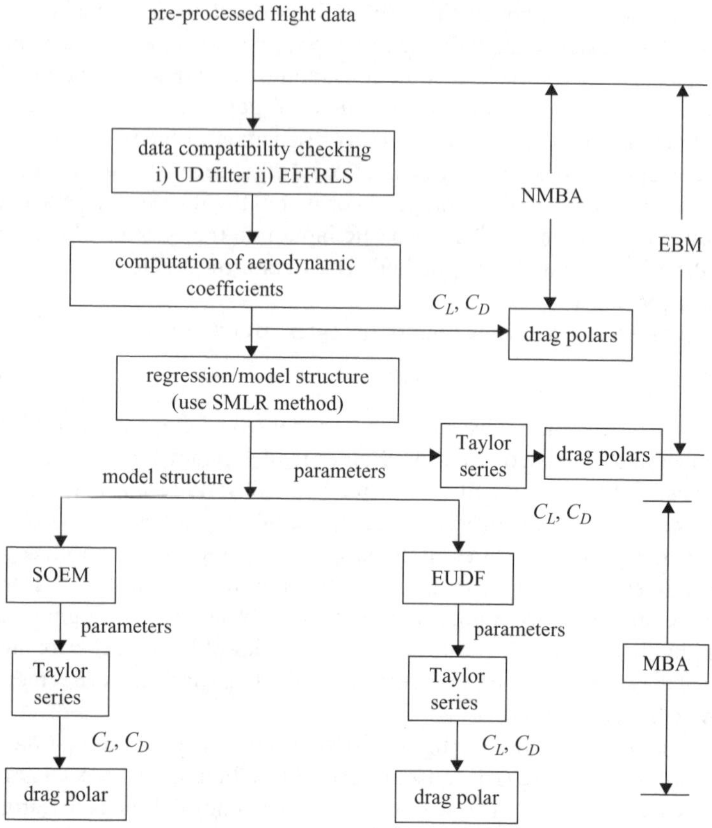

Figure 9.9 Relations between the four methods for drag polar estimation

involves performing dynamic flight manoeuvres on the aircraft, recording the relevant response variables and using the output error method for estimation of the drag polars. The demands of improved performance characteristics of modern flight vehicles have led to aerodynamically unstable configurations, which need to be highly augmented so they can be flown. For such an inherently unstable, augmented aircraft, parameter estimation and determination of performance characteristics would require special considerations.

For such aircraft, model based and non-model based approaches could be considered for determination of drag polar. The two approaches are linked as shown in Fig. 9.9. The estimation before modelling method is used for determination of the structure of the aerodynamic model to be used in the model based approach.

9.11.1 Model based approach for determination of drag polar

In this method, an explicit aerodynamic model for the lift and drag coefficients is formulated as shown below.

State model

$$\dot{V} = -\frac{\bar{q}S}{m}C_D + \frac{F_e}{m}\cos(\alpha + \sigma_T) + g\sin(\alpha - \theta)$$

$$\dot{\alpha} = -\frac{\bar{q}S}{mV}C_L - \frac{F_e}{mV}\sin(\alpha + \sigma_T) + q + \frac{g}{V}\cos(\alpha - \theta) \tag{9.137}$$

$$\dot{\theta} = q$$

Here, the C_L and C_D are modelled as

$$\begin{aligned} C_L &= C_{L_o} + C_{LV}\frac{V}{u_o} + C_{L_\alpha}\alpha + C_{L_q}\frac{q\bar{c}}{2u_o} + C_{L_{\delta_e}}\delta_e \\ C_D &= C_{D_o} + C_{DV}\frac{V}{u_o} + C_{D_\alpha}\alpha + C_{D_{\alpha^2}}\alpha^2 + C_{D_q}\frac{q\bar{c}}{2u_o} + C_{D_{\delta_e}}\delta_e \end{aligned} \tag{9.138}$$

Observation model

$$V_m = V$$

$$\alpha_m = \alpha$$

$$\theta_m = \theta$$

$$a_{x_m} = \frac{\bar{q}S}{m}(C_X) + \frac{F_e}{m}\cos\sigma_T \tag{9.139}$$

$$a_{z_m} = \frac{\bar{q}S}{m}(C_Z) - \frac{F_e}{m}\sin\sigma_T$$

$$C_Z = -C_L\cos\alpha - C_D\sin\alpha$$

$$C_X = C_L\sin\alpha - C_D\cos\alpha$$

The aerodynamic derivatives in the above equations could be estimated using the output error method (Chapter 3) for stable aircraft (stabilised output error method for unstable aircraft) or using an extended UD filter (Chapter 4). In the extended Kalman filter, the aerodynamic derivatives in eq. (9.138) would form part of the augmented state model (Examples 4.2 and 4.3). The estimated C_L and C_D are then used to generate the drag polar.

9.11.2 Non-model based approach for drag polar determination

This method does not require an explicit aerodynamic model to be formulated. The determination of drag polars is accomplished using the following two steps:

1 In the first step, sub-optimal smoothed states of aircraft are obtained using the procedure outlined in Chapter 7. Scale factors and bias errors in the sensors are estimated using the data compatibility checking procedure outlined in Appendix B (Example 7.1).

2 In the second step, the aerodynamic lift and drag coefficients are computed using the corrected measurements (from step 1) of the forward and normal accelerations using the following relations:

$$C_x = \frac{m}{\bar{q}S}\left(a_x - \frac{F_e}{m}\cos\sigma_T\right)$$
$$C_z = \frac{m}{\bar{q}S}\left(a_z + \frac{F_e}{m}\sin\sigma_T\right) \tag{9.140}$$

The lift and drag coefficients are computed from C_x and C_z using

$$C_L = -C_Z\cos\alpha + C_X\sin\alpha$$
$$C_D = -C_X\cos\alpha - C_Z\sin\alpha \tag{9.141}$$

C_D versus C_L is plotted to obtain the drag polar. The first step could be accomplished using the state and measurement models for kinematic consistency (Chapter 7 and Appendix B) and the extended UD filter (Chapter 4) or the extended forgetting factor recursive least squares method. A brief description of the latter is given below.

9.11.3 *Extended forgetting factor recursive least squares method*

The extended forgetting factor recursive least squares method does not require knowledge of process and measurement noise statistics, but requires a suitable choice of a forgetting factor λ [16]. Only one adjustable parameter λ is required to be selected as compared to several elements of Q and R required for tuning of a Kalman filter. The algorithm is given as

$$x(k+1/k) = \phi x(k/k)$$
$$x(k+1/k+1) = \phi[x(k/k) + L(y(k+1) - H\phi x(k/k)]$$
$$L = P(k/k)\phi^T H^T(\lambda I + H\phi P(k/k)\phi^T H^T)^{-1} \tag{9.142}$$
$$P(k+1/k+1) = \lambda^{-1}\phi[I - LH\phi]P(k/k)\phi^T$$

A simple explanation of the role of λ is given for the sake of completeness. The memory index of the filter can be defined as $\mathrm{MI} = 1/(1-\lambda)$. Thus if $\lambda = 1$, then MI is infinity – the filter is said to have infinite memory. This means that the entire data set is given equal weighing and the procedure gives an ordinary least squares solution. If λ is smaller then the MI will also be smaller (finite memory), thereby implying that the past data are given less weighting, since the weighting factor used in the least squares performance functional is given as [16]:

$$\lambda^{k-i}; \qquad i = 1, 2, \ldots, k$$

Choice of forgetting factor is based on the following considerations. If the process noise variance is expected to be large then the forgetting factor should be small, since the past data is not giving more information on the current state/parameter. If the process noise variance is relatively smaller than the measurement noise variance, then the forgetting factor should be of a large value. This implies that more data should be used to average out the effect of the noise on measurements. The forgetting factor can also be linked to the column rank of the observation model H. If this rank is larger, then there is more information (contained by the kth measurement data) on the present state. The forgetting factor can be also taken as inversely proportional to the condition number of the data matrix. If the condition number of the matrix is large then one would like to give less emphasis on the past data, and hence the forgetting factor should be smaller.

The above are general guidelines to choosing a forgetting factor. For a given application, specific evaluation study is generally required to arrive at a suitable forgetting factor. Thus, the forgetting factor can be chosen as

$$\lambda \propto \frac{\text{variance }(R)}{\text{variance }(Q)}\,\frac{1}{\text{condition no. (data matrix }P)}\,\frac{1}{\text{column rank }(H)}$$

From the above it is clear that the forgetting factor is intended to ensure that data in the distant past are 'forgotten' in order to afford the possibility of following the statistical variation of the measurement data.

The performance of the model based and non-model based approaches were evaluated by estimating the drag polars and comparing the same with the reference polars of an unstable/augmented aircraft using the data from a six degree of freedom fixed base flight simulator [17]. Roller coaster and windup turn manoeuvres (see Section B.6) were performed at a number of flight conditions to evaluate the methods outlined. It was found that the extended forgetting factor recursive least squares method with the non-model based approach (EFFRLS-NMBA) and the extended UD filter with the non-model based approach (EUDF-NMBA) performed better than the other two model based approaches. The stabilised output error method, being an iterative process, required more time for drag polar determination. The extended UD filter, being a recursive process, could be an attractive alternative to the stabilised output error method. However, it required proper choice of the process and measurement noise statistics. The estimation before modelling (EBM) helped in model selection based on statistical criteria. A non-model based approach could be preferred over a model based approach, as it would require less computation time and still give accurate results for drag polars from flight data. It is also a potential candidate for real-time on-line determination of drag polars.

9.12 Epilogue

Parameter estimation for inherently unstable/augmented (control) systems has found major applications in modelling of aerospace vehicles [1]. Many modern day high performance fighter aircraft are made inherently unstable or with relaxed static stability for gaining higher (lift/drag ratio) performance. However, such systems cannot fly without full authority control (laws) constantly working. Thus, the aircraft becomes a plant or system working within the closed loop control system. Several approaches for explicit parameter estimation of dynamic systems, in general, and aircraft in particular, have been elucidated in this chapter. A few other approaches for such applications are given in Reference 18. Frequency domain methods, as discussed in Chapter 11, could find increasing applications for such unstable/augmented systems/aircraft, if linear models are considered adequate.

9.13 References

1 KOEHLER, R., and WILHELM, K.: 'Closed loop aspects of aircraft identification', *AGARD LS*, 1979, **104**, pp. 10-1 to 10-25

2 KLEIN, V.: 'Estimation of aircraft aerodynamic parameters from flight data', *Prog. Aerospace Sciences*, 1989, **26**, pp. 1–77

3 HOU, D., and HSU, C. S.: 'State space model identification of unstable linear systems', *Control Theory and Advanced Technology*, 1992, **8**, (1), pp. 221–231

4 PREISSLER, H., and SCHAUFELE, H.: 'Equation decoupling – a new approach to the aerodynamic identification of unstable aircraft', *Journal of Aircraft*, 1991, **28**, (2), pp. 146–150

5 GIRIJA, G., and RAOL, J. R.: 'Analysis of stabilised output error methods', *IEE Proc. of Control Theory and Applications*, 1996, **143**, (2), pp. 209–216

6 LABAN, M., and MASUI, K. 'Total least squares estimation of aerodynamic model parameters from flight data', *Journal of Aircraft*, 1992, **30**, (1), pp. 150–152

7 GIRIJA, G., and RAOL, J. R.: 'Asymptotic and generalisation theory of equation de-coupling method for parameter estimation of dynamic systems', *Journal of the Inst. of Engrs. (Ind.)*, 1996, **77**, pp. 80–83

8 GIRIJA, G., and RAOL, J. R.: 'Controller information based identification methods'. Proceedings of *34th Aerospace Sciences Meeting and Exhibit (AIAA)*, Reno, NV, USA) paper no. 96-0900, January 15–18, 1996

9 GIRIJA, G., and RAOL, J. R.: 'An approach to parameter estimation of unstable systems', *Journal of Instn. of Engrs.*, 1995, **77**, pp 133–137

10 MAINE, R. E., and MURRAY, J. E.: 'Application of parameter estimation to highly unstable aircraft', *Journal of Guidance, Control and Dynamics*, 1988, **11**, (3), pp. 213–219

11 GIRIJA, G., and RAOL, J. R.: 'Estimation of parameters of unstable and augmented aircraft using recursive mixed estimation technique', *Journal of the Inst. of Engrs. (Ind.), Aerospace Division*, 1995, **76**, pp. 15–22

12 VAN DEN HOF, P. M. J., and SCHRAMA, R. J. P.: 'An indirect method for transfer function estimation from closed loop data', *Automatica*, 1993, **29**, (6), pp. 1523–1527

13 SINGH, J., and RAOL, J. R.: 'Improved estimation of lateral-directional derivatives of an augmented aircraft using filter error method', *Aeronautical Journal*, 2000, **14**, (1035), pp. 209–214

14 ILIFF, K. W.: 'Maximum likelihood estimates of lift and drag characteristics obtained from dynamic aircraft manoeuvres'. Mechanics Testing Conf. Proceedings, pp. 137–150, 1976

15 KNAUS, A.: 'A technique to determine lift and drag polars in flight', *Journal of Aircraft*, 1983, **20**, (7), pp. 587–592
16 ZHU, Y.: 'Efficient recursive state estimator for dynamic systems without knowledge of noise covariances', *IEEE Trans., AES*, 1999, **35**, (1), pp. 102–113
17 GIRIJA, G., BASAPPA, RAOL, J. R., and MADHURANATH, P.: 'Evaluation of methods for determination of drag polars of unstable/augmented aircraft'. Proceedings of *38th Aerospace Sciences Meeting and Exhibit (AIAA)*, Reno, NV, USA, paper no. 2000-0501, January 10–13, 2000
18 JATEGAONKAR, R. V., and THIELECKE, F.: 'Evaluation of parameter estimation methods for unstable aircraft', *Journal of Aircraft*, 1994, **31**, (3), pp. 510–519

9.14 Exercises

Exercise 9.1

Derive the expression for the system state equation for differential feedback (see Table 9.1):

$$u = Kx + L\dot{x} + \delta$$

Exercise 9.2

Derive the expression for the system state equation for integrating feedback (see Table 9.1):

$$\dot{u} + Fu = Kx + \delta$$

Exercise 9.3

Let the system be given by eq. (9.2) and the system responses be correlated as per eq. (9.5). Derive the expression for $\dot{x}$, eq. (9.6).

Exercise 9.4

Determine the observability matrix for the system of eq. (9.45), assuming that the linear system eq. (9.1) is without noise terms.

Exercise 9.5

Explain the significance of eq. (9.47), the mixed estimation solution.

Exercise 9.6

Let

$$x(k+1) = \phi x(k) + \psi Bu(k)$$

$$y(k) = Hx(k) + Du(k)$$

Obtain sensitivity equations with respect to β, the parameter vector containing elements of ϕ, ψ, B, H, D etc.

Exercise 9.7

What is the series expansion for ϕ and ψ given $\dot{x} = Ax + Bu$?

Exercise 9.8

Take

$$A = \begin{bmatrix} -1 & 0 \\ 0 & 2 \end{bmatrix}$$

Determine its eigenvalues and comment on the stability of the linear system governed by this matrix. Then choose a suitable value of δ to convert the system into a stable one.

Exercise 9.9

Determine the transition matrices for A and $\bar{A}$ of Exercise 9.8. Comment on equivalent δ between these matrices. Use $\phi = I + A\Delta t$ as an approximation for the transition matrix.

Exercise 9.10

Let

$$A = \begin{bmatrix} -1 & -2 \\ -3 & 4 \end{bmatrix}$$

Determine matrices A_d and A_{od} (see eq. (9.59)).

Exercise 9.11

Let A be as in Exercise 9.10. Determine A_s and A_{us} (see eq. (9.60)).

Exercise 9.12

What does the following expression signify if r is a white noise?

$$\frac{1}{N-1}\sum_{k=1}^{N} r(k)r(k-1)\Delta t$$

Exercise 9.13

Consider the expression given in Example 9.6 and show with details how the system could be made stable when it is unstable with $M_w = 0.2$?

Exercise 9.14

Determine the sensitivity function of eq. (9.128), for the closed loop system of Fig. 9.7.

Exercise 9.15

In eq. (9.130), why are $\hat{u}$ and v considered uncorrelated?

Chapter 10

Parameter estimation using artificial neural networks and genetic algorithms

10.1 Introduction

Research in the area of artificial neural networks has advanced at a rapid pace in recent times. The artificial neural network possesses a good ability to learn adaptively. The decision process in an artificial neural network is based on certain nonlinear operations. Such nonlinearities are useful: i) in improving the convergence speed (of the algorithm); ii) to provide more general nonlinear mapping between input-output signals; and iii) to reduce the effect of outliers in the measurements.

One of the most successful artificial neural networks is the so-called feed forward neural network. The feed forward neural network has found successful applications in pattern recognition, nonlinear curve fitting/mapping, flight data analysis, aircraft modelling, adaptive control and system identification [1–6]. An illustration and comparison of biological neuron and artificial neuron are given in Fig. 10.1 and Table 10.1 [7].

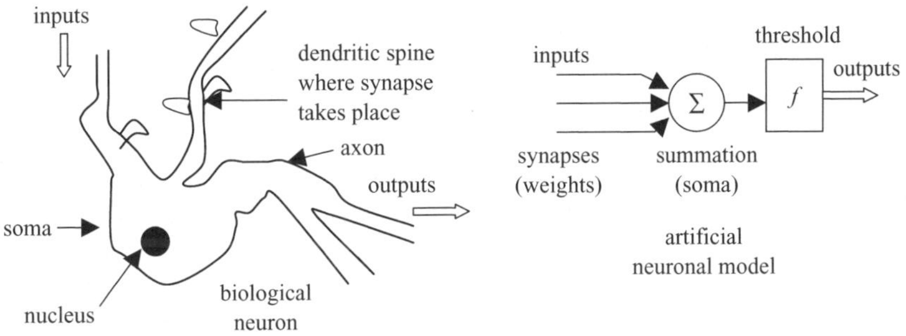

Figure 10.1 Artificial neuron imitates biological neuron in certain ways

Table 10.1 Comparison of neural systems

Biological neuron (of human brain)	Artificial neuron
Signals received by dendrites and passed on to neuron receptive surfaces	Data enter through input layer
Inputs are fed to the neurons through specialised contacts called synapses	Weights provide the connection between the nodes in the input and output layers
All logical functions of neurons are accomplished in soma	Nonlinear activation function operates upon the summation of the product of weights and inputs $f(\sum Wx_i)$
Output signal is delivered by the axon nerve fibre	The output layer produces the network's predicted response

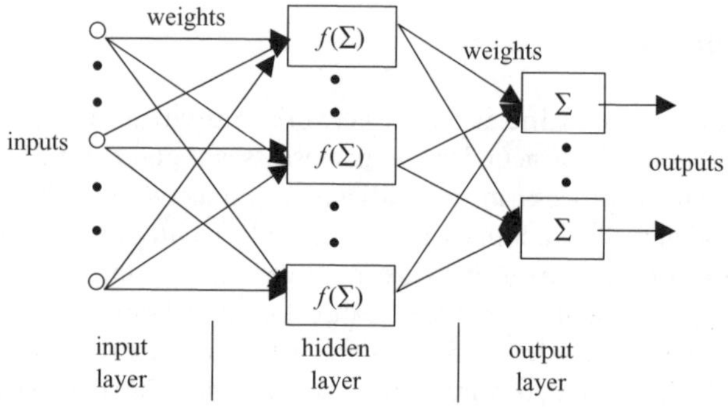

Figure 10.2 Feed forward neural network structure with one hidden layer

The artificial neural networks have some similarities to the biological neuron system, which has massive parallelism and consists of very simple processing elements. The feed forward neural network is an information processing system of a large number of simple processing elements (Fig. 10.2). These elements are called artificial neurons or nodes. These neurons are interconnected by links, which are represented by the so-called weights, and they cooperate to perform parallel distributed computing in order to carry out a desired computational task. The neural networks are so-called because the background of early researchers who were involved in the study of functioning of the human brain and modelling of the neuron system was in the area of biology, psychology or science [1]. Artificial neural networks have some resemblance to real neural networks. They should be more appropriately called massively parallel adaptive circuits or filters. This is because the artificial neural networks have technical roots in the area of analogue circuits, computing and signal processing. However, for the present, we continue to use the artificial neural network terminology keeping in mind that we are dealing with massively parallel adaptive circuits or filters.

Artificial neural networks are used for input-output subspace modelling because the basic neural network functions can adequately approximate the system behaviour in an overall sense. The feed forward neural networks can be thought of as nonlinear black-box model structures, the parameters (weights) of which can be estimated by conventional optimisation methods. These are more suitable for system identification, time-series modelling and prediction, pattern recognition/classification, sensor failure detection and estimation of aerodynamic coefficients [5, 6, 8]. Lately these have also been used for parameter estimation of dynamical system [9]. In this case, the feed forward neural network is used for predicting the time histories of aerodynamic coefficients and then some regression method is used to estimate the aerodynamic parameters (the aerodynamic stability and control derivatives, see Appendix B) from the predicted aerodynamic coefficients. This procedure parallels the so-called estimation before modelling approach discussed in Chapter 7.

In this chapter first the description of the feed forward neural network and its training algorithms is given. Next, parameter estimation using this approach is discussed. The presentation of training algorithms is such that it facilitates MATLAB implementation. Subsequently, recurrent neural networks are described. Several schemes based on recurrent neural networks are presented for parameter estimation of dynamical systems. Subsequently, the genetic algorithm is described and its application to parameter estimation considered.

10.2 Feed forward neural networks

The feed forward neural networks have a non-cyclic and layered topology and hence can be considered to have structure free (in the conventional polynomial model sense) nonlinear mapping between input-output signals of a system (see Fig. 10.2). The chosen network is first trained using the training set data and then it is used for prediction using a different input set, which belongs to the same class of the data. This is the validation set. The process is similar to the one used as cross-validation in system identification literature. The weights of the network are determined using the so-called back propagation/gradient-based procedure. Because of the layered disposition of weights of the feed forward neural network, the estimation of the weights requires propagation of the error of the output layer in a backward direction and hence the name back propagation. The estimation algorithms are described using the matrix/vector notation for the sake of clarity and ease of implementation in PC MATLAB. Even if one does not have the neural network toolbox of MATLAB, the simulation studies can be carried out easily and very efficiently using the available and newly formulated dot-em (.m) files.

The feed forward neural network has the following variables:

u_0 = input to (input layer of) the network;
n_i = number of input neurons = number of inputs u_0;

n_h = number of neurons of the hidden layer;
n_o = number of output neurons = number of outputs z;
$W_1 = n_h \times n_i$, weight matrix between input and hidden layer;
$W_{10} = n_h \times 1$, bias weight vector;
$W_2 = n_o \times n_h$, weight matrix between hidden and output layer;
$W_{20} = n_o \times 1$, bias weight vector;
μ = learning rate or step size.

10.2.1 *Back propagation algorithm for training*

This algorithm is based on the steepest descent optimisation method (see Section A.42) [10]. The forward pass signal computation is done using the following sets of equations, since u_0 is known and initial guesstimates of the weights are known.

$$y_1 = W_1 u_0 + W_{10} \tag{10.1}$$

$$u_1 = f(y_1) \tag{10.2}$$

Here y_1 is a vector of intermediate values and u_1 is the input to the hidden layer. The function $f(y_1)$ is a nonlinear sigmoidal activation function given by

$$f(y_i) = \frac{1 - e^{-\lambda y_i}}{1 + e^{-\lambda y_i}} \tag{10.3}$$

Next, the signal between the hidden and output layers is computed:

$$y_2 = W_2 u_1 + W_{20} \tag{10.4}$$

$$u_2 = f(y_2) \tag{10.5}$$

Here u_2 is the signal at the output layer. The learning rule is derived next.

Often, an unconstrained optimisation problem for parameter estimation is transformed into an equivalent system of differential equations, which in turn constitute a basic neural network algorithm to solve:

$$\frac{dW}{dt} = -\mu(t)\frac{\partial E(W)}{\partial W} \tag{10.6}$$

With the output error defined as $e = z - u_2$, and a suitable quadratic cost function based on it, the expression for the gradient is obtained as

$$\frac{\partial E}{\partial W_2} = -f'(y_2)(z - u_2)u_1^T \tag{10.7}$$

Here, u_1 is the gradient of y_2 with respect to W_2. The derivative f' of the node activation function f is given by

$$f'(y_i) = \frac{2\lambda_i e^{-\lambda y_i}}{(1 + e^{-\lambda y_i})^2} \tag{10.8}$$

The expression (10.7) follows directly from the quadratic function defined as $E = (1/2)(z - u_2)(z - u_2)^T$ and using eqs (10.4) and (10.5).

The modified error of the output layer can be expressed as

$$e_{2b} = f'(y_2)(z - u_2) \tag{10.9}$$

Thus, the recursive weight update rule for the output layer is given as

$$W_2(i+1) = W_2(i) + \mu e_{2b} u_1^T + \Omega[W_2(i) - W_2(i-1)] \tag{10.10}$$

The Ω is the momentum constant and is used to smooth out the weight changes and to accelerate the convergence of the algorithm.

The back propagation of the error and the weight update rule for W_1 are given as

$$e_{1b} = f'(y_1) W_2^T e_{2b} \tag{10.11}$$

$$W_1(i+1) = W_1(i) + \mu e_{1b} u_0^T + \Omega[W_1(i) - W_1(i-1)] \tag{10.12}$$

The data are presented to the network in a sequential manner and this process is called pattern learning in neural network literature. The data are presented again but with initial weights as the outputs from the previous cycle. This process is stopped when the convergence is reached. The entire process is called recursive-iterative. It must be noted here that the values of μ in eqs (10.10) and (10.12) need not be same. Similar observation applies to Ω.

10.2.2 Back propagation recursive least squares filtering algorithms

10.2.2.1 Algorithm with nonlinear output layer

During the forward pass training of the network, the signals y and u are computed for each layer as is done in the back propagation algorithm. The filter gains K_1 and K_2 are computed for both the layers and the forgetting factors f_1 and f_2 are chosen. The formulation is the usual scalar data processing scheme, as shown below.

For layer 1, the updates for filter gain K_1 and covariance matrix P_1 are given as [11]:

$$K_1 = P_1 u_0 (f_1 + u_0 P_1 u_0)^{-1} \tag{10.13}$$

$$P_1 = \frac{P_1 - K_1 u_0 P_1}{f_1} \tag{10.14}$$

For layer 2, the updates for filter gain K_2 and covariance matrix P_2 are given as

$$K_2 = P_2 u_1 (f_2 + u_1 P_2 u_1)^{-1} \tag{10.15}$$

$$P_2 = \frac{P_2 - K_2 u_1 P_2}{f_2} \tag{10.16}$$

The modified output error is given as

$$e_{2b} = f'(y_2)(z - u_2) \tag{10.17}$$

The back propagation of the output error to inner/hidden layer gives inner layer error as

$$e_{1b} = f'(y_1)W_2^T e_{2b} \tag{10.18}$$

And finally, the weight update rule for the output error is

$$W_2(i+1) = W_2(i) + (d - y_2)K_2^T \tag{10.19}$$

Here, d is given by

$$d_i = \frac{1}{\lambda} \ln\left[\frac{1+z_i}{1-z_i}\right]; \qquad z_i \neq 1 \tag{10.20}$$

For the hidden layer, the rule is

$$W_1(i+1) = W_1(i) + \mu e_{1b} K_1^T \tag{10.21}$$

Here, the additional computation of Kalman gains is needed, otherwise the procedure for training is similar to the back propagation algorithm. We note here that when the weight update rule of eq. (10.21) is used, the range of values of μ would not generally be the same as when the rule of eq. (10.12) is applied.

10.2.2.2 Algorithm with linear output layer

In this case, the output layer does not have nonlinearities. Only the inner layer has nonlinearities. The linear Kalman filter concept, therefore, is directly applicable in this case. Since the output layer block is linear, the output signal is computed as

$$u_2 = y_2 \tag{10.22}$$

The Kalman gain computations are as per the algorithm discussed in Section 10.2.2.1. Since the output layer has no nonlinearity, the error for the output layer is

$$e_{2b} = e_2 = (z - y_2) \tag{10.23}$$

The back propagation of the output error gives

$$e_{1b} = f(y_1)W_2^T e_{2b} \tag{10.24}$$

Finally, the weight update rules are

$$W_2(i+1) = W_2(i) + e_{2b} K_2^T \tag{10.25}$$

$$W_1(i+1) = W_1(i) + \mu e_{1b} K_1^T \tag{10.26}$$

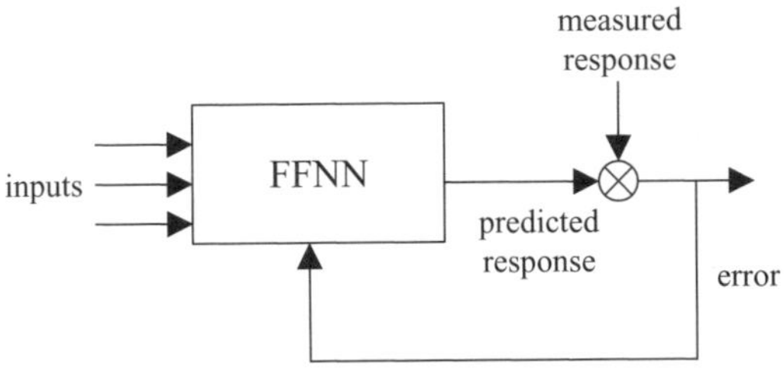

Figure 10.3 Parameter estimation with feed forward neural network

Once the data are scanned and the convergence achieved, the estimated weights of the last iteration are used as inputs and presented again to the network to predict the output. This output is compared with the desired/available output in order to judge the network's ability for prediction.

10.3 Parameter estimation using feed forward neural network

The very fact that the feed forward neural network (FFNN) provides for nonlinear mapping of the input-output data suggests that it should be possible to use it for system characterisation. We are aware how, based on *a priori* knowledge of the system and the underlying physics, mathematical models are developed and subjected to parameter estimation using conventional techniques like the equation error and output error methods. The feed forward neural network, however, works with a black-box model structure, which cannot be physically interpreted. The parameters of the network are the weights, which have no interpretation in terms of the actual system parameters.

The parameter estimation procedure using the feed forward neural network has two steps: i) the network is given the measured data and is trained to reproduce the clean/predicted responses which are compared with the system responses in the sense of minimisation of the output error (see Fig. 10.3); ii) these predicted responses are perturbed in turn for each parameter to be estimated and the changed predicted response is obtained. Assume that $z = \beta x$ and the network is trained to produce clean z. The trained network is used to produce $z + \Delta z$ and $z - \Delta z$ when x is changed to $x + \Delta x$ and $x - \Delta x$. Then β is obtained as $\beta = (z^+ - z^-)/(x^+ - x^-)$, and this method is called the Delta method. Since the variables are the signals, the parameter time histories are obtained and hence, the estimates are obtained by averaging these respective parameter time histories.

The above procedure is used for parameter estimation of Example 10.1.

10.3.1.1 Example 10.1

Generate the simulated data using the following equation:

$$z = a + bx_1 + cx_2 \tag{10.27}$$

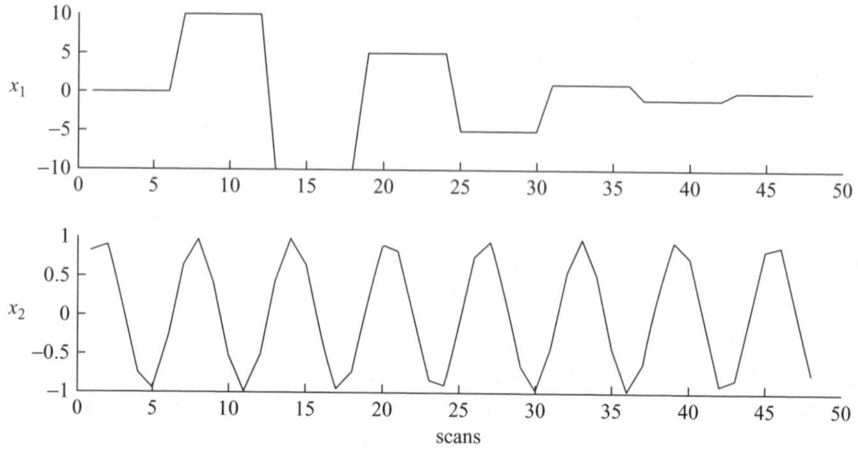

Figure 10.4 Time history of input signals (Example 10.1)

Here, parameters $a = 1$, $b = 2$, $c = 1$ and x_1, x_2 are the input to the model and z is the output of the model: i) train the neural network for the input variables x_1, x_2 and output variable z using the feed forward neural network with back propagation (FFNN-BPN); and ii) estimate a, b, and c using the Delta method with the help of a trained feed forward network for various levels of noise added to input signals.

10.3.1.2 Solution

The data generation is carried out using eq. (10.27) with constant value of parameters a, b, and c. The input signals x_1 and x_2 are shown in Fig. 10.4. The input signal x_2 is generated using the inbuilt MATLAB function 'sin(k)' with k varying from 1 to 48. The signal x_1 is generated as a periodic pulse with decreasing amplitude.

1 The simulated input and output signals are scaled and subsequently used to train FFNN using the back propagation algorithm. The training parameters were set to $\mu = 0.2, \Omega = 0.4$, in the feed forward neural network with the back propagation algorithm with four neurons in the hidden layer. The sigmoid slope parameters λ_1, λ_2 for hidden and output layers were taken as 0.8 and 0.75 respectively. The training was stopped after 10 000 iterations and the percentage fit error (PFE) of predicted data from the network w.r.t. true data was found to be 0.1. Figure 10.5 shows the time history match of predicted signal $\hat{z}$ to the true signal z. The training is done using file '**trainffnn.m**' residing in folder '**ch10FFNNex1**'.
2 After optimal training of the network, the Delta method is used for estimation of parameters a, b, and c. The estimated parameters and the parameter estimation error norm are given in Table 10.2. We see that with increase in noise, the parameter estimation error norm increases, but still the estimates are just

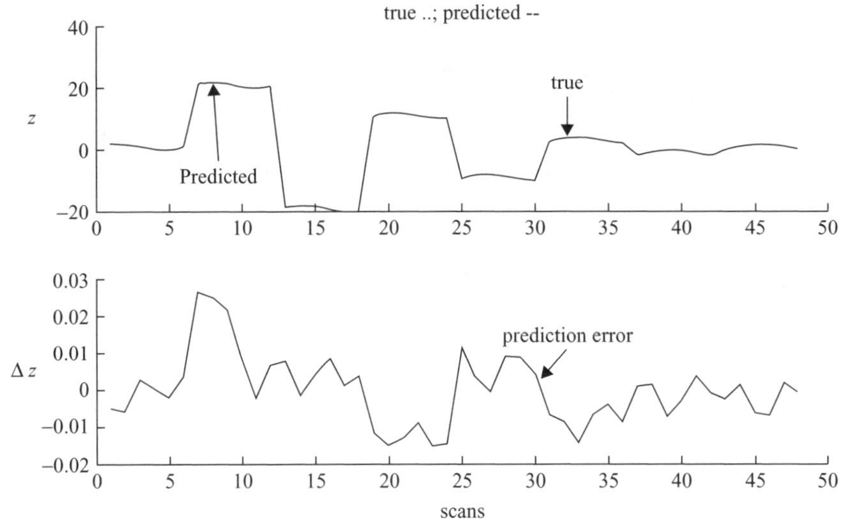

Figure 10.5 *FFNN-BPN algorithm time history match, prediction phase (Example 10.1)*

Table 10.2 *Parameter estimation with FFNN-BPN (Example 10.1)*

Parameters	True values	Estimated values using Delta method for different noise levels		
	–	SNR = ∞	SNR = 100	SNR = 10
a	1	0.9989	1.0272	1.1188
b	2	1.999	1.957	1.928
c	1	1.0004	0.9862	0.9441
PEEN	–	0.048	2.15	6.105

acceptable. The estimation is accomplished by using file ‘**peffnndm.m**’ placed in folder ‘**Ch10FFNNex1**’.

A Delta method that uses the generalisation properties of the feed forward neural network to estimate model parameters has been suggested [9]. The method, when applied to aircraft flight test data, was shown to yield the aircraft stability and control derivatives. The method makes use of the basic definition of derivative which states that a derivative represents the change in the aerodynamic force or moment caused by a small change in the motion or control variable about its nominal position. For example, the derivative C_{m_α} can be defined as the change in the aircraft pitching moment C_m due to a small change in the angle-of-attack α with all other motion and control variables held constant. To estimate aircraft stability and control derivatives, the input layer of the network contains the motion and control variables, such as angle-of-attack, sideslip angle, rates and control inputs. The output layer comprises the aerodynamic force and moment coefficients. In the following examples, the application of the Delta method and feed forward neural network to estimate aircraft derivatives from simulated flight test data is demonstrated for better understanding.

10.3.1.3 Example 10.2

Generate the simulated data using the following state and aerodynamic models of the aircraft dynamics (see Appendix B):

$$\begin{aligned}
\dot{V} &= -\frac{\bar{q}S}{m}C_D - g\sin(\theta - \alpha) \\
\dot{\alpha} &= -\frac{\bar{q}S}{mV}C_L + \frac{g}{V}\cos(\theta - \alpha) + q \\
\dot{q} &= \frac{\bar{q}S\bar{c}}{I_y}C_m \\
\dot{\theta} &= q
\end{aligned} \tag{10.28}$$

The aerodynamic model is

$$\begin{aligned}
C_D &= C_{D0} + C_{D\alpha}\alpha + C_{D\delta_e}\delta_e \\
C_L &= C_{L0} + C_{L\alpha}\alpha + C_{L\delta_e}\delta_e \\
C_m &= C_{m_0} + C_{m\alpha}\alpha + C_{m_q}\frac{q\bar{c}}{2V} + C_{m\delta_e}\delta_e
\end{aligned} \tag{10.29}$$

For a given set of parameter values (true values) do the following:

(i) Generate the time histories of variables $\dot{V}, \dot{\alpha}, \dot{q}, \dot{\theta}, V, \alpha, q, \theta, \delta_e$ and coefficients C_D, C_L, and C_m with sinusoidal input data.

(ii) Train the feed forward network for the variables $\dot{\alpha}, \dot{q}$ using
- Feed Forward Neural Network with Back Propagation (FFNN-BPN) and
- Feed Forward Neural Network with Back Propagation Recursive Least Square Filter algorithm with Linear output layer (FFNN-BPNRLSFL).

(iii) Train the feed forward network for the aerodynamic coefficients $C_D, C_L$and C_m using
- Feed Forward Neural Network with Back Propagation Recursive Least Squares Filter Algorithm with Nonlinear output layer (FFNN-BPNRLSFNL).

(iv) Use the Delta method to estimate the aerodynamic derivatives appearing in eq. (10.29), using the predicted time histories of the aerodynamic coefficients obtained by training the neural network for each of the aerodynamic coefficients individually and with different noise levels added to the variables V, α, q, θ.

10.3.1.4 Solution

(i) Time histories of variables $\dot{V}, \dot{\alpha}, \dot{q}, \dot{\theta}, V, \alpha, q, \theta, \delta_e$ and coefficients C_D, C_L and C_m are generated using eqs (10.28) and (10.29) with sinusoidal input $\delta_e = A\sin(\theta)$; $A = 1, \theta = 0 : \pi/8 : n\pi$ and $n = 25$. For the simulation, true values of aerodynamic coefficients are given in Table 10.5. The other parameters related to simulated aircraft are $\bar{c} = 10\,\text{m}$, $S = 23.0\,\text{m}^2$, $m = 7000\,\text{kg}$, $I_y = 50\,000\,\text{kg/m}^2$, $V = 100\,\text{m/s}$, $\bar{q} = 5000\,\text{kg/ms}^2$, and $g = 9.81\,\text{m/s}^2$. The initial values of α, q, and θ were taken as 0.1 rad, 0.0 rad/s, and 0.1 rad respectively. A total number of 200 data samples are simulated for analysis. The programs for data simulation, training, prediction and parameter estimation are contained in folder **Ch10FFNNex2**.

(ii) The following model is used for the purpose of training feed forward neural networks:

$$\dot{\alpha} = h_1(V, \alpha, q, \delta_e)$$

$$\dot{q} = h_2(V, \alpha, q, \delta_e)$$

Here h is a nonlinear functional relationship. The signals V, α, q, and δ_e are presented to the network as inputs and signals $\dot{\alpha}$ and $\dot{q}$ as outputs. The network was trained using both FFNN-BPN and FFNN-BPNRLSFL algorithms. The tuning parameters used for training the algorithms for $\dot{\alpha}$ and $\dot{q}$ signals are given in Table 10.3. Figures 10.6 and 10.7 show the time history match for prediction phase using FFNN-BPN and FFNN-BPNRLSL algorithms respectively, and we see that the latter gives somewhat better results.

(iii) Next, the FFNN-BPNRLSNL algorithm was used for prediction of the aerodynamic coefficients (time histories) C_D, C_L, and C_m as function of α, q, V and δ_e. The coefficient time histories are used as the outputs and α, q, V, δ_e as inputs to the network. The tuning parameters used for training are given in

Table 10.3 Tuning parameters used for feed forward neural network training for steps (ii) and (iii)

Tuning parameter	$\dot{\alpha}, \dot{q}$		C_D, C_L, C_m
	BPN	BPNRLSFL	BPNRLSFNL
Function slope of hidden layer λ_1	0.8	0.8	0.8
Function slope of hidden layer λ_2	0.75	0.75	0.75
Number of hidden layers	1	1	1
Number of nodes in the hidden layer	6	6	6
Data scaling range	±0.1	±0.1	±0.1
Learning rate parameter μ	0.2	0.2	0.2
Momentum parameter Ω	0.4	NA	NA
Training iterations	10 000	10 000	2000

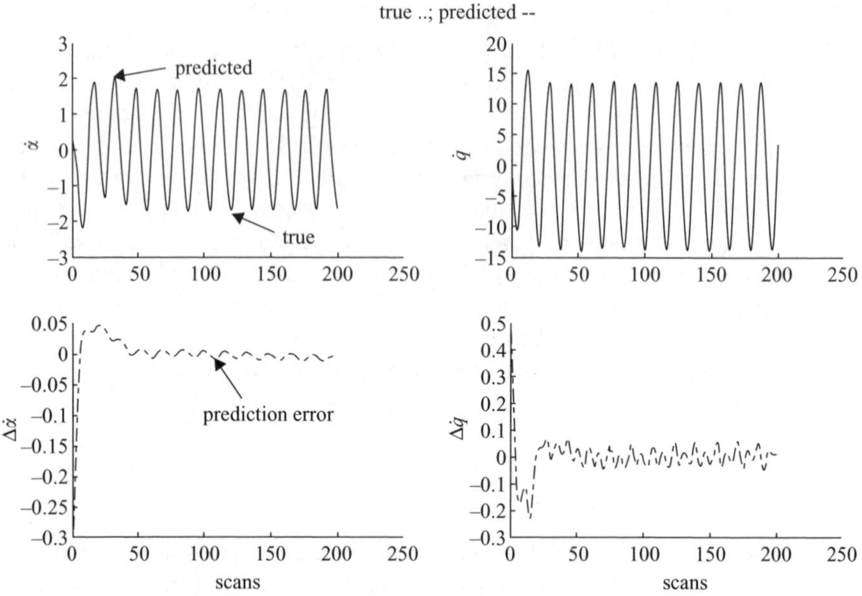

Figure 10.6 *Time history match and prediction error for $\dot{\alpha}$ and $\dot{q}$ using FFNN-BPN (Example 10.2)*

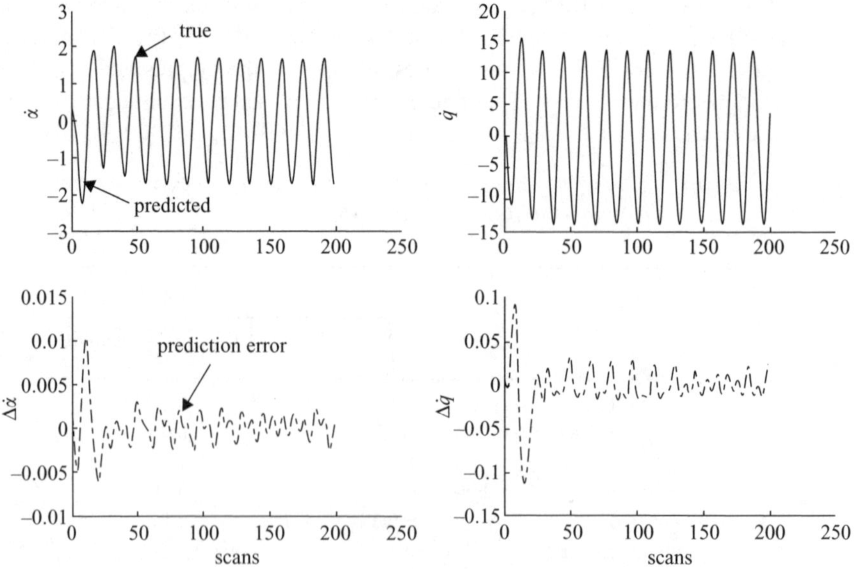

Figure 10.7 *Time history match and prediction error for $\dot{\alpha}$ and $\dot{q}$ using FFNN-BPNRLSFL (Example 10.2)*

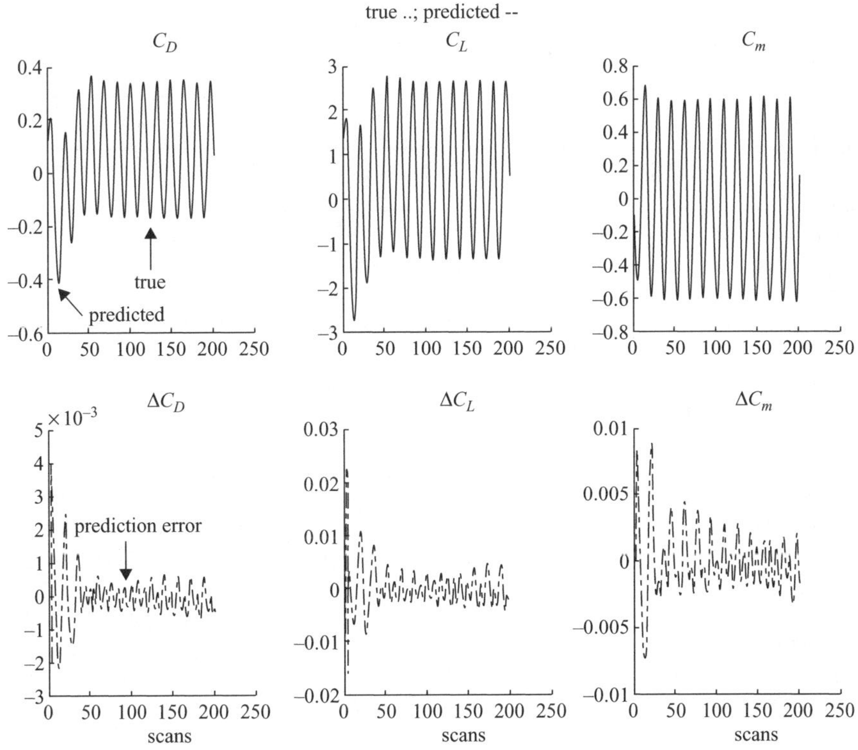

Figure 10.8 *Time history match and prediction error for C_D, C_L and C_m using FFNN-BPNRLSFNL (Example 10.2)*

Table 10.3 and the time history match for the coefficients C_D, C_L, and C_m are shown in Fig. 10.8.

(iv) FFNN-BPN and FFNN-BPNRLSFNL algorithms are used to train the network and predict the coefficients' time histories for C_D, C_L and C_m one at a time. The tuning parameters used for training the network are listed in Table 10.4. Once the feed forward network maps the input variables to output variables correctly, the Delta method is used for estimation of derivatives $C_{D0}, C_{D\alpha}, C_{D\delta_e}, C_{L0}, C_{L\alpha}, C_{L\delta_e}, C_{m0}, C_{m_\alpha}, C_{m_q}$ and $C_{m\delta_e}$. Having trained the network, any one variable in the input layer can be perturbed to cause a corresponding change in the output response. For example, with the weights in the network frozen after training, changing the value of α to $\alpha + \Delta\alpha$ at all points (other input variables remain unchanged) yields values of C_D^+ that are slightly different from C_D. Likewise, changing α to $\alpha - \Delta\alpha$ will yield the response C_D^-. Then $C_{D\alpha}$ derivative is given by $C_{D\alpha} = (C_D^+ - C_D^-)/2\Delta\alpha$. Following this procedure, the other derivatives can be determined. It is to be noted that the network produces as many estimates of the derivatives as the number of data points used to train the network. The final value of the

Table 10.4 Tuning parameters used for feed forward neural network training for step (iv)

Tuning parameter	C_D		C_L		C_m	
	BPN	BPNRLSFNL	BPN	BPNRLSFNL	BPN	BPNRLSFNL
Function slope of hidden layer λ_1	0.9	0.9	0.9	0.9	0.8	0.9
Function slope of hidden layer λ_2	0.85	0.85	0.85	0.85	0.75	0.85
Number of hidden layers	1	1	1	1	1	1
Number of nodes in the hidden layer	6	6	6	6	6	6
Data scaling range	±0.1	±0.1	±0.1	±0.1	±0.2	±0.1
Learning rate parameter μ	0.2	0.2	0.2	0.2	0.2	0.2
Momentum parameter Ω	0.4	NA	0.4	NA	0.2	NA
Training iterations	10 000	2000	10 000	2000	50 000	5000

derivative is obtained by taking the mean of these values for the corresponding derivative. After computing $C_{D\alpha}$,$C_{D\delta e}$ at all points, an estimate of C_{D0} can be obtained as: $C_{D0} = C_D - [C_{D\alpha}\alpha + C_{D\delta_e}\delta_e]$. The results of estimation are given in Table 10.5.

We see that the back propagation recursive least squares filter algorithm with nonlinear output layer gives somewhat better results compared to the back propagation with the steepest descent method in certain cases as can be seen from Table 10.5. Some improvement is surely possible.

10.3.1.5 Example 10.3

Consider the aircraft aerodynamic model:

$$\begin{aligned}
C_x &= C_{x0} + C_{x\alpha}\alpha + C_{x\alpha^2}\alpha^2 \\
C_z &= C_{z0} + C_{z\alpha}\alpha + C_{zq}q + C_{z\delta}\delta \\
C_m &= C_{m0} + C_{m\alpha}\alpha + C_{mq}q + C_{m\delta}\delta
\end{aligned} \tag{10.30}$$

Table 10.5 Parameter estimation with feed forward neural network (Example 10.2)

Parameters	True values	Estimated values using Delta method for different noise levels					
		SNR = ∞		SNR = 100		SNR = 10	
		BPN	BPNR-LSFNL	BPN	BPNR-LSFNL	BPN	BPNR-LSFNL
C_{D0}	0.046	0.0480	0.0465	0.0487	0.0472	0.0552	0.0534
$C_{D\alpha}$	0.543	0.5406	0.5467	0.5392	0.5456	0.5069	0.5121
$C_{D\delta_e}$	0.138	0.1383	0.1368	0.1284	0.1270	0.1160	0.1149
PEEN	–	**0.565**	**0.696**	**1.893**	**2.024**	**7.688**	**6.981**
C_{L0}	0.403	0.4177	0.4030	0.4279	0.4138	0.5002	0.4859
$C_{L\alpha}$	3.057	3.0475	3.0540	3.0708	3.0779	2.9733	2.9824
$C_{L\delta_e}$	1.354	1.3542	1.3530	1.2703	1.2690	1.1818	1.1804
PEEN	–	**0.520**	**0.094**	**2.625**	**2.619**	**6.375**	**6.127**
C_{m0}	0.010	−0.0175	−0.0383	−0.0170	−0.0377	−0.0132	−0.0321
$C_{m\alpha}$	−0.119	−0.1160	−0.1219	−0.1170	−0.1226	−0.1219	−0.1272
C_{mq}	−1.650	−1.6385	−1.6298	−1.6560	−1.6454	−1.6191	−1.6065
$C_{m\delta_e}$	−0.571	−0.5696	−0.5664	−0.5274	−0.5238	−0.5162	−0.5118
PEEN	–	**1.715**	**3.007**	**2.956**	**3.852**	**3.837**	**4.859**

For a given set of (true) parameter values, simulation is carried out to generate time histories consisting of 250 data samples for the variables α, α^2, q, δ and coefficients C_x, C_z and C_m. Using the feed forward neural network in conjunction with the Delta method, estimate the model parameters $C_{x0}, C_{x\alpha}, C_{x\alpha^2}, \ldots, C_{m\delta}$ appearing in eq. (10.30). Apply the regression method discussed in Chapter 2 to the simulated data and determine the parameter values. Compare the parameters values estimated using Delta and regression methods with true values.

10.3.1.6 Solution

The input layer of the feed forward neural network consists of the variables α, α^2, q and δ, and the output layer consists of the measured values of the non-dimensional force and moment coefficients C_x, C_z and C_m. The (FFNN-BPN) network can be trained using one of the two options: i) considering all the three measurements in the output layer; or ii) considering only one coefficient at a time in the output layer. In the present example, we adopt the second approach of training the network to predict only one coefficient at a time. Following this procedure gives the user more freedom to come up with a suitable set of tuning parameters that can lead to better prediction of C_x, C_z and C_m. Once the network maps the input variables to output variables, the Delta method is used to estimate derivatives $C_{x\alpha}, C_{x\alpha^2}, C_{z\alpha}, C_{zq}, C_{z\delta}, C_{m\alpha}, C_{mq}$

Table 10.6 Tuning parameters used for feed forward neural network training (Example 10.3)

Tuning parameter	Values of tuning parameters selected in FFNN to predict		
	C_x	C_z	C_m
Nonlinear function slope of hidden layer λ_1	0.8	0.8	0.8
Nonlinear function slope of hidden layer λ_2	0.75	0.75	0.75
Number of hidden layers	1	1	1
Number of nodes in the hidden layer	6	6	6
Data scaling range	−0.2 to 0.2	−0.1 to 0.1	−0.15 to 0.15
Learning ratc parameter μ	0.2	0.2	0.2
Momentum parameter Ω	0.4	0.4	0.4

and $C_{m\delta}$. After computing $C_{m\alpha}$, C_{mq} and $C_{m\delta}$ at all points, an estimate of C_{m0} can be obtained as

$$C_{m0} = C_m - [C_{m\alpha}\alpha + C_{mq}q + C_{m\delta}\delta]$$

The values of the tuning parameters used for network training are listed in Table 10.6. As seen from Table 10.6, the data scaling range selected for each of the coefficients for the feed forward neural network training is different. For this example, it is observed that the choice of the different scaling range for C_x, C_z and C_m leads to improved prediction of measured coefficients. The results of parameter estimation are provided in Table 10.7. Estimates obtained from applying the regression error method to the simulated data are also listed for comparison.

It is concluded that if one can tune the feed forward neural network to yield good prediction of training data, one can expect to achieve satisfactory values of the parameter estimates using the Delta method. The training and estimation are accomplished by using file '**trainffnn.m**' placed in folder '**Ch10FFNNex3**'.

We see from Table 10.7 that the Delta method gives estimates slightly different from the true values compared to the regression method. It is surprising that despite very low values of percentage fit error, the parameter estimation error norms are a bit high. We see that the feed forward neural network based parameter estimation approach offers an alternative method and could be made more robust and accurate by choosing the training parameters automatically and optimally. This requires further research.

Table 10.7 Parameter estimation with feed forward neural network BPN (Example 10.3)

Derivatives	True value+	Estimated values using		Comments
		Delta method	Regression	
C_{x0}	−0.054	−0.058	−0.0539	Fit error (PFE) after 10 000 iterations was 0.53%. Thereafter, change in PFE was < 0.011%
$C_{x\alpha}$	0.233	0.279	0.2318	
$C_{x\alpha^2}$	3.609	3.532	3.6129	
PEEN	–	2.475	0.11	
C_{z0}	−0.12	−0.121	−0.1188	Fit error (PFE) after 10 000 iterations was 0.11%. Thereafter, change in PFE was < 2.72e−6%
$C_{z\alpha}$	−5.68	−5.679	−5.6799	
C_{zq}	−4.32	−4.406	−4.1452	
$C_{z\delta}$	−0.407	−0.407	−0.3961	
PEEN	–	1.20	2.449	
C_{m0}	0.055	0.056	0.055	Training was stopped at 10 000 iterations and the PFE achieved was 0.95%. Subsequent change in PFE was of the order 0.001%
$C_{m\alpha}$	−0.729	−0.733	−0.729	
C_{mq}	−16.3	−16.61	−16.3	
$C_{m\delta}$	−1.94	−1.956	−1.94	
PEEN	–	1.887	0.00	

+ parameter values used to generate simulated data

10.4 Recurrent neural networks

Modelling of a system using artificial neural networks has recently become popular with application to signal processing, pattern recognition, system identification and control. Estimation of parameters using empirical data plays a crucial role in modelling and identification of dynamic systems. Often equation error and output error methods are used for parameter estimation of dynamic systems. These are generally batch iterative procedures where a set of data is processed to compute the gradient of a cost function and estimation error. The estimation of parameters is then refined using an iterative procedure based on the improved estimates of error and its gradients. Such methods can be termed as batch iterative. The artificial neural networks provide new/alternative paradigms to handle the problem of parameter estimation with potential application to on-line estimation. Especially recurrent neural networks are easily amenable to such possibilities due to their special structure-feed forward neural networks with feedback feature (see Fig. 10.9) [12–14]. In order to obtain fast solutions, a system of parallel computers can be used. This will require the parallelisation of the conventional parameter estimation algorithms. Since artificial neural networks have massively parallel processing capacity, they can be easily adapted to parameter estimation problems for on-line applications. In particular, the recurrent neural networks can be considered as more suitable for the problem of parameter estimation of linear dynamical systems, as compared with perhaps feed forward neural networks. The recurrent neural networks are dynamic neural networks, and hence amenable to explicit parameter estimation in state-space models.

10.4.1 Variants of recurrent neural networks

In this section, four variants of recurrent neural networks are studied from the point of view of explicit parameter estimation. In the literature, several variants of the basic Hopfield neural network structure are available. The three variants are related to each other by affine or linear transformation of their states. The variants are classified by the way in which the sigmoid nonlinearity operates: either on states, weighted states, residual of the network signal or forcing input [15].

10.4.1.1 RNN-S (HNN)

This network is known as the Hopfield neural network (HNN). The Hopfield neural network model has a number of mutually interconnected information processing units called neurons. In this configuration, the outputs of the network are nonlinear functions of the states of the network (and hence the 'S'). The dynamic representation of the network is given as (see Fig. 10.10)

$$\dot{x}_i(t) = -x_i(t)R^{-1} + \sum_{j=1}^{n} w_{ij}\beta_j(t) + b_i; \qquad j = 1, \ldots, n \tag{10.31}$$

Here, x is the internal state of the neurons, β the output state, $\beta_j(t) = f(x_j(t))$, w_{ij} are the neuron weights, b the bias input to the neurons and f the sigmoid nonlinearity. R is the neuron impedance and n is the dimension of the neuron state. The above

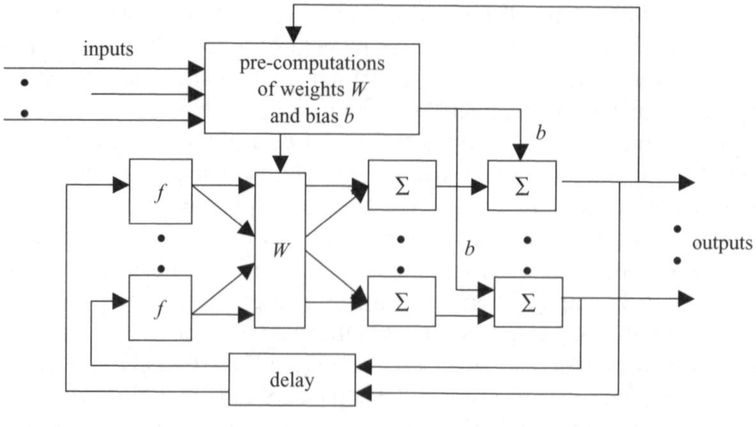

Figure 10.9 Typical block schematic of a recurrent neural network [13]

equation can also be written as

$$\dot{x}(t) = -x(t)R^{-1} + W\{f(x(t))\} + b \tag{10.32}$$

Equation (10.32) can be considered as a representation of 'classical' neuro-dynamics [16]. In comparison to biological neurons, the equation obtains a simple system retaining essential features: neuron as a transducer of input to output and a smooth sigmoidal response up to a maximum level of output, feedback nature of connections. Thus, the model retains two aspects: dynamics and nonlinearity.

10.4.1.2 RNN-FI

In this configuration of the recurrent neural networks, the nonlinearity operates on the forcing input: FI = weighted states + input to the networks → modified input = $f(Wx + b)$. The dynamics of this network can be given as (see Fig. 10.11)

$$\dot{x}_i(t) = -x_i(t)R^{-1} + f\left(\sum_{j=1}^{n} w_{ij}x_j(t) + b_i\right) \tag{10.33}$$

Here, $f(\cdot) = f(FI)$.

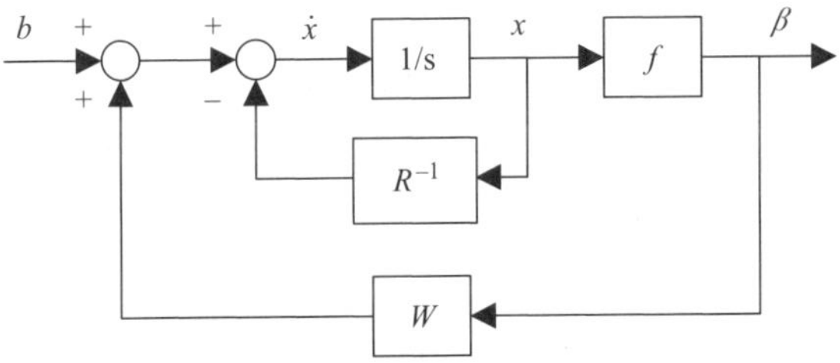

Figure 10.10 Schematic of RNN-S structure

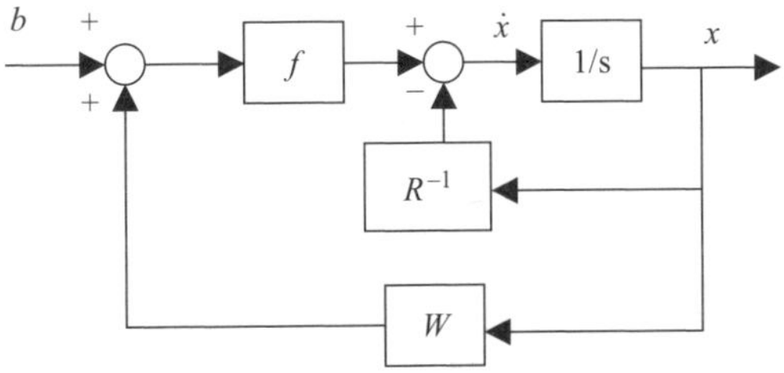

Figure 10.11 Schematic of RNN-FI structure

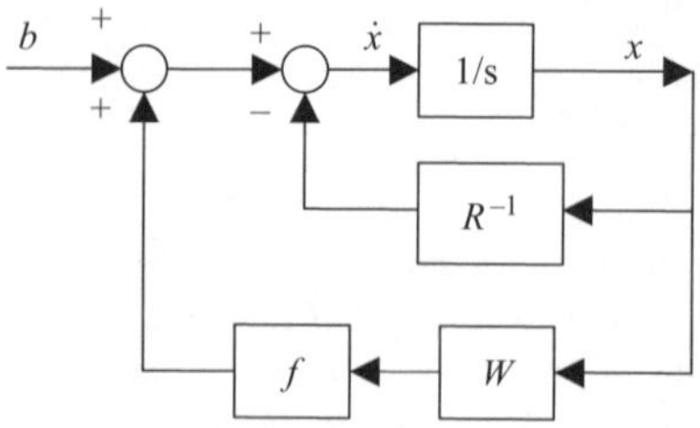

Figure 10.12 Schematic of RNN-WS structure

This network is related to the RNN-S by affine transformation. Use $x_H(t) = Wx + bR$ in eq. (10.32) to obtain the following equivalence:

$$\begin{aligned} W\dot{x} &= -(Wx + bR)R^{-1} + Wf(Wx + bR) + b \\ W\dot{x} &= -WxR^{-1} - b + Wf(Wx + bR) + b \\ \dot{x} &= -xR^{-1} + f(Wx + bR) \\ \dot{x} &= -xR^{-1} + f(FI) \end{aligned} \tag{10.34}$$

Here, *FI* is the modified input vector, due to the *bR* term. The invertibility of W is a necessary condition. We see that the above equation has exactly the same form as that of RNN-FI.

10.4.1.3 RNN-WS

In this configuration, the nonlinearity operates on the weighted states, hence WS. The dynamics of this neural network are described as (see Fig. 10.12)

$$\dot{x}_i(t) = -x_i(t)R^{-1} + f(s_i) + b_i \tag{10.35}$$

Here, $s_i = \sum_{j=1}^{n} w_{ij}x_j$.

It can be seen that the network is related to RNN-S by linear transformation. Substitute $x_H(t) = Wx$ in eq. (10.32) to obtain

$$\begin{aligned} W\dot{x} &= -(Wx)R^{-1} + Wf(Wx) + b \\ \dot{x} &= -xR^{-1} + f(s) + W^{-1}b \end{aligned} \tag{10.36}$$

Here, we have a modified input vector. The matrix W must be invertible.

10.4.1.4 RNN-E

In this type of configuration, the nonlinearity directly operates on the residual error or equation error. Hence, the function f or its derivative f' does not enter into the neuron dynamic equation. Yet, it does affect the residual by way of quantising them and thereby reducing the effect of measurement outliers. The dynamics are given by

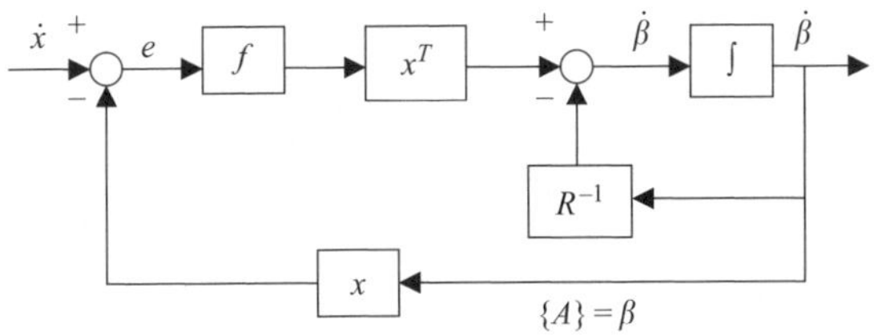

Figure 10.13 Schematic of RNN-E structure

(see Fig. 10.13)

$$\dot{x}_i(t) = -x_i(t)R^{-1} + \sum_{j=1}^{n} w_{ij}x_j(t) + b_i \tag{10.37}$$

In the case of RNN-E, we say that the internal state x_i is β_i, the parameters of the general dynamic system. In that case, the x_i of eq. (10.37) does not represent the state of this general dynamic system (see eq. (10.38)).

10.4.2 Parameter estimation with Hopfield neural networks

Consider the dynamic system

$$\dot{x} = Ax + Bu; \quad x(0) = x_0 \tag{10.38}$$

For parameter estimation using Hopfield neural networks, $\beta = \{A, B\}$ represents the parameter vector to be estimated and n is the number of parameters to be estimated. Based on the theory of Hopfield neural networks, a suitable functional can be associated with it, which iterates to a stable parameter estimation solution.

In this network, the neurons change their states x_i according to eq. (10.32). We can consider that the dynamics are affected by the nonlinear function f, i.e., $\beta_i = f(x_i)$.

Let the cost function be given as

$$E(\beta) = \frac{1}{2}\sum_{k=1}^{N} e^T(k)e(k) = \frac{1}{2}\sum_{k=1}^{N} (\dot{x} - Ax - Bu)^T(\dot{x} - Ax - Bu) \tag{10.39}$$

Here $e(k)$ is the equation error

$$e = \dot{x} - Ax - Bu \tag{10.40}$$

From optimisation theory we have:

$$\frac{d\beta}{dt} = -\frac{\partial E(\beta)}{\partial \beta} = -\frac{1}{2}\frac{\partial\{\sum_{k=1}^{N} e^T(k)e(k)\}}{\partial \beta} \tag{10.41}$$

Since β as a parameter vector contains the elements of A and B, we can obtain expressions $\partial E/\partial A$ and $\partial E/\partial B$ for A and B vectors, with $\sum(\cdot) = \sum_{k=1}^{N}(\cdot)$.

$$\frac{\partial E}{\partial A} = \sum(\dot{x} - Ax - Bu)(-x^T) = A\sum xx^T + B\sum ux^T - \sum \dot{x}x^T$$
$$\frac{\partial E}{\partial B} = \sum(\dot{x} - Ax - Bu)(-u) = A\sum xu + B\sum u^2 - \sum \dot{x}u \tag{10.42}$$

Expanding we get, for $A(2,2)$ and $B(2,1)$:

$$\begin{bmatrix} \dfrac{\partial E}{\partial a_{11}} & \dfrac{\partial E}{\partial a_{12}} \\ \dfrac{\partial E}{\partial a_{21}} & \dfrac{\partial E}{\partial a_{22}} \end{bmatrix} = \begin{bmatrix} a_{11} & a_{12} \\ a_{21} & a_{22} \end{bmatrix} \begin{bmatrix} \sum x_1^2 & \sum x_1x_2 \\ \sum x_2x_1 & \sum x_2^2 \end{bmatrix} + \begin{bmatrix} b_1 \\ b_2 \end{bmatrix} \begin{bmatrix} \sum ux_1 & \sum ux_2 \end{bmatrix} - \begin{bmatrix} \sum \dot{x}_1x_1 & \sum \dot{x}_1x_2 \\ \sum \dot{x}_2x_1 & \sum \dot{x}_2x_2 \end{bmatrix} \tag{10.43}$$

Simplifying, we get:

$$\begin{aligned}
\frac{\partial E}{\partial a_{11}} &= a_{11}\sum x_1^2 + a_{12}\sum x_2x_1 + b_1\sum x_1u - \sum \dot{x}_1x_1 \\
\frac{\partial E}{\partial a_{12}} &= a_{11}\sum x_1x_2 + a_{12}\sum x_2^2 + b_1\sum ux_2 - \sum \dot{x}_1x_2 \\
\frac{\partial E}{\partial a_{21}} &= a_{21}\sum x_1^2 + a_{22}\sum x_2x_1 + b_2\sum ux_1 - \sum \dot{x}_2x_1 \\
\frac{\partial E}{\partial a_{22}} &= a_{21}\sum x_1x_2 + a_{22}\sum x_2^2 + b_2\sum ux_2 - \sum \dot{x}_2x_2
\end{aligned} \tag{10.44}$$

In addition we have

$$\begin{aligned}
\frac{\partial E}{\partial b_1} &= a_{11}\sum x_1u + a_{12}\sum x_2u + b_1\sum u^2 - \sum \dot{x}_1u \\
\frac{\partial E}{\partial b_2} &= a_{21}\sum x_1u + a_{22}\sum x_2u + b_2\sum u^2 - \sum \dot{x}_2u
\end{aligned} \tag{10.45}$$

Next, assuming that the impedance R is very high, we describe the dynamics of RNN-S as

$$\dot{x}_i = \sum_{j=1}^{n} w_{ij}\beta_j + b_i \tag{10.46}$$

We also have $E = -(1/2)\sum_i \sum_j W_{ij}\beta_i\beta_j - \sum_i b_i\beta_i$ as the energy landscape of the recurrent neural network. Then, we get

$$\frac{\partial E}{\partial \beta_i} = -\sum_{j=1}^{n} w_{ij}\beta_j - b_i \tag{10.47}$$

or

$$\frac{\partial E}{\partial \beta_i} = -\left[\sum_{j=1}^{n} w_{ij}\beta_j + b_i\right] = -\dot{x}_i \tag{10.48}$$

or

$$\dot{x}_i = -\frac{\partial E}{\partial \beta_i}$$

Since

$$\beta_i = f(x_i), \quad \dot{x}_i = (f^{-1})'\dot{\beta}_i \tag{10.49}$$

Thus

$$(f^{-1})'\dot{\beta}_i = -\frac{\partial E}{\partial \beta_i}$$

Here $'$ denotes derivative w.r.t. β. Hence

$$\dot{\beta}_i = -\frac{1}{(f^{-1})'(\beta_i)}\frac{\partial E}{\partial \beta_i} = \frac{1}{(f^{-1})'(\beta_i)}\left[\sum_{j=1}^{n} w_{ij}\beta_j + b_i\right] \tag{10.50}$$

Now comparing expressions from eqs (10.44) and (10.45) to eq. (10.47), we get expressions for the weight matrix W and the bias vector b as:

$$W = -\begin{bmatrix} \sum x_1^2 & \sum x_2x_1 & 0 & 0 & \sum ux_1 & 0 \\ \sum x_1x_2 & \sum x_2^2 & 0 & 0 & \sum ux_2 & 0 \\ 0 & 0 & \sum x_1^2 & \sum x_2x_1 & 0 & \sum ux_1 \\ 0 & 0 & \sum x_1x_2 & \sum x_2^2 & 0 & \sum ux_2 \\ \sum x_1u & \sum x_2u & 0 & 0 & \sum u^2 & 0 \\ 0 & 0 & \sum x_1u & \sum x_2u & 0 & \sum u^2 \end{bmatrix} \tag{10.51}$$

$$b = -\begin{bmatrix} \sum \dot{x}_1x_1 \\ \sum \dot{x}_1x_2 \\ \sum \dot{x}_2x_1 \\ \sum \dot{x}_2x_2 \\ \sum \dot{x}_1u \\ \sum \dot{x}_2u \end{bmatrix} \tag{10.52}$$

Thus, the algorithm for parameter estimation of the dynamical system can be given as:

1 Compute W matrix, since the measurements of x, $\dot{x}$ and u are available (equation error formulation) for a certain time interval T, eq. (10.51).
2 Compute bias vector in a similar way from eq. (10.52).
3 Choose the initial values of β_i randomly.
4 Then solve the following differential equation.

Since $\beta_i = f(x_i)$ and since the sigmoid nonlinearity is a known function f, by differentiating and simplifying, we get

$$\frac{d\beta_i}{dt} = \frac{\lambda(\rho^2 - \beta_i^2)}{2\rho}\left[\sum_{j=1}^{n} w_{ij}\beta_j + b_i\right] \tag{10.53}$$

Here

$$f(x_i) = \rho\left(\frac{1 - e^{-\lambda x_i}}{1 + e^{-\lambda x_i}}\right) \tag{10.54}$$

Integration of eq. (10.53) yields the solution to the parameter estimation problem posed in the structure of the Hopfield neural network. For good convergence of the estimates to the true parameters, proper tuning of λ and ρ is essential. Often λ is chosen small, i.e., less than 1.0. The ρ is chosen such that when x_i (of recurrent neural network) approaches $\pm\infty$, the function f approaches $\pm\rho$. Equation (10.53) can be discretised to obtain the estimates by recursion. Also, it is possible to use the inverse of the weighting matrix W on the right hand side of eq. (10.53) to enhance the rate of convergence of the algorithm. The matrix W can be regarded as the information matrix for the parameter estimator defined by eq. (10.53). The foregoing scheme is termed as non-recursive, since the required computation of elements of W and b is performed by considering all the data. The discrete form of eq. (10.53) is given as

$$\beta_i(k+1) = \beta_i(k) + \frac{\lambda(\rho^2 - \beta_i^2(k))}{2\rho}\left[\sum_{j=1}^{n} w_{ij}\beta_j(k) + b_j\right] \tag{10.55}$$

The Δt can be absorbed in the constants of the 2nd term of eq. (10.55).

10.4.2.1 Example 10.4

Consider the second order system described by

$$\dot{x} = \begin{bmatrix} -0.7531 & 1 \\ -1.3760 & -1.1183 \end{bmatrix} x + \begin{bmatrix} 0 \\ -2.49 \end{bmatrix} u \tag{10.56}$$

1 obtain the response of the system to a doublet input; and
2 use x, $\dot{x}$, and u in the RNN-S algorithm to estimate all six parameters. Also comment on the accuracy of the results.

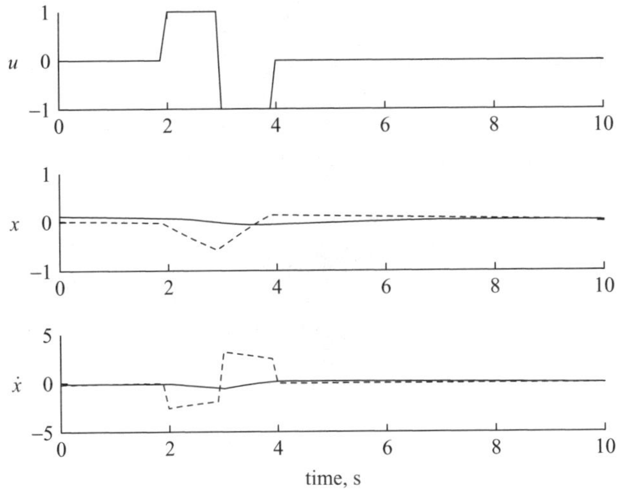

Figure 10.14 Doublet input and system states (Example 10.4)

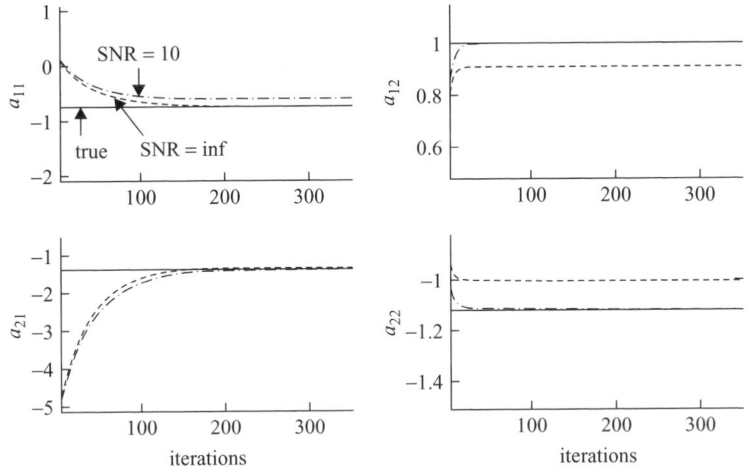

Figure 10.15 Estimated parameters for different SNR (Example 10.4)

The example is the same as in Reference 15, but the results are regenerated.

10.4.2.2 Solution

1 The 100 data samples are generated using a doublet input and initial state of the system $x(0) = [0.1 \quad 0.01]$. The input signal and system response are shown in Fig. 10.14.

Table 10.8 Parameter Estimation with RNN-S (Example 10.4)

Parameters	True values	Estimated values using RNN-S (HNN) method for different noise levels		
		SNR $= \infty$	SNR $= 100$	SNR $= 10$
a_{11}	−0.7531	−0.7531	−0.758	−0.707
a_{12}	1.0	1.0000	1.004	0.947
a_{21}	−1.376	−1.3760	−1.369	−1.276
a_{22}	−1.1183	−1.1183	−1.108	−1.017
b_{11}	0.0	−0.0000	−0.002	−0.011
b_{21}	−2.49	−2.4900	−2.485	−2.477
PEEN	–	0.0	0.451	4.840

2 The equation error formulation is used in RNN-S (Hopfield neural network) for parameter estimation. The estimation was carried out using noise free data and data with additive noise. The tuning parameters λ and ρ were kept at 0.1 and 100 respectively. It was noted that RNN-S took around 350 iterations before the convergence of estimated parameters to true values. Figure 10.15 shows the estimated parameters for noisy data with SNR $= 10$, and noise free data. It can be concluded from the figure that the convergence patterns for both cases are similar. Table 10.8 shows estimated parameters and PEENs for different SNRs. The system simulation and parameter estimation are accomplished by using file **'parestrnn1.m'** placed in folder **'Ch10RNNex4'**.

10.4.2.3 Example 10.5

Consider the second order unstable system described by

$$\dot{x} = \begin{bmatrix} -1.1 & 0.8 \\ 0.12 & -0.05 \end{bmatrix} x + \begin{bmatrix} -0.12 \\ 0.8 \end{bmatrix} u \qquad (10.57)$$

1 simulate the above system with doublet input using a sampling interval $\Delta t = 0.1$ s (number of data points $= 100$); and
2 use x, $\dot{x}$, and u in the RNN-S algorithm to estimate the parameters and comment on the accuracy of the results.

10.4.2.4 Solution

1 The above system is unstable (eigenvalues are $\lambda_1 = -1.18$ and $\lambda_1 = 0.03$) because one of the roots lies in right half of the s-plane. The system response is

obtained using doublet input with initial state of the system $x(0) = [0.5 \quad 0.002]$. The input signal and system response are shown in Fig. 10.16.

2 The equation error formulation is used in RNN-S (Hopfield neural network) for parameter estimation. The estimation was carried out using noise free data and data with additive noise. The tuning parameters λ and ρ were kept at 0.1 and 100 respectively. It was noted that RNN-S took around 350 iterations before the convergence of estimated parameters to true values. Figure 10.17 shows the

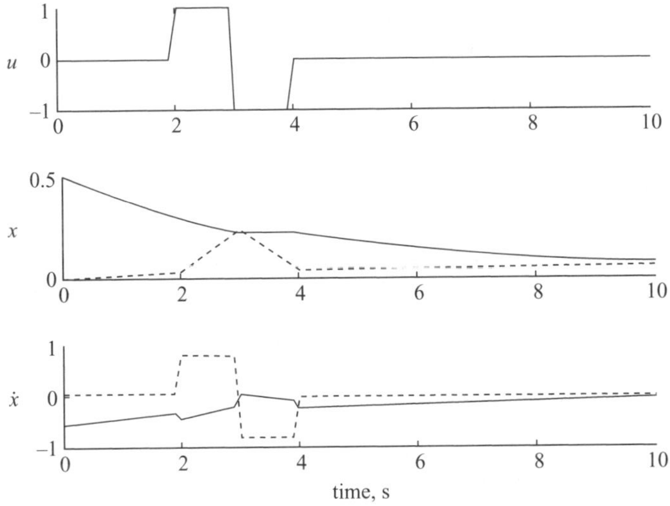

Figure 10.16 Doublet input and system states (Example 10.5)

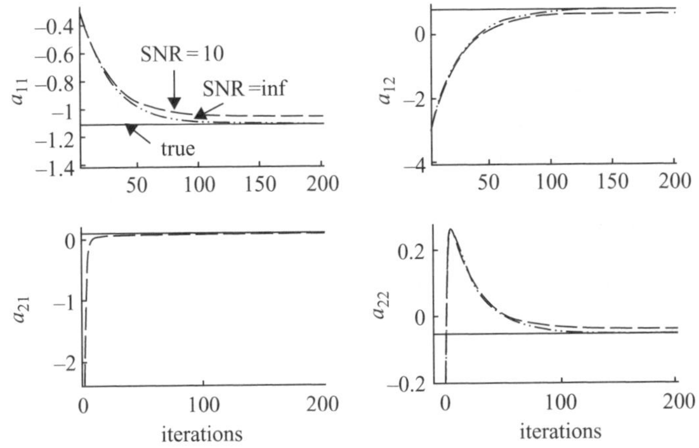

Figure 10.17 Estimated parameters for different SNR (Example 10.5)

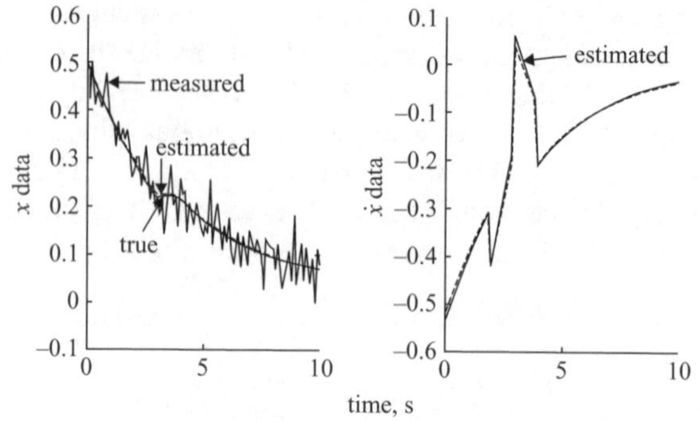

Figure 10.18 *True, measured and estimated system states for SNR = 10 (Example 10.5)*

Table 10.9 *Parameter estimation with RNN-S (Example 10.5)*

Parameters	True values	Estimated values using RNN-S (HNN) method for different noise levels		
		SNR = ∞	SNR = 100	SNR = 10
a_{11}	−1.1	−1.1	−1.10	−1.070
a_{12}	0.8	0.8	0.81	0.745
a_{21}	0.12	0.12	0.12	0.117
a_{22}	−0.05	−0.05	−0.05	−0.046
b_{11}	−0.12	−0.12	−0.12	−0.121
b_{21}	0.8	0.8	0.80	0.800
PEEN	–	0.0	0.710	4.067

estimated parameters for noisy data with SNR = 10, and noise free data. It can be concluded from the figure that the convergence patterns for both cases are similar. Figure 10.18 shows the true and estimated system state (x_1 and $\dot{x}_1$) for SNR = 10. Table 10.9 shows the estimated parameters and PEENs for different SNRs. The system simulation and parameter estimation are accomplished by using file '**parestrnn2.m**' placed in folder '**Ch10RNNex5**'.

Next, consider the following system [17]:

$$\begin{aligned}
\dot{x}_1 &= \beta_1 x_4 \\
\dot{x}_2 &= \beta_2 x_5 \\
\dot{x}_3 &= \beta_3 x_6 \\
\dot{x}_4 &= \beta_4 u \\
\dot{x}_5 &= \beta_5 u \\
\dot{x}_6 &= \beta_6 u
\end{aligned} \tag{10.58}$$

Here, $\beta_1, \beta_2, \beta_3, \beta_4, \beta_5$, and β_6 are the parameters to be estimated using HNN and u is the input to the system.

Cost function is defined as

$$J(\beta) = \frac{1}{2}\sum_{k=1}^{N} e^T(k)e(k) = \frac{1}{2}\sum_{k=1}^{N} (\dot{x} - f(x))^T(\dot{x} - f(x)) \tag{10.59}$$

Here, $\dot{x} = [\dot{x}_1 \quad \dot{x}_2 \quad \dot{x}_3 \quad \dot{x}_4 \quad \dot{x}_5 \quad \dot{x}_6]$,

$$f(x) = [\beta_1 x_4 \quad \beta_2 x_5 \quad \beta_3 x_6 \quad \beta_4 u \quad \beta_5 u \quad \beta_6 u]$$

For the optimal estimation, we have from eq. (10.59)

$$\frac{d\beta}{dt} = -\frac{\partial J(\beta)}{\partial \beta} = -\frac{1}{2}\frac{\partial\{\sum_{k=1}^{N} e^T(k)\, e(k)\}}{\partial \beta} = \sum_{k=1}^{N} \frac{\partial f(x)}{\partial \beta} \cdot e(k) \tag{10.60}$$

For simplification of expressions, let us assume $\sum(\cdot) = \sum_{k=1}^{N}(\cdot)$.

Now putting the value of $e(k)$ in eq. (10.60), we get

$$\frac{\partial J(\beta)}{\partial \beta} = -\sum \begin{bmatrix} \dot{x}_1 - \beta_1 x_4 \\ \dot{x}_2 - \beta_2 x_5 \\ \dot{x}_3 - \beta_3 x_6 \\ \dot{x}_4 - \beta_4 u \\ \dot{x}_5 - \beta_5 u \\ \dot{x}_6 - \beta_6 u \end{bmatrix}^T \begin{bmatrix} x_4 & 0 & 0 & 0 & 0 & 0 \\ 0 & x_5 & 0 & 0 & 0 & 0 \\ 0 & 0 & x_6 & 0 & 0 & 0 \\ 0 & 0 & 0 & u & 0 & 0 \\ 0 & 0 & 0 & 0 & u & 0 \\ 0 & 0 & 0 & 0 & 0 & u \end{bmatrix} \tag{10.61}$$

Dynamics of RNN-S are described as

$$\frac{\partial J(\beta)}{\partial \beta_i} = -\left[\sum_{j=1}^{n} w_{ij}\beta_j + b_i\right] \tag{10.62}$$

Here, n is the total number of parameters to be estimated. Now comparing the elements of eqs (10.61) and (10.62) we have:

Let us say $i = 1$, and then expanding eq. (10.62) we get

$$\frac{\partial J(\beta)}{\partial \beta_1} = -w_{11}\beta_1 - w_{12}\beta_2 - w_{13}\beta_3 - w_{14}\beta_4 - w_{15}\beta_5 - w_{16}\beta_6 - b_1 \tag{10.63}$$

Similarly by expanding eq. (10.61) for $i = 1$ we have

$$\frac{\partial J(\beta)}{\partial \beta_1} = -\sum \dot{x}_1 x_4 + \beta_1 \sum x_4^2 \tag{10.64}$$

By comparing expressions from eqs (10.63) and (10.64), we get expressions for 1st row elements of weight matrix W and bias vector b as

$$w_{11} = -\sum x_4^2, \quad w_{12} = w_{13} = w_{14} = w_{15} = w_{16} = 0$$

and

$$b_1 = \sum \dot{x}_1 x_4$$

One can get full expressions of W and b for $i = 2, \ldots, n$. After complete evaluation, we get W and b as

$$W = \begin{bmatrix} -\sum x_4^2 & 0 & 0 & 0 & 0 & 0 \\ 0 & -\sum x_5^2 & 0 & 0 & 0 & 0 \\ 0 & 0 & -\sum x_6^2 & 0 & 0 & 0 \\ 0 & 0 & 0 & -\sum u^2 & 0 & 0 \\ 0 & 0 & 0 & 0 & -\sum u^2 & 0 \\ 0 & 0 & 0 & 0 & 0 & -\sum u^2 \end{bmatrix}$$

$$b = \begin{bmatrix} \sum \dot{x}_1 x_4 \\ \sum \dot{x}_2 x_5 \\ \sum \dot{x}_3 x_6 \\ \sum \dot{x}_4 u \\ \sum \dot{x}_5 u \\ \sum \dot{x}_6 u \end{bmatrix}$$

These W and b can be used in eq. (10.50) and the parameters can be estimated.

10.4.2.5 Example 10.6

Consider the system below with all eigenvalues at the origin

$$\begin{aligned} \dot{x}_1 &= b_1 x_4; & \dot{x}_2 &= b_2 x_5 \\ \dot{x}_3 &= b_3 x_6; & \dot{x}_4 &= 0 \\ \dot{x}_5 &= 0; & \dot{x}_6 &= b_4 \end{aligned}$$

Here, true parameters are $b_1 = 1$, $b_2 = 1$, $b_3 = 1$, and $b_4 = -9.8$.

1 Simulate the above system with a unit step input signal and a sampling interval $\Delta t = 0.1$ s (number of data points $= 10$).
2 Use x, $\dot{x}$, and u in the RNN-S algorithm to estimate the parameters b_1, b_2, b_3, b_4.

10.4.2.6 Solution

The simulation of the system is carried out with the initial conditions as $x_1(0) = 10$ m, $x_2(0) = 3$ m, $x_3(0) = 0.1$ m, $x_4(0) = 0.5$ m/s, $x_5(0) = 0.1$ m/s, and $x_6(0) = 0.8$ m/s. The simulated data are generated for 1 s with 0.1 s sampling interval.

The parameter estimation was carried out using noise free data and data with additive noise. The tuning parameters λ and ρ were kept at 0.1 and 10 respectively. Figure 10.19 shows the true and estimated system state (x_1 and x_3) for SNR $= 10$. Table 10.10 shows the final value and PEEN of estimated parameters for different

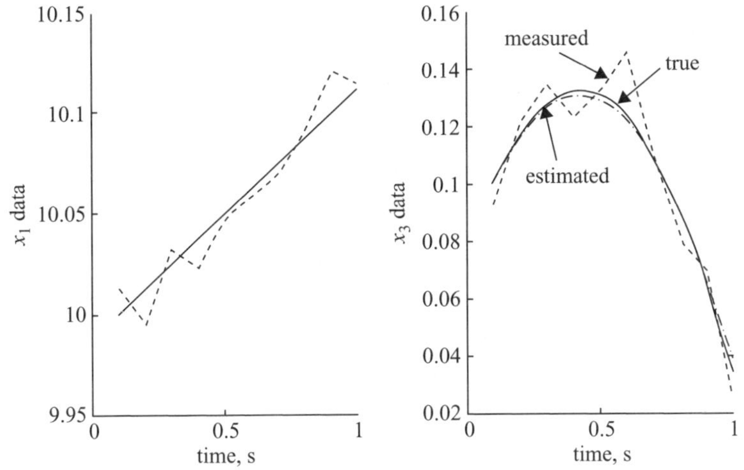

Figure 10.19 True, measured and estimated system states (Example 10.6)

Table 10.10 Parameter estimation with RNN-S (Example 10.6)

Parameters	True values	Estimated values using RNN-S (HNN) method for different noise levels		
		SNR = ∞	SNR = 10	SNR = 2
b_1	1	1.0000	1.0000	1.0000
b_2	1	1.0003	1.0003	1.0003
b_3	1	1.0000	0.9500	0.7272
b_4	−9.8	−9.799	−9.799	−9.799
PEEN	–	0.003	0.5	2.74

SNR levels. The system simulation and parameter estimation are accomplished by using file '**parestrnn3.m**' placed in folder '**Ch10RNNex6**'. Reasonably good estimation has been accomplished.

10.4.3 Relationship between various parameter estimation schemes

From Section 10.4.2, we have the following important relationships [13]:

(*a*)
$$\dot{\beta}_i = -\frac{\lambda}{(f^{-1})'(\beta_i)}\frac{\partial E}{\partial \beta_i} \tag{10.65}$$

(b) $$E(\beta) = \frac{1}{2}\sum_{k=1}^{N}(\dot{x} - Ax - Bu)^T(\dot{x} - Ax - Bu) \tag{10.66}$$

(c) $$\frac{\partial E}{\partial \beta_i} = -\left[\sum_{j=1}^{n} W_{ij}\beta_j + b_i\right] \tag{10.67}$$

From the above expressions, we have the following equivalence (assuming $B = 0$ in eq. (10.38)):

$$\frac{dx}{dt} = -\frac{\partial E}{\partial \beta_i} = \left[\sum_{j=1}^{n} w_{ij}\beta_j + b_i\right] = \sum_{k=1}^{N}[\dot{x}(k) - Ax(k)]x^T(k) \tag{10.68}$$

$$= \sum_{k=1}^{N}[-\{\beta\}x(k)x^T(k) + \dot{x}(k)x^T(k)] \tag{10.69}$$

Normally using the right hand side 3rd and 5th terms of the above, the explicit formulae for the matrix W and b have been derived in Section 10.4.2, since $\{\beta\}$ represents the elements of A. We note that for the discussion of this section only, the x of dx/dt in eq. (10.69) is not the same as $\dot{x}$, x, etc.

Alternatively, one can use equivalence of the 1st and 5th terms. With some initial parameters $\beta(0)$, integrate the following equation:

$$\frac{dx}{dt} = \sum_{k=1}^{N}[\dot{x}(k) - \{\beta(t)\}x(k)]x^T(k) \tag{10.70}$$

The complete information required for the evaluation of the right hand side is available for solving this equation. Then compute $\beta = f(x)$, since Hopfield neural network decision-making is nonlinear. Then use the new vector, β in eq. (10.69) for the next update. This procedure avoids explicit computation of the weight matrix and input vector. It can be further ascertained that the role played by the sigmoid nonlinearity is somewhat similar to that played by the damping parameter in some of the gradient-based parameter estimation methods.

We obtain from optimisation theory, that for the parameter vector the following holds true (for non-neural based methods):

$$\frac{d\beta}{dt} = \mu(t)\sum_{k=1}^{N}[\dot{x}(k) - Ax(k)]x^T(k) \tag{10.71}$$

or equivalently:

$$\beta(i+1) = \beta(i) + \mu\sum_{k=1}^{N}[\dot{x}(k) - Ax(k)]x^T(k) \tag{10.72}$$

For RNN-S (HNN), the discretisation approach leads to

$$\beta(i+1) = \beta(i) + \sum_{k=1}^{N} [\dot{x}(k) - f(x)x(k)]x^T(k); \qquad \beta = f(x) \tag{10.73}$$

Similarly for RNN-E, the parameter estimation rule is

$$\beta(i+1) = \beta(i) + \sum_{k=1}^{N} f[\dot{x}(k) - Ax(k)]x^T(k) \tag{10.74}$$

Here f could be 'tanh' nonlinearity. Next, from the theory of the Kalman filter, the following state estimation rule follows:

$$\hat{x}_a(k+1) = \tilde{x}_a(k+1) + K(z(k+1) - H\tilde{x}_a(k+1)) \tag{10.75}$$

Here, we presume that the state is an augmented state vector with unknown parameters β. The gradients of error w.r.t. states are implicit in the formulation of K. The Kalman filter is generally defined in the form of output error, which is also often known as the prediction error.

From the above development, the following facts emerge:

1 In the Hopfield neural network, the nonlinearity directly influences the parameter vector (the state of the Hopfield neural network).
2 In the case of RNN-E, the nonlinearity influences the residuals directly. It can also be viewed as affecting the parameter vector indirectly.
3 In the conventional parameter estimator, the factor μ affects the change in the parameter vector $\Delta\beta$, since from eq. (10.72), we get $\beta(i+1) = \beta(i) + \mu\Delta\beta$.
4 The Kalman filter gain operates on the residuals and optimally helps to determine the state estimate.

From the above equations and observations, we infer that nonlinearity f, μ or Kalman gain can affect the convergence of the parameter estimation algorithm. In eq. (10.72) the inherent decision-making process is linear. Thus, the distinction is in the way in which the nonlinear/linear element affects the convergence of the algorithm, measurement errors, states and parameters and hence overall accuracy of the estimates.

In principle, the recurrent neural network schemes developed in this chapter can be used for parameter estimation of stable or unstable/augmented dynamical systems [17,18]. The schemes are straightforward and require simple programming code. However, they require proper use of the sigmoid nonlinearities. When formulated using the equation error, the schemes need accurate measurements of states and their derivatives. It is also possible to incorporate measurement models and formulate them in the form of output error. This will automatically extend the application of the recurrent neural network based parameter scheme to general dynamic systems. Such a development can be found in Reference 18.

10.5 Genetic algorithms

First, a short description of genetic algorithms is given, and then the procedure of using them for parameter estimation is described.

Genetic algorithms are search methods inspired by nature's evolutionary systems [19]. They can be used to obtain global and robust solutions to many optimisation problems in science, engineering, economics, psychology and biology. Natural systems have evolved over millions of years. They have gone through iterations over many generations and in the process have become very robust, especially to their many different environments. Due to its strong evolutionary 'experience', the natural system offers good solutions whenever robustness is called for. Biological systems are generally more robust, efficient and flexible compared to the most sophisticated artificial systems. Artificial systems have to learn from biological systems to improve their performance and carry out their daily-required functions for a longer period of time and with greater efficiency. Genetic algorithms are based on some of the principles that govern the natural systems [20,21].

Genetic algorithms are computational optimisation schemes with an approach that seems rather unconventional. The algorithms solve optimisation problems imitating nature in the way it has been working for millions of years on the evolution of life forms. Inspired by the biological systems, genetic algorithms adopt the rules of natural selection and genetics to attain robustness. Acting on the premise of survival of the fittest, a population or sample of feasible solutions is combined in a manner similar to the combination of chromosomes in a natural genetic system. The fitter population members pass on their structures as genes in far greater measure than their less fit members do. As the generations evolve, the net effect is evolution of the population towards an optimum (species, solution, etc.). Genetic algorithms operate by combining the information present in different possible solutions so that a better solution is obtained in the next/future generations.

The terms used in the study of genetic algorithms are given in Table 10.11 [22].

Table 10.11 Comparison of genetic algorithm with natural genetic system

Natural genetic system	**Genetic algorithm**
Chromosomes	String of numbers
Gene	Feature or detection
Allele	Feature value
Locus	String position
Genotype	Structure
Phenotype	Parameter set, alternative form, a decoded structure

10.5.1 Operations in a typical genetic algorithm

10.5.1.1 Chromosomes

Chromosomes represent encoding of information in a string of finite length and each chromosome consists of a string of bits (binary digit; 0 or 1). Or it could be a symbol from a set of more than two elements. Generally, for function optimisation, chromosomes are constructed from binary strings as seen from the following table:

Parameter value	String
6	000110
34	100010

The long stretches of DNA that carry the genetic information needed to build an organism are called chromosomes. The chromosomes consist of genes. Each gene represents a unit of information and it takes different values. These values are called alleles at different locations called loci. The strings, composed of features or detectors, assume values such as 0 or 1, which are located at different positions in the string. The total package or system is called the genotype or structure. The phenotype results when interaction of genotype with environment takes place.

10.5.1.2 Population and fitness

Genetic algorithms operate on the population of possible solutions with chromosomes. The population members are known as individuals. Each individual is assigned a fitness value based on the objective function, or cost function. Better individuals (solutions) have higher fitness values and weaker ones have lower fitness values.

10.5.1.3 Initialisation and reproduction

By randomly selecting information from the search space and encoding it, a population of possible initial solutions is created. Reproduction is a process in which individual strings are copied as per their fitness values. Thus, the strings with a greater fitness value have a higher probability of contributing one or more offsprings to the next generation.

10.5.1.4 Crossover

In a crossover, a site is selected randomly along the length of the chromosomes, and each chromosome is split into two pieces at the crossover site. The new ones are formed by joining the top piece of one chromosome with the tailpiece of the other.

10.5.1.5 Mutation

Mutation is a small operation in which a bit in a string is changed at a random location. The main idea is to break monotony and add a bit of novelty. This operation would help gain information not available to the rest of the population. It lends diversity to the population.

10.5.1.6 Generation

Each iteration in the optimisation procedure is called a generation. In each generation pairs are chosen for crossover operation, fitness is determined, and mutation is carried out during the crossover operation (during or after has a subtle distinction). With these operations performed, a new population evolves that is carried forward.

10.5.1.7 Survival of the fittest

The individuals may be fitter or weaker than some other population members. So the members must be ranked as per their fitness value. In each generation, the weaker members are allowed to wither and the ones with good fitness values take part in the genetic operations. The net result is the evolution of the population towards the global optimum.

10.5.1.8 Cost function, decision variables and search space

In most practical optimisation problems, the goal is to find optimal parameters to increase the production and/or to reduce the expenditure/loss. That is, to get maximum profit by reorganising the system and its parameters that affect the cost function. Since, in effect, this reflects on the cost, it is represented by the cost function. A carefully devised and convergent computational algorithm would eventually find an optimum solution to the problem. The parameters of the system that decide the cost are termed decision variables. The search space is a Euclidean space in which parameters take different values and each point in the space is a probable solution.

10.5.2 Simple genetic algorithm illustration

A simple genetic algorithm is described, which will use the binary coding technique.

Step 1: Create population of N samples from a chosen search space – denoting the decision variables.
Step 2: Produce series of 0s and 1s to create chromosomes – i.e., encoding the decision variables.
Step 3: Calculate the cost function values and assign fitness (values) to each member.
Step 4: Sort the members accordingly to their respective fitness values.
Step 5: Carry out crossover operation taking two chromosomes at a time.
Step 6: Mutate the chromosomes with a given probability of mutation.
Step 7: Retain the best members of the population and remove the weaker members based on their fitness values.
Step 8: Replace the old generation by the new one and repeat steps 3 to 8.

Let us consider the problem of maximising the function [22]:

$$f(x) = x^2 - 64x + 100$$

Here, x varies from 0 to 63.

The function f has a maximum value of 100 at $x = 0$. The decision variables are coded in strings of finite length. We can encode the variables as a binary string of length 6. We create an initial population with 4 samples by randomly selecting them from the interval 0 to 63 and encode each sample. A binary string of length 6 can represent any value from 0 to 63; $(2^6 - 1)$. Four encoded samples in the initial population are: 5 (000101); 60 (111100); 33 (100001); 8 (001000). These individuals are sorted according to their fitness values and arranged in descending order of their fitness values. For simplicity, mutation is not used. Also, the problem that could have been solved using the conventional approach is used to illustrate GA operations for simplicity. For the present example, the fitness value is the same as the value of cost function and these individuals are sorted according to their fitness values:

No.	x	String	Fitness value
1	60	111100	−140
2	5	000101	−195
3	8	001000	−348
4	33	100001	−923

Next, the crossover is randomly selected, and in the first generation, the 1st and 2nd strings are crossed over at site 3 to get two new strings:

Crossover site	New strings	Fitness of new strings
111 ! 100	111101	−83
000 ! 101	000100	−140

Similarly, the 3rd and 4th strings are crossed over at site 2, to get:

Crossover site	New strings	Fitness of new strings
00 ! 1000	000001	37
10 ! 0001	101000	−860

Sorting these new individuals one gets:

No.	x	String	Fitness value
1	1	000001	37
2	61	111101	−83
3	4	000100	−140
4	40	101000	−860

It is now seen that in one generation fitness is improved from −140 to 37 ($f(1) > f(60)$). The weakest member of the population is replaced by the fittest member of the previous population; string 101000 that has fitness −860 is replaced by string

111100, whose fitness is −140. In the 2nd generation, the 1st and 2nd strings are crossed over at site 1 to obtain the following:

Crossover site	New strings	Fitness of new strings
0 ! 00001	011101	−915
1 ! 11101	100001	−923

Similarly, the 3rd and 4th strings are crossed over at site 3 to obtain:

Crossover site	New strings	Fitness of new strings
000 ! 100	000100	−140
111 ! 100	111100	−140

We replace the weakest member by the fittest member of the previous population (string 100001 with fitness value of −923 is replaced by the string 000001 with fitness value of 37). The sorting results in:

No.	x	String	Fitness value
1	1	000001	37
2	4	000100	−140
3	60	111100	−140
4	29	011101	−915

In the 3rd generation, the process of crossover at site 4 is carried out (not shown here). The new set of strings in the population, after replacement of the weakest by the fittest member is give as:

No.	x	String	Fitness value
1	0	000000	100
2	1	000001	37
3	61	111101	−83
4	5	000101	−195

We see that as the genetic algorithm progresses from one generation to the next, the improved solutions evolve. At $x = 0$, $f(x) = 100$, the desired result.

10.5.2.1 Stopping strategies for genetic algorithms

One needs to know where and when to stop the genetic algorithm iterations. If the population size is fixed, then more generations might be needed for the convergence of a genetic algorithm to an optimal solution. One way is to track the fitness value for no further improvement. As the algorithmic steps progress, a situation would occur where we need a large number of generations to bring about a small improvement in the fitness value. One can define a predetermined number of generation/iterations to solve the problem. Also, insignificant change in the norm of estimated parameters can be tracked for a few consecutive iterations before stopping the search. It must be possible to do an effective search if one exploits some important similarities in the coding used in genetic algorithms. Another way is to evaluate the gradient of the cost function and use the conventional approaches for assessing the quality of the estimates for their convergence to true values. It is possible to use GA with a gradient-based approach for evaluating the estimation accuracy as is done for OEM (Chapter 3). Again, as is true with all the other parameter estimation methods, the matching of time histories of the measured data and model responses is necessary but not a sufficient condition. An increase in the number of samples would generally increase the success rate.

10.5.2.2 Genetic algorithms without coding of parameters

Genetic algorithms become more complex because of coding the chromosomes, especially for more complex problems. In the problems of science and engineering, we come across real numbers. Thus, we need to use real numbers and still use genetic algorithms on these numbers for solving optimisation problems. A major change is in the crossover and mutation operations. Averaging the two samples, for instance, the two sets of parameter values can perform the crossover operation. After the crossover, the best individual is mutated. In mutation, a small noise is added. Assume that two individuals have β_1 and β_2 as numerical values of the parameters. Then after crossover, we obtain the new individual as $(\beta_1 + \beta_2)/2$. For mutation we have $\beta_3 = \beta_1 + \varepsilon * v$, where d is a constant and v is a number chosen randomly between -1 and 1.

Thus, all the genetic algorithm operations can be performed by using real numbers like 4.8904, etc., without coding the samples. This feature is extremely well suited for several engineering applications: parameter estimation, control, optimisation and signal processing [23].

10.5.2.3 Parallelisation of genetic algorithms

The genetic algorithms are powerful and yet very simple strategies for optimisation problems. They can be used for multi-modal, multi-dimensional and multi-objective optimisation problems, not only in science and engineering, but also in business and related fields. However, despite the fact that the computations required in genetic algorithm operations are very simple, they become complex as the number of iterations grows. This will put heavy demand on the computational power. Often, the procedures can be parallelised and the power of the parallel computers can be used. Since genetic algorithms can work on population samples simultaneously, their natural parallelism can be exploited to implement them on parallel computers.

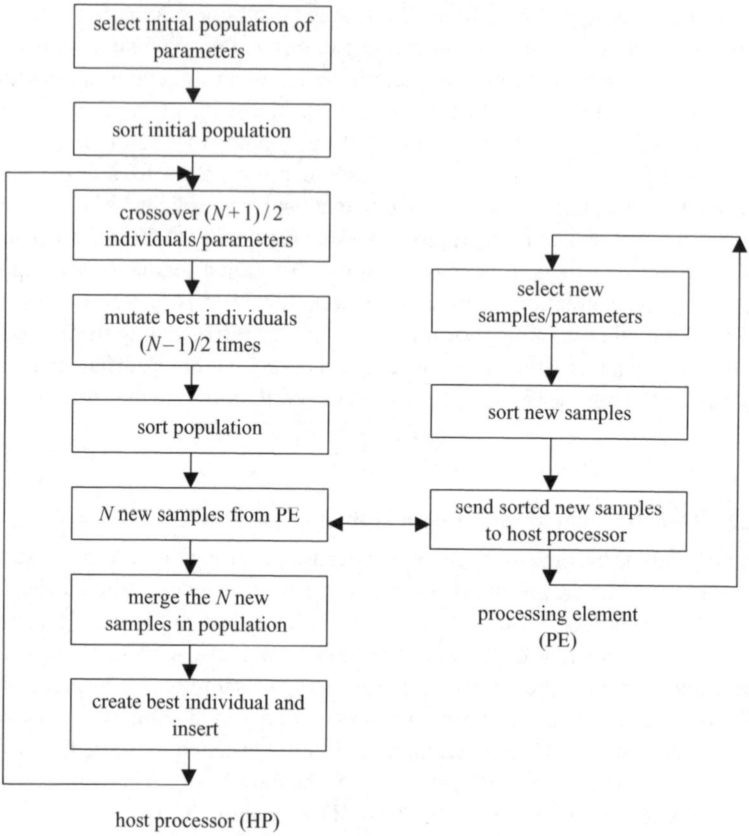

Figure 10.20 A schematic of the parallel genetic algorithm [24]

10.5.2.4 Scheme for parallel genetic algorithm

One scheme is shown in Fig. 10.20. The sorting is split between two processors. In this scheme, the host processor does the job of crossover, mutation, etc.

10.5.3 Parameter estimation using genetic algorithms

As we have seen in previous chapters, most of the parameter estimation methods are based on the minimisation of the cost function resulting in utilisation of the gradient of the cost function. The application of the genetic algorithm to the parameter estimation problem does not need utilisation of the gradient of the cost function.

Consider the problem of parameter estimation as follows:

$$z = H\beta + v; \quad \hat{z} = H\hat{\beta} \tag{10.76}$$

The cost function is formulated as

$$E = \frac{1}{2}\sum (z - \hat{z})^T (z - \hat{z}) = \frac{1}{2}\sum (z - H\hat{\beta})^T (z - H\hat{\beta}) \qquad (10.77)$$

Now in the gradient-based method, the minimum is obtained by $\partial E/\partial\beta$ and the result will be eq. (2.4). However, instead we can use the genetic algorithm as explained in steps 1 to 8 in Section 10.5.2 of this chapter.

10.5.3.1 Example 10.7

Consider the third order system described by

$$\begin{bmatrix} \dot{x}_1 \\ \dot{x}_2 \\ \dot{x}_3 \end{bmatrix} = \begin{bmatrix} -2 & 0 & 1 \\ 1 & -2 & 0 \\ 1 & 1 & -1 \end{bmatrix} \begin{bmatrix} x_1 \\ x_2 \\ x_3 \end{bmatrix} + \begin{bmatrix} 1 \\ 0 \\ 1 \end{bmatrix} u \qquad (10.78)$$

Here, u is the doublet input to the system. The output is described by

$$z = [2 \quad 1 \quad -1] \begin{bmatrix} x_1 \\ x_2 \\ x_3 \end{bmatrix} \qquad (10.79)$$

Obtain the doublet response of the system and use u and z in the genetic algorithm to estimate all the 15 parameters.

10.5.3.2 Solution

The system is simulated with a doublet input and has total simulation time of 20 s (sampling interval $\Delta t = 0.1$ s; number of data points = 200). Figure 10.21 shows the doublet input u and system response z. Figure 10.22 shows the response error for the case with no noise. The estimation of parameters is accomplished by using

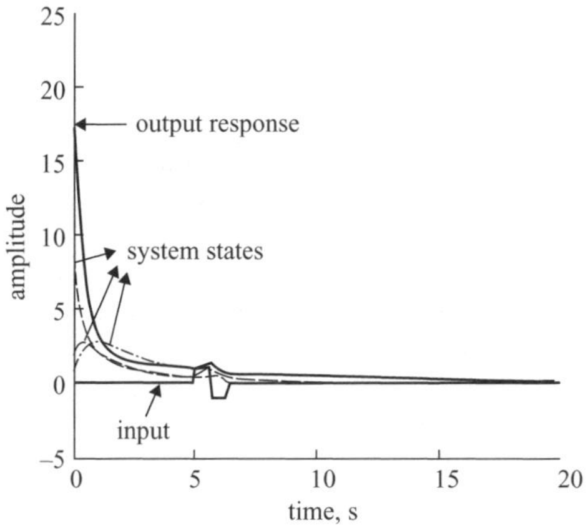

Figure 10.21 System response and doublet input (Example 10.7)

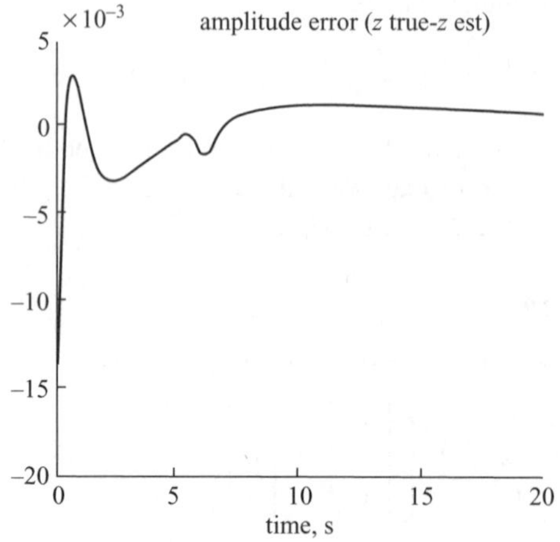

Figure 10.22 Outpur error w.r.t. true data (SNR = ∞) (Example 10.7)

file '**parestga.m**' placed in folder '**Ch10GAex7**'. The initial state of the system, $x(0) = [10 \quad 1 \quad 0.1]$.

- POPSIZE = 100 (sets of parameters/population size)
- MAXITER = 100 (number of GA iterations)

The initial population of parameters and fitness values are given in Table 10.12 and the estimated parameters for various noise levels are given in Table 10.13.

10.5.3.3 Example 10.8

Find the minimum of the function $f(b) = b^2 - 64b + 1025$ using a genetic algorithm, where b varies from 0 to 63 (see Fig. 10.23).

10.5.3.4 Solution

From Fig. 10.23 the minimum for $f(b)$ is at $b = 32$. Using a genetic algorithm the minimum was found at $b = 32$. Figure 10.24 shows the plot of b versus genetic algorithm iterations. The estimation of parameter b is accomplished by using file '**parestga.m**' placed in folder '**Ch10GAex8**'.

- POPSIZE = 10 (sets of parameter/population size)
- MAXITER = 20 (number of iterations)

We see that the convergence is reached in less than 10 iterations.

10.5.3.5 Example 10.9

Find the global minimum of the function (see Fig. 10.25) $f(b) = b^3 - 45b^2 + 600b + v$ using genetic algorithm, where b varies from 1 to 25, and v is the measurement noise.

Table 10.12 Initial population of parameters and fitness (Example 10.7)

	Initial 10 populations of parameters									
	1	2	3	4	5	6	7	8	9	10
a_{11}	−1.1	−2.19	−2.97	−1.99	−2.01	−1.60	−2.73	−2.55	−2.17	−2.18
a_{12}	0	0	0	0	0	0	0	0	0	0
a_{13}	1.11	1.42	0.95	0.93	1.32	0.95	1.39	1.26	1.37	0.77
a_{21}	0.99	0.91	1.43	0.81	1.15	1.20	0.70	1.03	0.52	0.94
a_{22}	−1.22	−1.21	−2.07	−2.62	−1.36	−1.76	−2.40	−1.72	−1.46	−1.13
a_{23}	0	0	0	0	0	0	0	0	0	0
a_{31}	0.96	0.85	1.35	1.18	0.84	1.46	0.78	0.88	1.49	0.71
a_{32}	0.52	1.31	1.03	0.80	0.79	1.02	0.97	1.28	1.29	1.34
a_{33}	−0.68	−1.49	−1.30	−0.96	−1.16	−0.62	−1.44	−0.82	−1.06	−0.87
b_{11}	0.95	0.64	1.17	0.65	1.03	0.67	1.49	0.96	1.0	0.63
b_{21}	0	0	0	0	0	0	0	0	0	0
b_{31}	1.29	0.70	0.52	0.88	0.81	0.77	0.92	1.29	1.14	1.11
c_{11}	2.84	2.21	2.36	2.72	2.68	1.51	2.03	1.12	1.64	2.26
c_{12}	1.24	0.77	0.88	1.35	1.07	1.38	0.83	1.10	1.46	0.87
c_{13}	−1.32	−1.30	−0.67	−0.91	−1.13	−0.76	−1.07	−1.45	−0.77	−0.93
Fitness*	0.0006	−0.015	−0.026	0.0082	0.0035	0.0007	0.024	0.0028	0.0018	−0.008

* Fitness value $= [(1/2)\sum_{k=1}^{N}(z(k)-\hat{z}(k))^T\hat{R}^{-1}(z(k)-\hat{z}(k)) + (N/2)\ln(|\hat{R}|)]^{-1}$

$\hat{R} = (1/N)\sum_{k=1}^{N}(z(k)-\hat{z}(k))(z(k)-\hat{z}(k))^T$

Table 10.13 Parameter estimation with GA (Example 10.7)

Parameters	True values	Estimated values	
		SNR = ∞	SNR = 10
a_{11}	−2	−2.0055	−2.0401
a_{12}	0	0.0000	0.0000
a_{13}	1	1.0012	1.0208
a_{21}	1	1.0033	1.0235
a_{22}	−2	−2.0121	−2.0459
a_{23}	0	0.0000	0.0000
a_{31}	1	1.0028	1.0185
a_{32}	1	1.0027	1.0194
a_{33}	−1	−1.0009	−1.0215
b_{11}	1	1.0011	1.0198
b_{21}	0	0.0000	0.0000
b_{31}	1	1.0078	1.0202
c_{11}	2	2.0015	2.0505
c_{12}	1	1.0043	1.0246
c_{13}	−1	−0.9979	−1.0145
PEEN	–	0.3730	2.1879

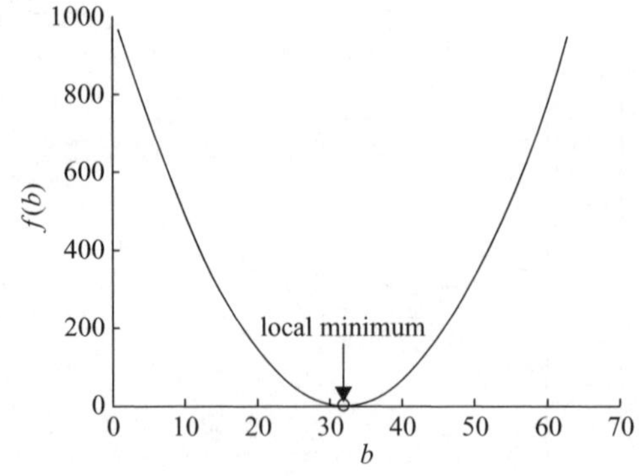

Figure 10.23 Cost function $f(b)$ w.r.t. parameter b (Example 10.8)

10.5.3.6 Solution

The data simulation is carried out using function $f(b)$ with v as an additive white Gaussian noise. In the Fig. 10.25 the global minimum for $f(b)$ is at $b = 1$. Using the genetic algorithm the global minimum was found to be at $b = 1.005$. Figure 10.26

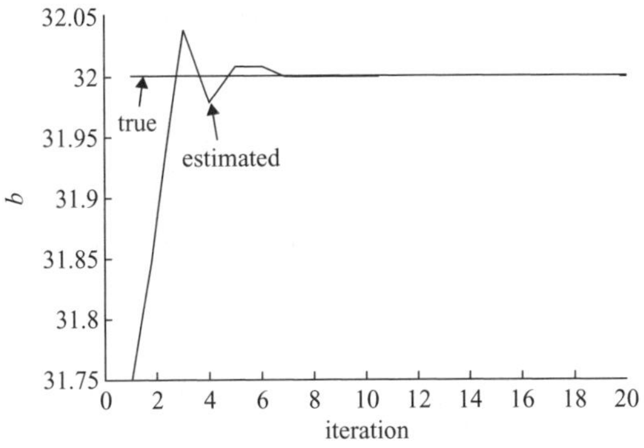

Figure 10.24 Estimation of parameter b versus iteration (Example 10.8)

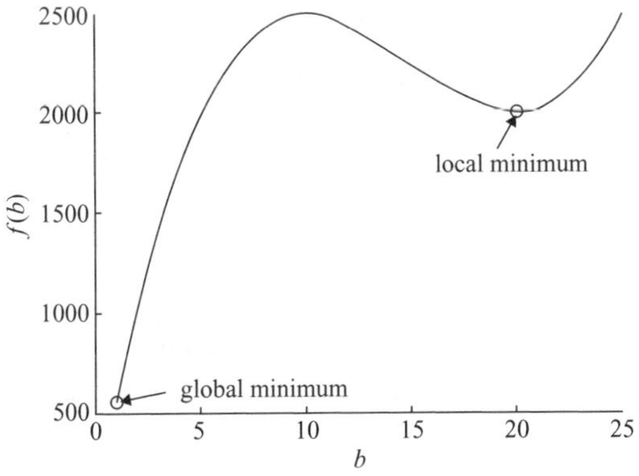

Figure 10.25 Cost function f (b) w.r.t. parameter b (Example 10.9)

shows the plot of b versus genetic algorithm iterations. The estimation of parameter b is accomplished by using file '**parestga.m**' placed in folder '**Ch10GAex9**'.

- POPSIZE = 100 (sets of parameter/population size)
- MAXITER = 250 (number of GA iterations)

The estimates of parameter b are presented in Table 10.14.

10.6 Epilogue

Certain circuit architectures of simple neuron-like analogue processors were given for on-line applications [12]. The recurrent neural network architectures can be used for solving linear systems, pseudo inversion of matrices and quadratic programming problems. These architectures can be made suitable for implementation on VLSI chips. System identification and control aspects of nonlinear systems have

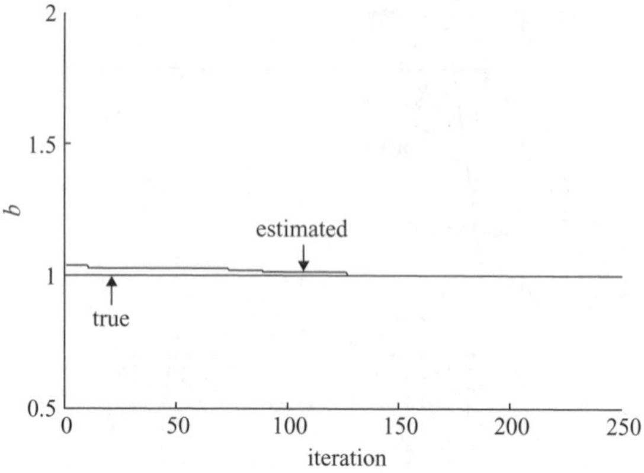

Figure 10.26 Parameter b versus iteration for SNR = 100 *(Example 10.9)*

Table 10.14 Parameter estimation with genetic algorithm

(Example 10.9)

Parameters	True values	Estimated values		
		SNR = ∞	SNR = 100	SNR = 50
b	1	1.00081	1.00015	1.0045

been treated [6], based on mainly recurrent neural networks. Several schemes were evaluated with simulated data. In Reference 3, a review of development in feed forward neural networks is given. Several algorithms for supervised training of the neural networks are presented. A concept of 'minimal disturbance' is adopted. It suggests that the already stored information is disturbed minimally when new information is incorporated into the network while training. Initial work on parameter estimation using recurrent neural networks can be found in Reference 14. As such, literature on recurrent neural network based explicit parameter estimation is limited [15, 17, 18]. In Reference 18, several architectures for parameter estimation using recurrent neural networks are presented: gradient-, weight (W) and bias (b)-, information matrix- and output error-based. Comprehensive treatment of artificial neural networks can be found in References 25 to 27. An extensive survey of artificial neural networks is provided in Reference 28, where various formulations of discrete and continuous-time recurrent neural networks are also considered. Some of these formulations [28] were further studied in this chapter from the parameter estimation point of view.

Work on parameter estimation using genetic algorithms is also limited. More research applications of artificial neural networks and genetic algorithms for parameter estimation to real life systems would be highly desirable.

10.7 References

1 EBERHART, R. C., and DOBBINS, R. W.: 'Neural network PC tools – a practical guide' (Academic Press, New York, 1993)

2 IRWIN, G. W., WARWICK, K., and HUNT, K. J. (Eds.): 'Neural network applications in control', IEE Control Engineering Series 53 (The IEE, London, 1995)

3 WIDROW, B., and LEHR, M. A.: 'Thirty years of adaptive neural networks: perceptron, madaline and back propagation', *Proc. of the IEEE*, 1998, **78**, (9), pp. 1415–1442

4 CICHOCKI, A., and UNBEHANEN, R.: 'Neural networks for optimisation and signal processing' (John Wiley and Sons, N.Y., 1993)

5 LINSE, D. J., and STENGEL, R. F.: 'Identification of aerodynamic coefficients using computational neural networks', *Journal of Guidance, Control and Dynamics*, 1993, **16**, (6), pp. 1018–1025

6 NARENDRA, K. S., and PARTHASARTHY, K.: 'Identification and control of dynamical systems using neural networks', *IEEE Trans. on Neural Networks*, 1990, **1**, (1), pp. 4–27

7 RAOL, J. R., and MANEKAME, S.: 'Artificial neural networks – a brief introduction', *Journal of Science Education*, 1996, **1**, (2), pp. 47–54

8 RAOL, J. R.: 'Feed forward neural networks for aerodynamic modelling and sensor failure detection', *Journal of Aero. Soc. of India*, 1995, **47**, (4), pp. 193–199

9 RAISINGHANI, S. C., GHOSH, A. K., and KALRA, P. K.: 'Two new techniques for aircraft parameter estimation using neural networks', *Aeronautical Journal*, 1998, **102**, (1011), pp. 25–29

10 WERBOS, P. J.: 'Back propagation through time: what it does and how to do it', *Proc. of the IEEE*, 1990, **78**, (10), pp. 1550–1560

11 SCALERO, R. S., and TEPEDELENLIOGH, N.: 'A fast new algorithm for training feed forward neural networks', *IEEE Trans. on Signal Processing*, 1992, **40**, (1), pp. 202–210

12 CICHOCKI, A., and UNBEHANEN, R.: 'Neural networks for solving systems of linear equations and related problems', *IEEE Trans. on Circuits and Systems – I: Fundamental theory and applications*, 1992, **39**, (2), pp. 124–138

13 RAOL, J. R., and JATEGAONKAR, R. V.: 'Aircraft parameter estimation using recurrent neural networks – a critical appraisal', AIAA Atm. Flight Mechanics Conference, Baltimore, Maryland, August 7–9, 1995 (AIAA-95-3504-CP)

14 CHU, S. R., and TENORIO, M.: 'Neural networks for system identification', *IEEE Control System Magazine*, 1990, pp. 31–35

15 RAOL, J. R.: 'Parameter estimation of state-space models by recurrent neural networks', *IEE Proc. Control Theory and Applications (U.K.)*, 1995, **142**, (2), pp. 114–118

16 HOPFIELD, J. J., and TANK, D. W.: 'Computing with neural circuits; a model', *Science*, 1986, pp. 625–633

17 RAOL, J. R.: 'Neural network based parameter estimation of unstable aerospace dynamic systems', *IEE Proc. Control Theory and Applications (U.K.)*, 1994, **141**, (6), pp. 385–388

18 RAOL, J. R., and HIMESH, M.: 'Neural network architectures for parameter estimation of dynamical systems', *IEE Proc. Control Theory and Applications (U.K.)*, 1996, **143**, (4), pp. 387–394

19 GOLDBERG, D. E.: 'Genetic algorithms in search, optimisation and machine learning' (Addison-Wesley Publishing Company, Reading, MA, 1989)

20 SINHA, N. K., and GUPTA, M. M.: 'Soft computing and intelligent systems – theory and applications' (Academic Press, New York, 2000)

21 MITCHELL, M.: 'An introduction to genetic algorithms' (Prentice Hall of India, New Delhi, 1998)

22 RAOL, J. R., and JALISATGI, A.: 'From genetics to genetic algorithms', *Resonance, The Indian Academy of Sciences*, 1996, **2**, (8), pp. 43–54

23 PATTON, R. J., and LIU, G. P.: 'Robust control design via eigenstructure assignment, genetic algorithms and gradient-based optimisation', *IEE Proc. Control Theory Applications*, 1994, **141**, (3), pp. 202–207

24 RAOL, J. R., JALISATGI, A. M., and JOSE, J.: 'Parallel implementation of genetic and adaptive portioned random search algorithms', *Institution of Engineers (India)*, 2000, **80**, pp. 49–54

25 ZURADA, J. M.: 'Introduction to artificial neural system' (West Publishing Company, New York, 1992)

26 HAYKIN, S.: 'Neural networks – a comprehensive foundation' (IEEE, New York, 1994)

27 KOSKO, B.: 'Neural networks and fuzzy systems – a dynamical systems approach to machine intelligence' (Prentice Hall, Englewood Cliffs, 1992)

28 HUSH, D. R., and HORNE, B. G.: 'Progress in supervised neural networks – what is new since Lippmann?' *IEEE Signal Processing Magazine*, 1993, pp. 8–39

10.8 Exercises

Exercise 10.1

Let the cost function be given as $E = (1/2)(z - u_2)^T(z - u_2)$ for the output layer of the feed forward neural network. Obtain a learning rule for weights W_2. (Hint: use $(dW_2/dt) = -\partial E/\partial W_2$.)

Exercise 10.2

Derive the weight update rule for W_1 of the feed forward neural network. (Hint: use $(dW_1/dt) = -\partial E/\partial W_1$.)

Exercise 10.3

In eq. (10.20), if $z_i = 1$, then what artifice will you use in your program code to avoid ill-conditioning, since with $z_i = 1$, the expression will be infinity?

Exercise 10.4

Why will the range of values of μ for eqs (10.12) and (10.21) be quite different? (Hint: Look at the relevant terms in the corresponding weight update rules and compare.)

Exercise 10.5

Compare and contrast eqs (10.15) and (10.16) of the recursive weight update rules, with somewhat similar equations in Chapter 4 for the Kalman filter.

Exercise 10.6

Consider eq. (10.12), use Δt as the time interval and convert the rule to the 'weight-derivative' update rule.

Exercise 10.7

What is signified by the expanded structure/elements of the weight matrix W and bias vector b? (Hint: these are computed as squares of certain variables.)

Exercise 10.8

Let $\beta_i = f(x_i)$ and $f = \rho[(1 - e^{-\lambda x_i})/(1 + e^{-\lambda x_i})]$. Obtain expression for x_i. (Hint: $x_i = f^{-1}(\beta_i)$.)

Exercise 10.9

Given the logistic sigmoid function $f(x_i) = 1/(1 + e^{-x_i})$, obtain its first derivative w.r.t. x_i.

Exercise 10.10

If for training the feed forward neural network, an extended Kalman filter is to be used, formulate the state-space model for the same.

Exercise 10.11

Compare recurrent neural network dynamic equations with the linear system state equations ($\dot{x} = Ax + Bu$) and comment.

Exercise 10.12

Obtain the gradient of the cost function

$$E = \frac{\rho}{2\lambda} \sum_{k=1}^{N} \ln(\cosh(\lambda \dot{e}(k))); \qquad \dot{e} = \dot{x}(k) - Ax(k).$$

Exercise 10.13

Given $(d\beta_1/dt) = -\mu(\partial E/\partial \beta_1)$, where β_1 is a parameter vector, obtain various parameter estimation rules if μ is a linear constant and μ is some nonlinear function f.

Exercise 10.14

Derive expressions for individual steps of recurrent neural network architecture based on direct gradient computation, given

$$E(v) = \frac{1}{2}\sum_{k=1}^{N}(\dot{x}(k) - Ax(k))^T(\dot{x}(k) - Ax(k))$$

Draw the block diagram. (Hint: use $(d\beta/dt) = -\partial E/\partial\beta$; with $\beta =$ (elements of A and B).)

Exercise 10.15

Explain the significance of momentum constant in the weight update rule of the feed forward neural network. (Hint: ponder on the weight-difference term.)

Chapter 11
Real-time parameter estimation

11.1 Introduction

In previous chapters, we have discussed several parameter estimation techniques for linear and nonlinear dynamic systems. It was stated often that the Kalman filter, being a recursive algorithm, is more suitable for real-time applications. Many other approaches like estimation before modelling and model error estimation algorithms can be used in a recursive manner for parameter estimation. However, they put a heavy burden on computation.

Modern day systems are complex and they generate extensive data, which puts a heavy burden on post-processing data analysis requirements. Many times, simple results of system identification and parameter estimation are required quickly. Often, it is viable to send data to a ground station by telemetry for 'real-time' analysis.

There are situations where on-line estimation could be very useful: a) model-based approach to sensor failure detection and identification; b) reconfigurable control system; c) adaptive control; and d) determination of lift and drag characteristics of an aircraft from its dynamic manoeuvres.

For the on-line/real-time parameter estimation problem, several aspects are important: i) the estimation algorithm should be robust; ii) it should converge to an estimate close to the true value; iii) its computational requirements should be moderately low or very low; and iv) the algorithm should be numerically reliable and stable so that condition (i) is assured.

It is possible to apply on-line techniques to an industrial process as long as transient responses prevail, since when these responses die out or subside, there is no activity and all input-output signals of the process (for identification) have attained the steady state and hence these signals are not useful at all for parameter estimation. Only the steady state gain of the plant/system can be determined.

Also, other considerations are important: i) too much uncertainty of the basic model of the system; and ii) system process and measurement noise will further degrade the estimation performance.

In this chapter, some parameter estimation approaches, which are suitable for on-line/real-time application, are discussed [1, 2].

11.2 UD filter

The UD filtering algorithm is a feasible approach for such a purpose. It is computationally very efficient, numerically reliable and stable. For parameter estimation, it has to be used in the extended Kalman filter/UD filter mode. What it means is that since the unknown parameters are considered as additional states, the original Kalman filter form will become the extended Kalman filter problem, for which the extended UD filter can be used. In that case, the time propagation and measurement data updates can be in the form of the nonlinear functions f and h, but the gain and covariance propagation/update recursions can be processed using UD factorisation formulation (see Section 4.3). The nonlinear system model f and h functions are linearised and discretised in real-time, using the finite difference method.

Alternatively, one can use the UD filter/extended UD filter for state estimation only and then use a recursive least squares method for parameter estimation. In that case, one can follow the procedure outlined in Chapter 7. However, the computations should be kept as simple as possible. Even for the recursive least squares method, the factorisation scheme can be used because for real-time implementation, numerical reliability and stability of algorithms are very essential. Here, it is also possible to put these two steps on separate parallel processors. Several approaches to recursive least squares and related methods have been discussed [2, 3, 4]. Since the UD filter, as presented in Section 4.3, can be used for real-time parameter estimation with trivial modification (of appending the parameters as additional states), it is not repeated here.

11.3 Recursive information processing scheme

In Chapter 10, we studied parameter estimation schemes based on recurrent neural networks. In the present scheme, the information on states and input is processed in a sequential manner. It should be feasible to use this scheme for on-line applications. In this scheme, the data x, $\dot{x}$ and u are processed as soon as they are available to obtain the elements of W and b without waiting to receive the complete set of the data. Thus, the scheme uses the current data (x, $\dot{x}$ and u in a cumulative manner). It is not necessary to store the previous data until the estimation process is completed. This is because the previous data has been already incorporated in the computation of W and b. However, in the start W and b are based on partial information. The solution of eq. (10.53) is also attempted immediately at each sampling instant. Such an algorithm is given below [5]:

Step 1: choose initial values of β randomly.
Step 2: compute W and b based on currently available data (at time index k)

$$\begin{aligned} W(k) &= \frac{k-1}{k}\left[W(k-1) - \frac{1}{k-1}P(k)\Delta t\right] \\ b(k) &= \frac{k-1}{k}\left[b(k-1) + \frac{1}{k-1}Q(k)\Delta t\right] \end{aligned} \tag{11.1}$$

with $W(1) = -W_w(1)\Delta t$ and $b(1) = -b_b(1)\Delta t$.

Step 3: integrate the following equation one-time step ahead

$$\frac{d\beta_i}{dt} = \frac{\lambda(\rho^2 - \beta_i^2(k))}{2\rho}\left[\sum_{j=1}^{n} w_{ij}(k)\beta_j(k) + b_i(k)\right] \tag{11.2}$$

Step 4: recursively cycle through steps 2 and 3 until convergence is reached or no more data are available.

It can be readily seen that the scheme has the following recursive form for information processing:

$$I_{Wb}(k) = h\left(I_{Wb}(k-1), x(k), \dot{x}(k), u(k)\right) \tag{11.3}$$

In the above expressions, W_w and b_b are essentially the correlation elements computed by using x, $\dot{x}$, u etc., as shown in eqs (10.51) and (10.52).

Here, h is some functional relationship between present and past information. Thus, the utilisation of data, computation of W and b and the solution of eq. (11.2) for the estimation of parameters are carried out in a recursive manner within the Hopfield neural network structure. Proper tuning and some regularisation in the parameter estimation rule of eq. (11.2) would be very desirable. In addition, it is felt that use of an inverse of $W^T W$ (or its norm) in eq. (11.2) will speed up the algorithm. A relation between cost function, tuning parameter and settling time has been given [6]. A similar relation for the present recursive information processing scheme can be evolved.

11.3.1.1 Example 11.1

Consider the second order system described by

$$\dot{x} = \begin{bmatrix} -1.43 & -1.5 \\ 0.22 & -3.25 \end{bmatrix} x + \begin{bmatrix} -6.27 \\ -12.9 \end{bmatrix} u$$

1 obtain the doublet response of the system and generate 100 data points using a sampling interval $\Delta t = 0.1$ s; and
2 use x, $\dot{x}$ and u in the recursive RNN-S (Hopfield neural network) algorithm to estimate parameters.

11.3.1.2 Solution

1 The system response is generated for doublet input with initial state of the system $x(0) = [0.0 \quad 0.0]$.

2 The recursive scheme is used in RNN-S (Hopfield neural network) for parameter estimation. The estimation was carried out using noise free data and data with additive noise. The tuning parameters λ and ρ were kept at 0.1 and 100 respectively. For the sake of faster and smoother convergence of estimated parameters to true values, internal local iterations for each data point in RNN-S were set to 200. This means that computed weight (W) and bias (b) values for each data point are used in eq. (11.2) to carry out local iterations by using the estimated β and the same W and b. These W and b are then upgraded when new data are received at the next time point. As long as these iterations can be finished within the sampling time (much ahead of the new data arrival), there should not be any problem of computer time overheads. It was noted that RNN-S took around 50 data samples before the convergence of estimated parameters to true values. Figure 11.1 shows the estimated parameters for data with SNR = 100, and noise free data. Table 11.1 shows estimated parameters for different SNR levels. Reasonably good estimation has been achieved. The system simulation and parameter estimation are accomplished by using file '**parestrnn4.m**' placed in folder '**Ch11RNNex1**'.

We see from the above example that 'local iterations' are required for the algorithm to avoid more transients during the process. This aspect of using local tuning is a disadvantage of the scheme and it requires further research.

11.4 Frequency domain technique

Time-domain methods have several advantages: i) the strings of data from an experiment are available in discrete form in time-domain from the data recording

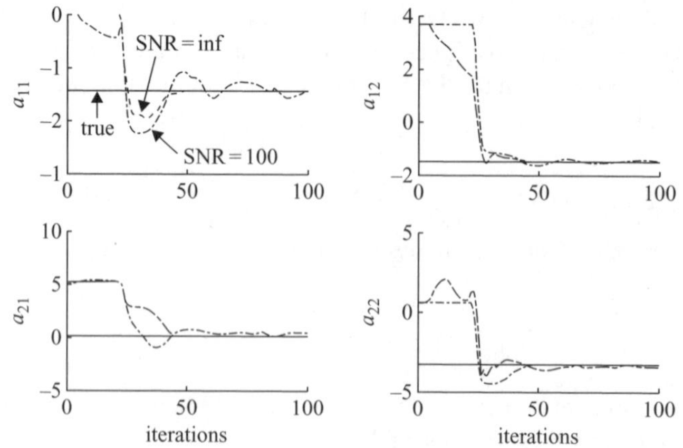

Figure 11.1 Estimated parameters for different SNR (Example 11.1)

Table 11.1 *Parameter estimation with recursive RNN-S (Example 11.1)*

Parameters	True values	Estimated values using RNN-S (HNN) method for different noise levels	
		SNR $= \infty$	SNR $= 100$
a_{11}	-1.43	-1.43	-1.34
a_{12}	-1.50	-1.50	-1.51
a_{21}	0.22	0.22	0.58
a_{22}	-3.25	-3.25	-3.38
b_1	-6.27	-6.27	-6.14
b_2	-12.9	-12.9	-12.63
PEEN	–	0.00	3.35

systems; ii) state-space models can be used as the models required in the estimation process; iii) the model parameters will have direct physical interpretation; iv) time-domain analysis of estimation results, like residuals, etc. is very well established and can be used for judging the statistical significance of the parameters and states; and v) many time-domain methods for parameter estimation are available in open literature.

However, based on the problem or experimental situation, time-domain methods can have certain limitations [7, 8]: i) measurement and process noise in the data systems; ii) in a closed loop control system, the independent input to plant is not available (as we have seen in Chapter 9); iii) the plant instability such that the data will have definite increasing trends; and iv) difficulty in assessing the performance of the method on-line.

Frequency domain parameter estimation methods overcome some of the limitations of the time-domain methods.

11.4.1 Technique based on the Fourier transform

In this subsection, the first offline scheme [7, 8] is described. Let the dynamical system be described by

$$\begin{aligned} \dot{x} &= Ax + Bu \\ z &= Cx \end{aligned} \tag{11.4}$$

The finite Fourier transform of signal $x(t)$ is given by

$$x(\omega) = \int_0^T x(t)e^{-j\omega t}\,dt \tag{11.5}$$

or its discrete domain approximation is given as

$$x(\omega) = \sum_{0}^{N-1} x(k)e^{-j\omega t_k} \tag{11.6}$$

Here, $t_k = k\Delta t$.

If the sampling rate is very high compared to the frequency range of our interest, then this discrete time approximation will be very accurate [7]. Applying the Fourier transform to eq. (11.4), we obtain

$$\begin{aligned} j\omega x(\omega) &= Ax(\omega) + Bu(\omega) \\ z(\omega) &= Cx(\omega) \end{aligned} \tag{11.7}$$

Our aim is to estimate the parameters, which are the elements of matrices A, B and C.

Expanding the above expressions, eq. (11.7), we get at

$$\omega = \omega_1, \quad \omega = \omega_2, \ldots, \omega = \omega_n \qquad \text{for } A = 2 \times 2 \text{ and } B = 2 \times 1$$

$$\begin{aligned}
\omega = \omega_1 \quad & \begin{aligned} j\omega_1 x_1(\omega_1) &= a_{11}x_1(\omega_1) + a_{12}x_2(\omega_1) + b_1 u(\omega_1) \\ j\omega_1 x_2(\omega_1) &= a_{21}x_1(\omega_1) + a_{22}x_2(\omega_1) + b_2 u(\omega_1) \end{aligned} \\
\omega = \omega_2 \quad & \begin{aligned} j\omega_2 x_1(\omega_2) &= a_{11}x_1(\omega_2) + a_{12}x_2(\omega_2) + b_1 u(\omega_2) \\ j\omega_2 x_2(\omega_2) &= a_{21}x_1(\omega_2) + a_{22}x_2(\omega_2) + b_2 u(\omega_2) \end{aligned} \\
& \vdots \\
\omega = \omega_m \quad & \vdots
\end{aligned} \tag{11.8}$$

Collating the above terms in particular order, we obtain

$$\begin{bmatrix} j\omega_1 \left(x_1(\omega_1) + x_2(\omega_1)\right) \\ j\omega_2 \left(x_1(\omega_2) + x_2(\omega_2)\right) \\ \vdots \\ \vdots \end{bmatrix}_{m\times 1}
= \begin{bmatrix} x_1(\omega_1) & x_2(\omega_1) & u(\omega_1) & x_1(\omega_1) & x_2(\omega_1) & u(\omega_1) \\ x_1(\omega_2) & x_2(\omega_2) & u(\omega_2) & x_1(\omega_2) & x_2(\omega_2) & u(\omega_2) \\ \vdots & \vdots & \vdots & \vdots & \vdots & \vdots \\ \vdots & \vdots & \vdots & \vdots & \vdots & \vdots \end{bmatrix}_{m\times 6} \begin{bmatrix} a_{11} \\ a_{12} \\ b_1 \\ a_{21} \\ a_{22} \\ b_2 \end{bmatrix}_{6\times 1} \tag{11.9}$$

The above equation has a general form given by

$$Z = H\beta + v \tag{11.10}$$

Here, $\beta = [a_{11}a_{12}b_1a_{21}a_{22}b_2]^T$ as the parameter vector. Then we obviously get the least squares solution (see Chapter 2) as

$$\beta = \left(\text{Re}(H^T H)\right)^{-1} \text{Re}(H^T z) \tag{11.11}$$

Here T indicates complex conjugate transpose and 'Re' indicates taking only the real part of the elements of matrices. Actually, other frequency domain data arrangements of the above expressions could be possible.

We note that v is the complex (domain) equation error. The equation error variance can be estimated as [7]:

$$\sigma_r^2 = \frac{1}{m-n}[(Z - H\hat{\beta})^T (Z - H\hat{\beta})] \tag{11.12}$$

Then covariance of estimates β can be obtained as:

$$\text{cov}(\tilde{\beta}) = \sigma_r^2 [\text{Re}(H^T H)]^{-1} \tag{11.13}$$

11.4.1.1 Example 11.2

Generate simulated data using the following equation:

$$\dot{x} = \begin{bmatrix} -1.43 & -1.5 \\ 0.22 & -3.75 \end{bmatrix} x + \begin{bmatrix} -6.27 \\ -12.9 \end{bmatrix} u \tag{11.14}$$

Using two doublet inputs and a sampling interval $\Delta t = 0.1$ s, obtain time histories of x consisting of 100 data samples. Estimate the parameters using the frequency domain least squares method (based on the discrete Fourier transform) in a batch/offline mode.

11.4.1.2 Solution

The data generation is carried out using eq. (11.14) and is implemented in the file '**Ch11fdsids.m**'. The signals u and x (x_1 and x_2) are shown in Fig. 11.2. The respective Fourier transforms as in eq. (11.9) are computed using '**Ch11fdsidft.m**' and are shown in Fig. 11.3. The matrices Z and H as per eq. (11.10) are computed. The unknown parameters in β are estimated using '**Ch11fdsidls.m**'. The estimated parameters are shown in Table 11.2. The program files for data generation and estimation are in folder '**Ch11FDex2**'.

Figure 11.4 demonstrates the model validation procedure with the aim to check the predictive capabilities of the model. If the system parameters are well estimated, then for any arbitrary input, the response from the estimated model and the actual system should show good match. The parameters in Table 11.2 are estimated from the data generated using two doublet inputs. For model validation, we use a different control input form (3211; see Appendix B) to generate the true system responses x_1 and x_2 from eq. (11.14). Next, the estimated parameters from Table 11.2 are used in eq. (11.14) and the 3211 input is used to obtain the model predicted responses $\hat{x}_1$ and

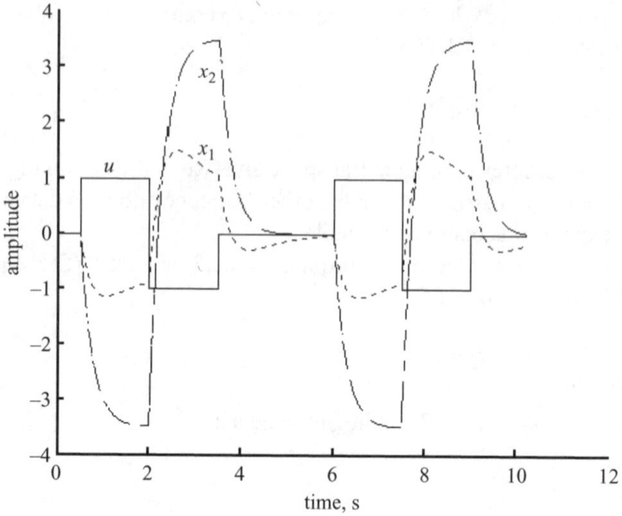

Figure 11.2 Time history of input signals (Example 11.2)

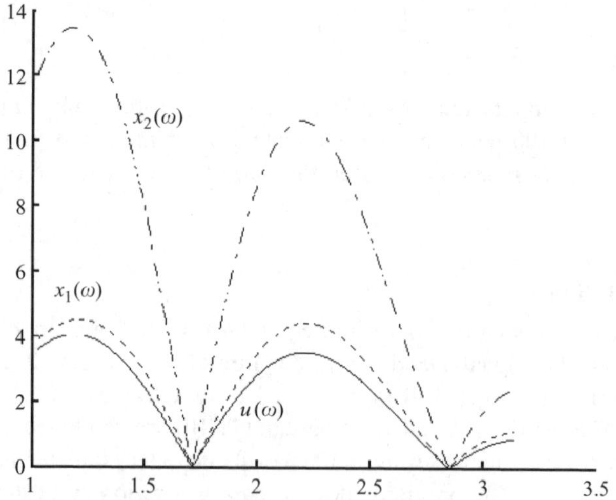

Figure 11.3 Fourier transform of the signals (Example 11.2)

$\hat{x}_2$. A comparison of the true and the model predicted responses in Fig. 11.4 shows that the estimated model has excellent predictive capabilities.

Model validation is necessary in parameter estimation studies, particularly when there are no reference parameter values available for comparison.

Table 11.2 Parameter estimation in the frequency domain (Example 11.2)

Parameter	True	Estimated
a_{11}	−1.43	−1.3979
a_{12}	−1.5	−1.48
a_{21}	0.22	0.2165
a_{22}	−3.75	−3.7522
b_1	−6.27	−6.1958
b_2	−12.9	−12.9081
PEEN	–	0.5596

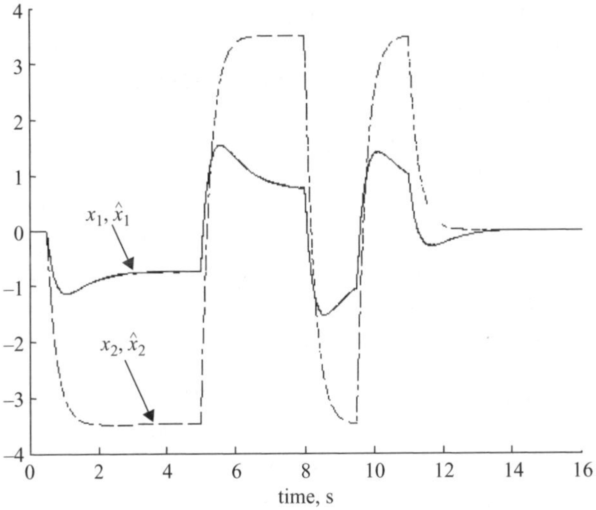

Figure 11.4 Model validation (Example 11.2)

11.4.2 Recursive Fourier transform

From eq. (11.6), we see that it should be possible to derive a recursive scheme for parameter estimation using discrete recursive Fourier transform updates. We see that the following relation holds [7]:

$$X_k(\omega) = X_{k-1}(\omega) + x_k e^{-j\omega k \Delta t} \tag{11.15}$$

with $x(\omega) \cong X(\omega)\Delta t$.

The above relationship, eq. (11.15), shows that the discrete Fourier transform at sample time k is related to that at sample time $k-1$. We also have the following equivalence:

$$e^{-j\omega k\Delta t} = e^{-j\omega\Delta t}e^{-j\omega(k-1)\Delta t} \tag{11.16}$$

The first term on the right hand side of eq. (11.16), for a given frequency and constant sampling interval, is constant.

From the foregoing, it can be seen that the discrete Fourier transform computations can be done in a recursive manner as and when the time-domain discrete data are available, thereby avoiding the storage of such data. It means that each sampled data is processed immediately. Based on the recursive discrete Fourier transform, the parameter estimation now can be accomplished in real-time fashion in frequency domain.

The major advantages of processing the data in the frequency domain are for unstable systems, and systems with noise, drift etc. The frequency domain technique allows one to choose a band of frequencies (ω_0 to ω_f) that covers the range of interest, i.e., approximately sufficiently more than the bandwidth of the dynamical system. This allows one to eliminate the transformed data outside the band. This reduces the effect of slow drift in data (at low frequencies) and the high frequency noise effect. If, for example, the band of interest is 0.1 to 10 rad/s, then we can closely space data with $\Delta\omega = 0.1$ rad/s to get about 100 points. In addition, another advantage is that since very low frequencies (say, below 0.1 rad/s) are eliminated, the effect of bias will be highly minimised and hence it is not necessary to estimate these bias parameters. It also removes the effect of trim values in the data. At the higher end (ω_f), other high frequency noise effects (like structural frequency interactions in aircraft, helicopters and spacecraft) are also eliminated, which occur beyond, say, 10 rad/s.

Thus, the frequency domain real-time parameter estimation has several such advantages as highlighted above. However, one major disadvantage is that it is not applicable directly to nonlinear system parameter estimation. Perhaps, it should be applicable to linearised nonlinear system problems. The problems with models linear-in-parameters can be also handled. However, the approach requires the measurement of all the states and measurement variables, since it is an equation error based method (see Chapter 2). This is now possible for systems with automatic control, since many internal states would also be measured.

Some other merits of the frequency domain approaches are:

1 It does not require the starting values of the parameters.
2 No tuning parameters are required like those in UD filter and recursive information processing schemes.
3 The scheme could be relatively faster than the UD filter and recurrent neural network based schemes.

However, it is felt that since recursive discrete Fourier transform computations are used, initially the information content used will be limited, and this might cause some transients. Some regularisation mechanism of bounding of the parameters would be required. One approach is to use a constraint condition on the parameters. This can be included in the cost function.

Table 11.3 Parameter estimation in the frequency domain (Example 11.3)

Parameter	True	Estimated at 6th s
a_{11}	−1.43	−1.433
a_{12}	−1.5	−1.4953
a_{21}	0.22	0.2189
a_{22}	−3.75	−3.7497
b_1	−6.27	−6.253
b_2	−12.9	−12.8995
PEEN	–	0.1198

11.4.2.1 Example 11.3

Repeat Example 11.2 and estimate the parameters using the frequency domain least squares method (based on the recursive discrete Fourier transform).

11.4.2.2 Solution

The data generation is carried out using eq. (11.14) and is implemented in '**Ch11fdsidsr.m**'. The signals u and x (x_1 and x_2) are shown in Fig. 11.2. The respective Fourier transforms as in eq. (11.9) are computed recursively at each instant of time. The matrices Z and H as per eq. (11.10) are updated accordingly. The unknown parameters in β are estimated using '**Ch11fdsidlsr.m**' at each instant. The estimated parameters at the 6th s are shown in Table 11.3. The true and recursively estimated parameters are shown in Fig. 11.5 (the initial transient effect is not shown). All programs are in folder '**Ch11FDRex3**'.

11.5 Implementation aspects of real-time estimation algorithms

With the advent of microprocessors/fast computers, the real-time implementation of the estimation algorithm has become greatly feasible and viable. In addition, parallel computers play a very important role in this direction. Several aspects need to be kept in mind for real-time implementation:

1. More reliable and stable algorithms should be used. The UD filter is one such algorithm.
2. One main aspect is to keep the algorithm structure as simple as possible. The system models used should not be too complex, otherwise, they will put a heavy burden on computation. Uncertainties in the model will cause additional errors in the estimation.
3. As much knowledge as possible on the system and data should be gathered for use in filter design (tuning, etc.), based on the previous experiments.

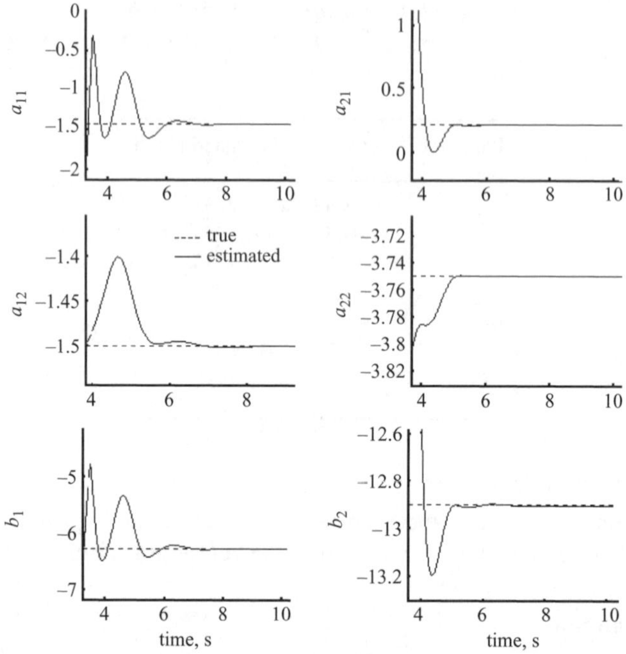

Figure 11.5 True and the recursively-estimated parameters (Example 11.3)

4 Necessary noise characterisation modules can be included or used.
5 Due to the availability of measurement data from multiple sensors, the demand on computer time will increase.
6 It may be necessary to split the data processing tasks and program on two or more individual (parallel) processors, which can have inter-processor communication links for transfer of data or results of state/parameter estimation. This calls for use of multi-programming concepts.
7 In the Kalman filter, gain/covariance computation are actually time consuming. UD filter will be more suitable here.

11.6 Need for real-time parameter estimation for atmospheric vehicles

The need for real-time parameter estimation for aircraft is becoming more realistic. The aerodynamic coefficients/parameters are required for various reasons and a variety of vehicles [7–13]:

1 Re-entry bodies.
2 To do reconfiguration control of fly-by-wire aircraft, with changing dynamics.
3 To save flight test time and fuel, since near-real time feedback of results will be available.

4 For having rapid analysis of data.
5 For aircraft development program – saving in cost and time are very important.
6 On-line failure detection and accommodation.
7 Adaptive flight control – would need changing dynamics to be taken into account.
8 Restructurable control systems, in case there is battle damage to a control surface.
9 To help expand the aircraft flight envelop.
10 For adaptive controller for in-flight simulators.
11 To take decisions on continuation of flight tests the next day – based on the results of real-time parameter estimation.

If the parameters are time varying, then we need rapid adaptation and hence the use of a short span of data. However, this requirement contradicts the need to have a longer span of data in order to avoid the correlation of data (closed loop system identification).

Specific reasons for real-time parameter estimation are as follows:

- Parameter adaptive control methods are very useful for inflight simulation to track and compensate for system parameter variations [10].
- To rapidly estimate the parameters of an aircraft's changing dynamics during a variety of flight-test manoeuvres.
- To formulate the (fault) accommodation control laws using on-line/real-time estimation of aircraft parameters in a restructurable control system.

The new generation and high performance aircraft have a highly integrated and software-intensive avionics, e.g., aircraft stall warning system, which is based on a stall warning algorithm amongst many other systems. There is a need for fault accommodation procedures for actuator and battle damage of control surface faults. These procedures can be designed based on real-time parameter estimation capability.

Major elements in real-time analysis process are:

- data acquisition in real-time at the flight test centre;
- data editing and pre-processing;
- collation of the data worthy of further analysis;
- modelling and parameter estimation;
- display of time histories and parameters.

The real-time schemes are also very useful and applicable to many industrial plants/processes, e.g., chemical plants. Quick modelling to obtain reasonably accurate models could be used in such cases to save costs by reducing the losses in the plant/process.

11.7 Epilogue

In Reference 9, a six-degree of freedom model of the aircraft is presented which accurately estimates the ratios of the aerodynamic coefficients or of derivatives. It also deals with determination of static stability margins. The approach used does not depend upon the assumptions about altitude measurements and atmospheric modelling. In Reference 8, need and methods for real-time parameters are considered for restructurable flight control systems, whereas elsewhere a computationally efficient real-time parameter scheme for reconfigurable control has been considered [13]. Although, the recursive estimation techniques have been around for more than a decade, their applications to aircraft parameter estimation are accountably small.

11.8 References

1 HSIA, T. C.: 'System identification – least squares methods' (Lexington Books, Lexington, Massachusetts, 1977)
2 SINHA, N. K., and KUSZTA, B.: 'Modelling and identification of dynamic system' (Van Nostrand, New York, 1983)
3 HAYKIN, S.: 'Adaptive filter theory' (Prentice-Hall, Englewood Cliffs, 1986)
4 LJUNG, L., and SODERSTROM, T.: 'Theory and practice of recursive identification' (MIT Press, Boston, 1983)
5 RAOL, J. R.: 'Parameter estimation of state-space models by recurrent neural networks', *IEE Proc. Control Theory and Applications (U.K.)*, 1995, **142**, (2), pp. 114–118
6 RAOL, J. R., and HIMESH, M.: 'Neural network architectures for parameter estimation of dynamical systems', *IEE Proc. Control Theory and Applications (U.K.)*, **143**, (4), pp. 387–394
7 MORELLI, E. A.: 'Real-time parameter estimation in frequency domain', AIAA-99-4043, 1999
8 NAPOLITANO, M. R., SONG, Y., and SEANOR, B.: 'In-line parameter estimation for restructurable flight control systems', *Aircraft Design*, 2001, **4**, pp. 19–50
9 QUANWEI, J., and QIONGKANG, C.: 'Dynamic model for real-time estimation of aerodynamic characteristics', *Journal of Aircraft*, 1989, **26**, (4), pp. 315–321
10 PINEIRO, L. A.: 'Real-time parameter identification applied to flight simulation', *IEEE Trans. on Aerospace and Electronic Systems*, 1993, **29**, (2), pp. 290–300
11 HARRIS, J. W., HINES, D. O., and RHEA, D. C.: 'Migrating traditional post test data analysis into real-time flight data analysis', AIAA-94-2149-CP, 1994
12 SMITH, T. D.: 'The use of in flight analysis techniques for model validation on advanced combat aircraft', AIAA-96-3355-CP, 1996
13 WARD, D. G., and MONACO, J. F.: 'Development and flight testing of a parameter identification algorithm for reconfigurable control', *Journal of Guidance, Control and Dynamics*, 1998, **21**, (6), pp. 1022–1028

11.9 Exercises

Exercise 11.1

Let $X = A + jB$. Obtain the real part of the matrix $X^T X$ where T represents the conjugate transpose.

Exercise 11.2

Obtain the inversion of a complex matrix $X = A + jB$ by 'real' operation.

Exercise 11.3

If $\hat{\beta} = [\mathrm{Re}(X^T X)]^{-1}\mathrm{Re}(X^T Y)$ simplify this expression to the extent possible by assuming $X = A + jB$ and $Y = C + jD$.

Bibliography

An additional list of books and papers related to parameter estimation is provided here.

BAKER, FRANK: 'Item response theory: parameter estimation techniques' (Assessment Systems Corporation, 1992)

NASH, JOHN C.: 'Nonlinear parameter estimation: an integrated system in BASIC' (Marcel Dekker, New York, 1987)

SINGH, V. P.: 'Entropy-based parameter estimation in hydrology' (Kluwer Academic Publishers, 1998)

KHOO, M. C. K.: 'Modelling and parameter estimation in respiratory control' (Kluwer Academic Publishers, 1990)

SODERSTROM, T.: 'Discrete-time stochastic systems: estimation and control' (Prentice Hall International Series in Systems and Control Engineering, 1995)

ENGLEZOS, P., and KALOGERAKIS, N.: 'Applied parameter estimation for chemical engineers' (Marcel-Dekker, New York, 2001)

BUZZI, H., and POOR, H. V.: 'On parameter estimation in long-code DS/CDMA systems: Cramer-Rao bounds and least-squares algorithms', *IEEE Transactions on Signal Processing*, 2003, **51**, (2), pp. 545–559

OBER, R. J.: 'The fisher information matrix for linear systems', *Systems and Control Letters*, 2002, **47**, (3), pp. 221–226

HOSIMIN THILAGAR, S., and SRIDHARA RAO, G.: 'Parameter estimation of three-winding transformers using genetic algorithm', *Eng. Applications of Artificial Intelligence: The International Journal of Intelligent Real-Time Automation*, 2002, **15**, (5), pp. 429–437

BEN MRAD, R., and FARAG, E.: 'Identification of ARMAX models with time dependent coefficients', *Journal of Dynamic Systems, Measurement and Control*, 2002, **124**, (3), pp. 464–467

VAN DER AUWERAER, H., GUILLAUME, P., VERBOVEN, P., and VANALANDUIT, S.: 'Application of a fast-stabilizing frequency domain parameter estimation method', *Journal of Dynamic Systems, Measurement and Control*, 2001, **123**, (4), pp. 651–658

STOICA, P., and MARZETTA, T. L.: 'Parameter estimation problems with singular information matrices', *IEEE Transactions on Signal Processing*, 2001, **49**, (1), pp. 87–90

JATEGAONKAR, R., and THIELECKE, F.: 'ESTIMA – an integrated software tool for nonlinear parameter estimation', *Aerospace Science and Technology*, 2002, **6**, (8), pp. 565–578

GHOSH, A. K., and RAISINGHANI, S. C.: 'Parameter estimation from flight data of an unstable aircraft using neural networks', *Journal of Aircraft*, 2002, **39**, (5) pp. 889–892

SONG, Y., CAMPA, G., NAPOLITANO, M., SEANOR, B., and PERHINSCHI, M. G.: 'On-line parameter estimation techniques – comparison within a fault tolerant flight control system', *Journal of Guidance, Control and Dynamics*, 2002, **25**, (3), pp. 528–537

NAPOLITANO, M. R., SONG, Y., and SEANOR, B.: 'On-line parameter estimation for restructurable flight control systems', *Aircraft Design: An International Journal of Theory, Technology, Applications*, 2001, **4**, (1), pp. 19–50

Appendix A

Properties of signals, matrices, estimators and estimates

A good estimator should possess certain properties in terms of errors in parameter estimation and/or errors in the predicted measurements or responses of the mathematical model thus determined. Since the measured data used in the estimation process are noisy, the parameter estimates can be considered to have some random nature. In fact, the estimates that we would have are the mean of the probability distribution, and hence the estimation error would have some associated covariance matrices. Thus, due to the stochastic nature of the errors, one would want the probability of the estimate being equal to the true value to be 1. We expect an estimator to be unbiased, efficient and consistent – not all of which might be achievable. In this appendix, we collect several properties of signals, matrices, estimators and estimates that would be useful in judging the properties and 'goodness of fit' of the parameter/state estimates and interpreting the results [1–4]. Many of these definitions, properties and other useful aspects [1–10] are used or indicated in the various chapters of the book and are compiled in this appendix.

A.1 Autocorrelation

For a random signal $x(t)$, it is defined as

$$R_{xx}(\tau) = E\{x(t)x(t+\tau)\}; \qquad \tau \text{ is the 'time-lag'}$$

Here E stands for a mathematical expectation operator.

For the stationary process, R_{xx} is dependent on τ and x only and not on t. Its value is maximum when $\tau = 0$, then it is the variance of the signal x (assuming the mean of the signal is removed). As the time tends to be large, if the R_{xx} shrinks then physically it means that the nearby values of the process x are not correlated and hence not dependent on each other. Autocorrelation of the white noise/process is an impulse function. Autocorrelation of discrete-time residuals is given as

$$R_{rr}(\tau) = \frac{1}{N-\tau}\sum_{k=1}^{N-\tau} r(k)r(k+\tau); \qquad \tau = 0, \ldots, \tau_{\max} \text{ are the discrete-time lag}$$

In order that residuals are white, the normalised values R_{rr} should lie within $\pm 1.97/\sqrt{N}$ band; only 5 per cent of R_{rr} are allowed out of the band. This property is used for checking the performance of state/parameter estimation algorithms. In practice, about 30 to 50 autocorrelation values are obtained and checked if at least 95 per cent of these values fall within the band. Then it is assumed that practically these autocorrelation values are zero and hence the residuals are white, thereby signifying that they are not 'autocorrelated'. This means that complete information has been extracted out of the measurement data for parameter estimation.

A.2 Aliasing or frequency folding

According to Shannon's sampling theorem, if the continuous time signal is sampled at more than twice the Nyquist frequency, the information content in the signal is preserved and the original continuous-time signal can be recovered from the sampled signal by reverse process. Now usually, the measured signal contains noise, which is believed to be of high frequency. For a white noise, the frequency spectrum is flat of constant (power) magnitude. For a band-limited noise, it extends up to a certain frequency. If such a continuous-time measurement is sampled, then aliasing or frequency folding is likely to occur. Let ω_N be the Nyquist or cut off frequency, ω_s the sampling frequency and Δt the sampling interval. For any frequency in the range $0 \le f \le f_N$, the higher frequencies that are aliased with f are

$$(2f_N \pm f), (4f_N \pm f), \ldots, (2nf_N \pm f)$$

Let

$$t = \frac{1}{2f_N} = \frac{1}{f_s}$$

Then

$$\cos(2\pi ft) \cong \cos\left\{2\pi(2nf_N \pm f)\frac{1}{2f_N}\right\}$$

$$\cos\left(2\pi f\left(\frac{1}{2f_N}\right)\right) \cong \cos\left\{2\pi(n) \pm \frac{\pi f}{f_N}\right\}$$

$$\cos\left(\pi\frac{f}{f_N}\right) = \cos\left\{\pi\frac{f}{f_N}\right\}$$

This shows that the noise spectra would alias with the signal spectra under certain conditions. This means that all data at frequencies $(2nf_N \pm f)$ will have the same cosine function as the data at the frequency f when sampled at points $1/f_N$ apart.

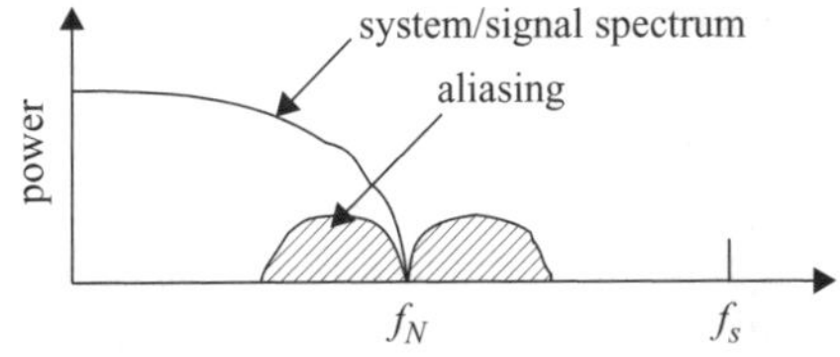

Figure A.1 Effect of aliasing

If $f_N = 100$ Hz, then data at $f = 30$ Hz would be aliased with data at frequencies 170, 230, etc. Similarly, power would also be aliased.

There are two approaches to overcome the problem of aliasing:

1 Sample the original signal at 4 to 6 times the Nyquist frequency. Then apparently, the (new) Nyquist frequency will be $f'_N = (1/2)f_s$ where the $f_s = 6f_N$, and hence we get

$$f'_N = \frac{1}{2}f_s = \frac{1}{2}(6f_N) = 3f_N$$

Now, the frequency folding will occur around $f'_N = 3f_N$ and not around f_N. This pushes the folding further away from the actual f_N, and hence, essentially minimising the aliasing of the power spectrum below f_N (thereby not affecting the frequency range of interest (see Figure A.1)).

2 Filter the continuous-time signal to reduce substantially the effect of noise. However, this will introduce time lag in the signal because of the low pass filter (lag).

Often the signals are collected at 200 samples/s and then digitally filtered down to 50 samples/s.

A.3 Bias and property of unbiased estimates

This is the difference between the true value of the parameter β and expectation value of its estimate: bias $(\beta) = \beta - E(\hat{\beta})$.

Bias, in general, cannot be determined since it depends on the true value of the parameter that is in practice unknown! Often the estimates would be biased, if the noise were not zero mean. We use a large amount of data to estimate a parameter, then we expect an estimate to centre closely on the true value. The estimate is called unbiased if $E\{\hat{\beta} - \beta\} = 0$. This property means that on the average the expected value of the estimate is the same as the true parameter. One would expect the bias to be small. Unbiased estimates are always sought and preferable. Unbiased estimate may not exist for certain problems. If an estimate is unbiased as the number of data points tends to infinity, then it is called an asymptotically unbiased estimate.

A.4 Central limit property/theorem

Assume a collection of random variables that are distributed individually according to some different distributions. Let $y = x_1 + x_2 + \cdots + x_n$; then the central limit theorem [5] states that the random variable y is approximately Gaussian (normally) distributed, if $n \to \infty$ and x should have finite expectations and variance. Often n is even 6 or 10 and the distribution of y would be almost similar to the theoretical normal distribution. This property helps in making a general assumption that noise processes are Gaussian, since one can say that they arise due to the sum of various individual noise processes of different types.

A.5 Centrally pivoted five-point algorithm

This is a numerical differentiation scheme, which uses the past and future values of the sampled data to obtain differentiated values of the variables. For example, if the past values of data y are denoted by $y_1, y_2, y_3, \ldots$, and the future values are denoted by $y_{-1}, y_{-2}, y_{-3}, \ldots$, with τ being the sampling interval, then the derivative $\dot{y}$ of y, evaluated at y_0 (pivotal point) is given by the expression [6]:

$$\text{Pivotal point} \qquad \dot{y} = \frac{1}{12\tau}[-8y_1 + y_2 - y_{-2} + 8y_{-1}]$$

with the derivative at other points expressed as

$$\text{Initial point} \qquad \dot{y} = \frac{1}{12\tau}[-25y_0 + 48y_{-1} - 36y_{-2} + 16y_{-3} - 3y_{-4}]$$

$$\text{Second point} \qquad \dot{y} = \frac{1}{12\tau}[-3y_1 - 10y_0 + 18y_{-1} - 6y_{-2} + y_{-3}]$$

$$\text{Penultimate point} \qquad \dot{y} = \frac{1}{12\tau}[3y_{-1} + 10y_0 - 18y_1 + 6y_2 - y_3]$$

$$\text{Final point} \qquad \dot{y} = \frac{1}{12\tau}[25y_0 - 48y_{-1} + 36y_2 - 16y_3 + 3y_4]$$

The estimated values are most accurate when the pivot is centrally located.

A.6 Chi-square distribution [3]

Let x_i be the normally distributed variables with zero mean and unit variance. Let

$$\chi^2 = x_1^2 + x_2^2 + \cdots + x_n^2$$

Then the random variable χ^2 has the pdf (probability density function) with n degrees of freedom:

$$p(\chi^2) = 2^{-n/2}\Gamma\left(\frac{n}{2}\right)^{-1}(\chi^2)^{(n/2)-1}\exp\left(-\frac{\chi^2}{2}\right)$$

Here, $\Gamma(n/2)$ is Euler's gamma function.

We also have $E(\chi^2) = n; \sigma^2(\chi^2) = 2n$.

Thus in the limit the χ^2 distribution approximates the Gaussian distribution with mean n and variance $2n$. If the probability density function is numerically computed from the random signal (data), then the χ^2 test can be used to determine if the computed probability density function is Gaussian or not.

A.7 Chi-square test [3]

Let x_i be normally distributed and mutually uncorrelated variables around mean m_i and with variance σ_i. Form the normalised sum of squares:

$$s = \sum_{i=1}^{n} \frac{(x_i - m_i)^2}{\sigma_i^2}$$

Then s follows the χ^2 distribution with n DOF. Often, in estimation practice, the χ^2 test is used for hypothesis testing.

A.8 Confidence level

In parameter/state estimation, requirement of high confidence in the estimated parameters/states is imperative without which the results cannot be trusted. Often this information is available from the estimation results. A statistical approach and judgment are used to define the confidence interval within which the true parameters/states are assumed to lie with 95 per cent of confidence, signifying the high probability with which truth lies within the upper and lower intervals. This signifies that the estimation error, e.g., $\tilde{\beta}_{LS}$, should be within a certain interval band. In that case, one can define:

$$P\{l < \beta < u\} = \alpha$$

It means that α is the probability that β is constrained in the interval (l, u). In other words, the probability that the true value, β, is between l (the lower bound) and u (the upper bound) is α. As the interval becomes smaller, the estimated value $\hat{\beta}$ can be taken, more confidently, as the value of the true parameter.

A.9 Consistency of estimates

One can study the behaviour of an estimator with an increased amount of data. An estimator is called asymptotically unbiased, if the bias approaches zero as the number of data tends to infinity. An asymptotically efficient estimator is obtained if the equality in CRI (Chapter 3) is approached as the number of data tends to infinity (see definition of an efficient estimator). It is very reasonable to postulate that as the number of data used increases, the estimate tends to the true value. This property is called 'consistency'. This is a stronger property than asymptotic unbiasedness, since it has to be satisfied for single realisation of estimates and not 'on the average' behaviour. It means that the strong consistency is defined in terms of the convergence of the individual realisations of the estimates and not in terms of the average properties of the estimates. Hence, all the consistent estimates are unbiased asymptotically.

The convergence is required to be with probability 1 (one) and is expressed as

$$\lim_{N\to\infty} P\{|\hat{\beta}(z_1, z_2, \ldots, z_n) - \beta| < \delta\} = 1 \qquad \forall\delta > 0$$

This means that the probability that the error in estimates (w.r.t. the true values) is less than a certain small positive value is one, as the number of data used in the estimation process tends to infinity.

A.10 Correlation coefficient

$$\rho_{ij} = \frac{\text{cov}(x_i, x_j)}{\sigma_{x_i}\sigma_{x_j}}; \qquad -1 \leq \rho_{ij} \leq 1$$

Here, $\rho_{ij} = 0$ for independent variables x_i and x_j.

For the certainly correlated process, $\rho = 1$. Thus ρ defines the degree of correlation between two random variables. This test is used in the model error method for parameter estimation. For example, in KF theory, often the assumption is made that the state error and measurement error or residuals are uncorrelated.

If a variable d is dependent on several x_i, then the correlation coefficient for each of x_i can be utilised to determine the degree (extent) of this correlation with d as

$$\rho(d, x_i) = \frac{\sum_{k=1}^{N}(d(k) - \underline{d})(x_i(k) - \underline{x}_i)}{\sqrt{\sum_{k=1}^{N}(d(k) - \underline{d})^2}\sqrt{\sum_{k=1}^{N}(x_i(k) - \underline{x}_i)^2}}$$

Here, the 'under bar' represents the mean of the variable.

If $|\rho(d, x_i)|$ is nearly equal to 1, then d can be considered to be linearly related to particular x_i. In that case, the x_i terms with the higher correlation coefficient can be included in the model (see Chapter 8).

A.11 Covariance

This is defined as

$$\text{cov}(x_i, x_j) = E\{[x_i - E(x_i)][x_j - E(x_j)]\}$$

For the independent variables x_i and x_j, the covariance matrix is null. But if the matrix is zero, it does not mean that x_i and x_j are independent. The covariance matrix is supposed to be symmetric and positive semi-definite by definition. However, in practice, when the estimation (iteration) proceeds the matrix may not retain these properties (Chapter 4). The covariance matrix plays a very important role in Kalman filter time-propagation and measurement data update equations. It provides theoretical prediction of the state-error variance and the covariance-matching concept can be used for judging the performance/consistency of the filter (tuning) (Chapter 4). A similar concept is also used in the method of model error for tuning the deterministic state estimator (see Chapter 8). The square roots of the diagonal elements of this matrix give standard deviations of the errors in estimation.

It must be also emphasised that the inverse of the covariance matrix gives the indication of the information content in the signals about the parameters. Thus, the large covariance matrix signifies higher uncertainty and low information and low confidence in the state/parameter estimation results.

A.12 Editing of data

The measured data could contain varieties of unwanted things: noise, spikes, etc. Therefore, it would be desirable to edit the raw data to get rid of noise and spikes. Since noise spectra is broadband from low frequency to high frequency, the best one can do is to filter out the high frequency component effectively. By editing the data for spikes, one removes the spikes or wild points and replaces them with suitable values. One approach is to remove the spikes and replace the data by taking the average of the nearby values of the samples. For judging the wild points, one can use the finite difference method to determine the slope. Any point exhibiting a higher slope than the allowable slope can be deleted. For filtering out the noise, one can use a Fourier transform or digital filtering methods.

A.13 Ergodicity

Assume a number of realisations of a random process are present. For an ergodic process, any statistic computed by averaging over all the members of this ensemble (realisations) at a fixed time point can also be calculated (and will be identical) by averaging over all times on a single representative member of the ensemble. Ergodicity implies stationarity, but stationary processes need not be ergodic. Often the assumption of ergodicity is implicit in the parameter estimation process. This assumption allows one to handle only one realisation of the process, e.g., data collected from only one experiment. However, from the point of view of consistency of results, it will be desirable to have at least three repeat experiments at the same operating condition. Then these data sets can be used for system identification and parameter estimation purposes, either by averaging the data or by using two sets of data for estimation and the third for model validation purposes.

A.14 Efficiency of an estimator

We have seen in Chapter 2 that we can obtain covariance of the estimation error. This covariance, which is theoretical in nature, can be used as a measure of the quality of an estimator. Assume that $\hat{\beta}_1$ and $\hat{\beta}_2$ are the unbiased estimates of the parameter vector β. We compare these estimates in terms of error covariance matrices. We form the inequality:

$$E\{(\beta - \hat{\beta}_1)(\beta - \hat{\beta}_1)^T\} \leq E\{(\beta - \hat{\beta}_2)(\beta - \hat{\beta}_2)^T\}$$

From this, we notice that the estimator $\hat{\beta}_1$ is said to be superior to $\hat{\beta}_2$ if the inequality is satisfied. If it is satisfied for any other unbiased estimator, then it is called an efficient estimator. Another useful measure is the mean square error. Since, the mean square error and the variance are identical for unbiased estimators, such optimal estimators are also called minimum variance unbiased estimators.

As we have seen in Chapter 3, the efficiency of an estimator can be defined in terms of the so-called Cramer-Rao inequality. It obtains a theoretical limit to the achievable accuracy, irrespective of the estimator used:

$$E\{[\hat{\beta}(z) - \beta][\hat{\beta}(z) - \beta]^T\} \geq M^{-1}(\beta)$$

The matrix M is the Fisher information matrix I_m (see eq. (3.44) of Chapter 3). The inverse of M is a theoretical covariance limit. It is assumed that the estimator is unbiased. Such an estimator with equality valid is called an efficient estimator. Thus, the Cramer-Rao inequality means that for an unbiased estimator, the variance of parameter estimates cannot be lower than its theoretical bound $M^{-1}(\beta)$. However, one can get an estimator with lower variance, but it would be the biased estimate. Therefore, a compromise has to be struck between acceptable bias and variance. The $M^{-1}(\beta)$ gives Cramer-Rao lower bounds for the estimates and is very useful in judging the quality of the estimates. Mostly these Cramer-Rao bounds are used in defining uncertainty levels around the estimates obtained by using a maximum likelihood/output error method (see Chapter 3).

A.15 Eigenvalues/eigenvector

The eigen (German word) values are the characteristics values of matrix A. Let $Ax = \lambda x$.

This operation means that a matrix operation on vector x simply upgrades the vector x by scalar λ. We formulate the eigenvalues/eigenvector problem as

$$(\lambda x - Ax) = 0 \quad \Rightarrow \quad (\lambda I - A)x = 0$$

Since we need a solution of x, $|\lambda I - A| = 0$ and λ_i are the so-called eigenvalues of the matrix A. If λ_i are distinct, then $A = T\Lambda T^{-1}$ and Λ is the diagonal matrix, with its elements as eigenvalues, and T is the modal matrix with its columns as eigenvectors (corresponding to each eigenvalue). A real symmetric matrix has distinct eigenvalues.

Also

$$\lambda(A) = \frac{1}{\lambda(A^{-1})}$$

Now consider a closed loop system shown in Fig. A.2.

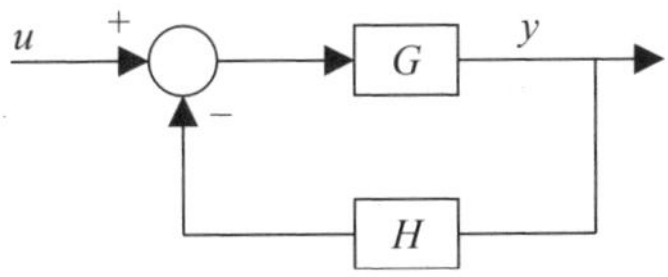

Figure A.2 Closed loop system

We have the transfer function as

$$\frac{y(s)}{u(s)} = \frac{G(s)}{1 + G(s)H(s)}$$

Here, $s = \sigma + j\omega$ is a complex frequency and $GH(s) + 1 = 0$ is the characteristic equation. Its roots are the poles of the closed loop transfer function. We also have

$$\dot{x} = Ax + Bu$$
$$y = Cx$$

Then, taking the Laplace transform, we get

$$sx(s) = Ax(s) + Bu(s)$$
$$y(s) = Cx(s)$$

By rearranging, we get

$$\frac{y(s)}{u(s)} = C(sI - A)^{-1}B = \frac{C\mathrm{adj}(sI - A)B}{|sI - A|}$$

We see the following similarities:

$$|\lambda I - A| = 0 \quad \text{and} \quad |sI - A| = 0$$

The latter will give the solution for s and they are the poles of the system $y(s)/u(s)$. We also get poles of the system from $GH(s) + 1 = 0$.

Due to the first similarity, we say that the system has 'eigenvalues' and 'poles' that are as such the same things, except that there could be cancellation of some 'poles' due to 'zeros' of $G(s)/(1 + G(s)H(s))$. Thus, in general a system will have more eigenvalues than poles. It means that all the poles are eigenvalues but all eigenvalues are not poles. However, for a system with minimal realisation, poles and eigenvalues are the same. For multi-input multi-output systems, there are specialised definitions for zeros (and poles).

Eigenvalues are very useful in control theory, however they have certain limitations when smallness or largeness of a matrix is defined. These limitations are avoided if, instead, the concept of singular values is used.

A.16 Entropy

This is a measure of some disorder in the system. Here, the system could be a plant or some industrial process. Always in a system, there could be some disorder and if the disorder is reduced, some regularisation will set in the system. Let P be the probability of the state of a system, then

$$E_s = k \log(P) + k_0$$

Let each state of the system be characterised by probability p_i, then

$$E_s = -\sum_{i=1}^{n} p_i \log p_i$$

In information theory concept, if new measurements are obtained, then there is a gain in information about the system's state and the entropy is reduced. The concept of entropy is used in model order/structure (Chapter 6) determination criteria. The idea here is that first a low order model is fitted to the data. The entropy is evaluated. Then a higher order model is fitted in succession and a reduction in the entropy is sought. Physical interpretation is when a better model is fitted to the data, the model is the refined one and the fit error is substantially reduced. The disorder is reduced, and hence the entropy.

A.17 Expectation value

Let x_i be the random variables, then the mathematical expectation E is given as

$$E(x) = \sum_{i=1}^{n} x_i P(x = x_i)$$

$$E(x) = \int_{-\infty}^{\infty} x p(x)\, dx$$

Here, P is the probability distribution of variables x, and p the pdf of variable x. Usual definition of mean of a variable does not take into account the probability of (favourable) occurrence of the variables and just gives the conventional average value of the variables. The expectation concept plays an important role in many parameter estimation methods. It can be considered as a weighted mean, where the weights are individual probabilities. In general, it can also be used to get average properties of squared quantities or two variables like x_i, y_i.

A.18 Euler-Lagrange equation [10]

Let

$$J = \int_{0}^{t_f} \phi(\dot{x}, x, t)\, dt$$

be the cost function to be minimised. We assume that the function ϕ is differentiable twice with respect to $\dot{x}$, x and t.

Let the variables be perturbed as

$$x(t) \to x(t) + \varepsilon\eta(t); \quad \dot{x}(t) \to \dot{x}(t) + \varepsilon\dot{\eta}(t); \qquad \varepsilon \text{ is a small quantity}$$

Then we get

$$\phi(\dot{x} + \varepsilon\dot{\eta}, x + \varepsilon\eta, t) = \phi(\dot{x}, x, t) + \varepsilon\left[\eta\frac{\partial\phi}{\partial x} + \dot{\eta}\frac{\partial\phi}{\partial\dot{x}}\right] + \text{higher order terms}$$

Then the differential in ϕ is obtained as

$$\Delta\phi = \int_0^{t_f} \varepsilon\left[\eta\frac{\partial\phi}{\partial x} + \dot{\eta}\frac{\partial\phi}{\partial\dot{x}}\right] dt$$

We note here that $\varepsilon \to 0$, the perturbed trajectory $\to x(t)$ and the cost function $J \to$ extremum, leading to the condition

$$\frac{\Delta\phi}{\varepsilon} \to 0 \quad \Rightarrow \quad \int_0^{t_f} \left[\eta\frac{\partial\phi}{\partial x} + \dot{\eta}\frac{\partial\phi}{\partial\dot{x}}\right] dt = 0$$

Performing integration by parts, of the second term, we get

$$\int_0^{t_f} \dot{\eta}\frac{\partial\phi}{\partial\dot{x}}\, dt = \left[\eta\frac{\partial\phi}{\partial\dot{x}}\right]_0^{t_f} - \int_0^{t_f} \eta\frac{d}{dt}\left(\frac{\partial\phi}{\partial\dot{x}}\right) dt$$

Combining the last two equations, we obtain

$$\int_0^{t_f} \eta\left[\frac{\partial\phi}{\partial x} - \frac{d}{dt}\left(\frac{\partial\phi}{\partial\dot{x}}\right)\right] dt + \left[\eta\frac{\partial\phi}{\partial\dot{x}}\right]_0^{t_f} = 0$$

Since $\eta(0) = \eta(t_f) = 0$ as $x(0)$ and $x(t_f)$ are fixed, we obtain (since η is arbitrary):

$$\frac{\partial\phi}{\partial x} - \frac{d}{dt}\left(\frac{\partial\phi}{\partial\dot{x}}\right) = 0$$

This is known as the Euler-Lagrange equation or Euler-Lagrange condition. This is applicable also to function ϕ of more variables, e.g., $\phi(\dot{x}, x, \dot{\lambda}, \lambda, \ldots, t)$, etc.

The 'integration by parts' rule used in deriving the above condition is as follows. Assume there are two variables $\dot{u}$ and v as integrand. Then, we have

$$\int_0^{t} \dot{u}v\, dt = (uv)|_0^t - \int_0^{t} \left(u\frac{dv}{dt}\right) dt$$

A.19 Fit error

Several related definitions can be found in Chapter 6.

A.20 F-distribution

See Chapter 6. Let x_1 and x_2 be normally distributed random variables with arbitrary means and variances as σ_1^2 and σ_2^2.

Let

$$s_1^2 = \frac{1}{N_1 - 1}\sum_{i=1}^{N_1}(x_{1i} - \bar{x}_1)^2 \quad \text{and} \quad s_2^2 = \frac{1}{N_2 - 1}\sum_{i=1}^{N_2}(x_{2i} - \bar{x}_2)^2$$

Now these s_1^2 and s_2^2 are the unbiased estimates of the variances, and x_{1i} and x_{2i} are the samples from the Gaussian distribution. Then

$$x_1^2 = \frac{(N_1 - 1)s_1^2}{\sigma_{x_1}^2} \quad \text{and} \quad x_2^2 = \frac{(N_2 - 1)s_2^2}{\sigma_{x_2}^2}$$

are χ^2 distributed variables with DOF $h_1 = N_1 - 1$ and $h_2 = N_2 - 1$. The ratio

$$F = \left(\frac{h_2}{h_1}\right)\left(\frac{x_1^2}{x_2^2}\right) = \frac{s_1^2\sigma_{x_2}^2}{s_2^2\sigma_{x_1}^2}$$

can be described by F-distribution with (h_1, h_2) degrees of freedom. The F-distribution is used in the F-test.

A.21 F-test

The F-test provides a measure for the probability that the two independent samples of variables of sizes n_1 and n_2 have the same variance. Let s_1^2 and s_2^2 be estimates of these variances. Then the ratio $t = s_1^2/s_2^2$ follows F-distribution with h_1 and h_2 degree of freedom. Then hypotheses are formulated as follows and tested for making decisions on the truth (which of course is unknown):

$$H_1(\sigma_1^2 > \sigma_2^2): t > F_{1-\alpha}$$
$$H_2(\sigma_1^2 < \sigma_2^2): t < F_{\alpha}$$

at the level of $1 - \alpha$ or α. The F-test is used in selecting an adequate order or structure in time-series and transfer function models. A model with lower variance of residuals is selected and a search for a better and better model is made.

A.22 Fuzzy logic/system

Uncertainty abounds in nature. Our interest is to model this uncertainty. One way is to use crisp logic and classical set theoretic based probability concepts. Uncertainties affect our systems and data. A set consists of a finite no. of elements that belong to some specified set called the universe of discourse. The crisp logic concerns itself with binary or bilinear decisions: Yes or No; 0 or 1; -1 or 1. Examples are: i) the light in a room is off or on; ii) an event A has occurred or not occurred. The real life experience shows that some extension of the crisp logic is needed. Events or occurrences leading to fuzzy logic are: i) the light could be dim; ii) day could be bright with a certain degree of brightness; iii) day could be cloudy to a certain degree; and iv) weather could be warm, cold or hazy. Thus, the idea is to allow for a degree of uncertainty with the truth and falsity (1 or 0) being at the extremes of a continuous spectrum of this uncertainty. This leads to multi-valued logic and to fuzzy set theory [7, 8].

Since 1970, fuzzy logic has seen applications in the process control industry, traffic, etc. Fuzziness is based on the theory of sets if the characteristic function is generalised to take an infinite number of values between 0 and 1. mA(x) is a membership function of x on the set A and is a mapping of the universe of discourse x on the closed interval [0,1] (see Figure A.3).

The membership function gives a measure of the degree to which x belongs to the set A: mA(x): X $\rightarrow$ [0,1]. Fuzzy variable low is described in terms of a set of positive integers in the range [0,100] $\rightarrow$ A $=$ {low}. This set expresses the degree to which the temperature is considered low over the range of all possible temperatures.

The rule based fuzzy systems can model any continuous function or system and the quality of the approximation depends on the quality of rules. These rules can be formed by the experts who have a great experience in dealing with the classical systems, which are designed/developed or maintained by them. Alternatively, the artificial neural networks can be used to learn these rules from the data. The fuzzy engineering deals with function approximations. Application to a washing machine might save the energy and wear and tear on the clothes. This approximation actually does not depend on words, cognitive theory or linguistic paradigm. It rests on the mathematics of function approximation and statistical learning theory. Since much of this mathematics is well known, there is no magic in fuzzy systems. The fuzzy system is a natural way to turn speech and measured action into functions that approximate the hard tasks.

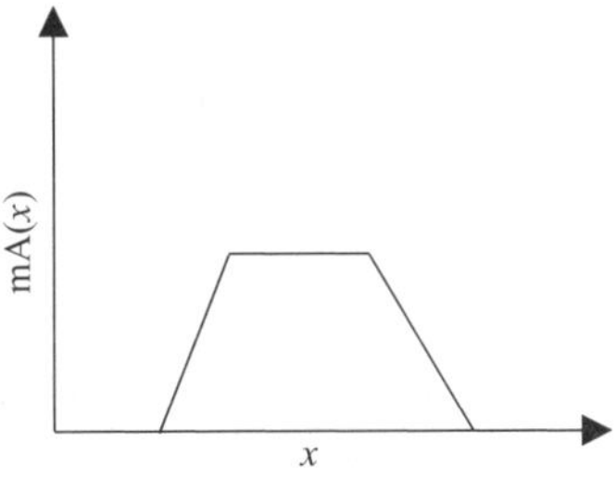

Figure A.3 Fuzzy membership

The basic unit of fuzzy approximation is the 'If…Then…' rule. As an example: If the wash water (in the washing machine) is dirty then add more detergent powder. Thus, the fuzzy system is a set of such well-defined and composed If…Then… rules that map input sets to output sets as in the previous example. The overlapping rules define polynomials and richer functions. Each input partially fires all the rules in parallel and the system acts as an associative processor as it computes the output function. The system then combines these partially fired Then part fuzzy sets in a sum and converts this sum to a scalar or vector output. These additive fuzzy systems are proven universal approximators for rules that use fuzzy sets of any shape and are computationally simple. A fuzzy variable is one whose values can be considered labels of fuzzy sets: temperature → fuzzy variable → linguistic values such as low, medium, normal, high, very high, etc. leading to membership values (on the universe of discourse – degree C). The no. of rules could be large, say 30. For a complex process control plant, one might need 60 to 80 rules and for a small task 5 to 10 rules might be sufficient, e.g., for a washing machine. A combination of 2 or 3 fuzzy conditional statements will form a fuzzy algorithm (see Chapter 4). A linguistic variable can take on values that are statements of a natural language such as: primary terms that are a label of fuzzy sets, such as high, low, small, medium, zero; negative NOT and connective AND and OR; 'hedges' like very, nearly, almost; and parenthesis. These primary terms may have either continuous or discrete membership functions. The continuous membership functions are defined by analytical functions.

The core of every fuzzy controller is the inference engine, which is a computation mechanism with which a decision can be inferred even though the knowledge may be incomplete. This mechanism gives the linguistic controller the power to reason by being able to extrapolate knowledge and search for rules, which only partially fit for any given situation for which a rule does not exist. The inference engine performs an exhaustive search of the rules in the knowledge base to determine the degree of fit for each rule for a given set of causes. A number of rules contribute to the final result to a varying degree. A fuzzy propositional implication defines the relationship between the linguistic variables of a fuzzy controller:

- Given two fuzzy sets A and B that belong to the universe of discourse X and Y respectively, then the fuzzy propositional implication is:
- R: If A then $B = A \rightarrow B = A \times B$ where $A \times B$ is the Cartesian product of the two fuzzy sets A and B.

The knowledge necessary to control a plant is usually expressed as a set of linguistic rules of the form: If (cause) then (effect). These are the rules with which new operators are trained to control a plant and they constitute the knowledge base of the system. All the rules necessary to control a plant might not be elicited, or known, and hence it is necessary to use some technique capable of inferring the control action from available rules. The fuzzy systems are suited to control of nonlinear systems and multi-valued nonlinear processes. The measurements of plant variables (even if contaminated by noise) and control actions to the plant actuators are crisp. First, fuzzify the measured plant variables, then apply fuzzy algorithm (rules/inferences) and finally de-fuzzify the results.

In Chapter 4 the fuzzy logic based adaptive Kalman filter is studied, for which the universe of the discourse is $U_{rs} = [0.0\ 0.4]$ and the universe of discourse $U_{\psi} = [0.1\ 1.5]$. Both the input and output universe spaces have been discretised into five segments. The fuzzy sets are defined by assigning triangular membership functions to each of the discretised universe. Then fuzzy implication inference leads to fuzzy output subsets. Finally, the adaptive estimation algorithm requires crisp values. A defuzzification procedure is applied using the *centre of area* method and to realise the fuzzy rule base, the fuzzy system toolbox of PC MATLAB was used for generating the results of Section 4.5.3.

Defuzzification of the output arising from the fuzzy controller is done using either the centre of gravity or centre of area method. In the centre of area method, the area under the composite membership function of the output of the fuzzy controller is taken as the final output [7].

A.23 Gaussian probability density function (pdf)

The Gaussian pdf is given as

$$p(x) = \frac{1}{\sqrt{2\pi}\sigma} \exp\left(-\frac{(x-m)^2}{2\sigma^2}\right)$$

Here, m is the mean and σ^2 is the variance of the distribution. For the measurements, given the state x (or parameters), the pdf is given by

$$p(z|x) = \frac{1}{(2\pi)^{n/2}|R|^{1/2}} \exp\left(-\frac{1}{2}(z - Hx)^T R^{-1}(z - Hx)\right)$$

In the above, R is the covariance matrix of measurement noise. The variable x can be replaced by β, the parameter vector. The maximisation of $p(z|x)$ is equivalent to minimisation of the term in the parenthesis.

A.24 Gauss-Markov process

Assume a lumped parameter linear system of first order driven by a white Gaussian noise. Then the output will be Gauss-Markov process of first order. This assumption is used in KF theory. A continuous process $x(t)$ is first order Markov if, for every k and $t_1 < t_2 < \cdots < t_k$,

$$P\{x(t_k)|x(t_{k-1}), \ldots, x(t_1)\} = P\{x(t_k)|x(t_{k-1})\}$$

This means that the probability distribution of $x(t_k)$ is dependent on the value at point $k-1$ only.

A.25 Hessian

The symmetric matrix of dimension $n \times n$ of second partial derivatives of a cost function f is termed as Hessian of the cost function. Let the cost function be dependent on the components of β, then

$$H_f = \begin{bmatrix} \dfrac{\partial f}{\partial \beta_1 \beta_1} & \dfrac{\partial f}{\partial \beta_1 \beta_2} & \cdots \\ \vdots & & \\ \dfrac{\partial f}{\partial \beta_n \beta_1} & \cdots & \dfrac{\partial f}{\partial \beta_n \beta_n} \end{bmatrix}$$

The positive Hessian indicates the minimum of the function f and the negative Hessian indicates the maximum of the cost function f. This property is useful in optimisation/estimation problems. For the LS method, $H_f = H^T H$ (see Chapter 2), and it indicates minimum of the cost function.

A.26 H-infinity based filtering

In the KF, the signal generating system is assumed a state-space model driven by a white noise process with known statistical properties. The sensor measurements are always corrupted by (white) noise process, the statistical properties of which are assumed known. Then the aim of the filter is to minimise the variance of the state estimation error.

The H-infinity problem differs from the KF, specifically in the following aspects [9]:

1 The white noise is replaced by unknown deterministic disturbance of finite energy. This is a major difference because white noise has a constant (and infinite length) spectrum – its energy is spread over the entire frequency band.
2 A specified positive real number, say γ^2, (a scalar parameter) is defined. Then the aim of the H_∞ filter is to ensure that the energy gain from the disturbance to the estimation error is less than the scalar parameter.

We know that in an estimation problem, the effect of input disturbance on the output of the estimator should be minimised, and the filter should produce the estimates of the state very close to the true states. In the H_∞ filter, this is explicitly stated and any gain from input disturbance energy to the output state error energy is to be minimised.

In the limit as $\gamma \to \infty$, the KF should emerge as a special case of the H_∞ filter.

The H_∞ philosophy has emerged from the optimal control synthesis paradigm in the frequency domain. The theory addresses the question of modelling errors and treats the worst-case scenario. The idea is to plan the worst and then optimise. Thus, we get the capability of handling plant modelling errors as well as unknown disturbances. It then also has a natural extension to the existing KF theory. The H_∞-based concept is amenable to the optimisation process and is applicable to multivariate problems.

The H_∞ concept involves a metric of signal or its error (from estimated signal), which should reflect the average size of the RMS value. In the H_∞ filtering process, the following norm is used:

$$H_\infty = \frac{\sum_{k=0}^{N}(x(k)-\hat{x}(k))^T(x(k)-\hat{x}(k))}{(x(0)-\hat{x}(0))^T P_0(x(0)-\hat{x}(0)) + \sum_{k=0}^{N} w^T(k)w(k) + \sum_{i=1}^{m}\sum_{k=0}^{N} v_i^T(k)v_i(k)}$$

We see from the structure of the H_∞ norm that input is the collection of energies from: i) the initial condition errors; ii) state disturbance; and iii) measurement noise. The output energy is directly related to state or parameter estimation error. Here, m denotes the number of sensors with independent measurement noises.

A.27 Identifiability

Given the input-output data of a system and the chosen form of the model (which when operated upon by the input, produces the output), one must be able to identify the coefficients/parameters of the model, with some statistical assumptions on the noise processes (acting on measurements). The identification methods (e.g., least squares) then yield the numerical values of these coefficients. The term 'system identification' is used in the context of identification of transfer function and time-series models. One important assumption is that the input should be persistently exciting, in order to be able to capture the modes of the system from its output. This roughly means that the spectrum of the input signal should be broader than the bandwidth of the system (that generates a time-series).

A.28 Lagrange multiplier [10]

Let the function to be optimised be given as

$$J = f(\beta_1, \beta_2)$$

subject to the constraint $e(\beta_1, \beta_2) = 0$.

From the constraint, we see that β_1 and β_2 are not independent. We form a composite cost function as

$$J_a = f(\beta_1, \beta_2) + \lambda e(\beta_1, \beta_2)$$

The above is identical to J because of the constraint equation. In J_a, λ is an arbitrary parameter. Now J_a is a function of the three variables β_1, β_2 and λ. The extremum of J_a can be obtained by solving the following equations:

$$\frac{\partial J_a}{\partial \beta_1} = \frac{\partial f}{\partial \beta_1} + \lambda \frac{\partial e}{\partial \beta_1} = 0$$

$$\frac{\partial J_a}{\partial \beta_2} = \frac{\partial f}{\partial \beta_2} + \lambda \frac{\partial e}{\partial \beta_2} = 0$$

$$\frac{\partial J_a}{\partial \lambda} = e(\beta_1, \beta_2) = 0$$

Assuming $(\partial e/\partial \beta_2) \neq 0$, we solve the second equation for λ and substitute λ in the first equation. We need to ensure that

$$\left(\frac{\partial e}{\partial \beta_1}\right)^2 + \left(\frac{\partial e}{\partial \beta_2}\right)^2 \neq 0$$

The parameter λ is called the 'Lagrange Multiplier' and it facilitates the incorporation of the constraint in the original cost function.

A.29 Measurement noise covariance matrix

This matrix for discrete-time noise, given as $R(k)$, is called the noise covariance matrix. For continuous-time measurement noise, the covariance matrix $R(t)$ is called the spectral density matrix.

In the limit $\Delta t \to 0$, $R(k) = R(t)/\Delta t$, such that the discrete noise sequence tends to the infinite valued pulses of zero duration. This ensures that the area under the 'impulse' autocorrelation function $R_k \Delta t =$ the area R under the continuous white noise impulse autocorrelation function.

A.30 Mode

In parameter estimation, we use data affected by random noise, etc. Hence, the estimate of the parameter vector is some measure or quantity related to the probability distribution. It could be mode, median or mean of the distribution. The mode of the distribution defines the value of x (here x could be a parameter vector) for which the probability of observing the random variable is a maximum. Thus mode signifies the argument (i.e. x or parameter vector) that gives the maximum of the probability distribution. The distribution could be unimodal or multi-modal. In practical situations multi-modal distribution could occur.

A.31 Monte-Carlo method

For a dynamic system, assume that the simulated data are used for parameter estimation. Therefore, for one set of data, we get one set of estimated parameters.

Next, we change the seed number for the random number generator, add these data as noise to measurements, and again estimate the parameters with the new data set. In the new data set, the original signal remains the same. Thus, we can formulate a number of such data sets with different seed nos. and obtain parameters to see the variability of the estimates across different realisations of the data, mimicking the practical real life situation. Then we can obtain the mean value and the variance of the parameter estimates using all the individual estimates from different realisations. This will help in judging the performance of the estimation method. The mean of the parameters should converge to the true values. If we take two estimation procedures/methods, then the one that gives estimates (mean value) closer to the true value and less variance will bc thc better choice. This approach can be used for linear or nonlinear systems. A similar procedure can be used for state estimation methods also. This procedure is numerical and could become computationally intensive. Depending upon the problem and its complexity often 400 or 500 simulation runs are required. However, as little as 20 runs are also often used to generate average results.

A.32 Norm of a vector

We need to have a measure of a vector, or matrix (of a signal) in order to have knowledge of their magnitudes and strengths. This will also help in judging the magnitude of state error or measurement error or residuals. Let x be a vector. Then the distance measure or norm is defined as

$$L_p = \|x\|_p = \left(\sum_{i=1}^{n} |x_i|^p\right)^{1/p}; \qquad p \geq 1$$

We have three possibilities [3]:

1 If $p = 1$ then the length of vector x is $\|x\|_1 = |x_1| + |x_2| + \cdots + |x_n|$. Then the centre of a probability distribution estimated using L_1 norm is the median of the distribution.

2 If $p = 2$, then it is called the Euclidean norm and gives a length of the vector. We see that it is the square root of the inner product of the vector x with itself. In addition, it is equal to the square root of the sum of the squares of the components of x. This leads to the Schwarz inequality:

$$|x^T y| \leq \|x\| \cdot \|y\|$$

Here y is another vector. Also for $p = 2$, the centre of a distribution estimated using L_2 norm is the mean of the distribution and is the chi-square estimator. This norm is used in many state/parameter estimation problems to define the cost functions in terms of state error or measurement error. The minimisation problems with this norm are mathematically highly tractable. This leads to the least squares or maximum likelihood estimator as the case may be.

3 If $p = \infty$, then it gives the Chebyshev norm. It signifies the maximum of the absolute value of x_i

$$\|x\|_{p=\infty} = \max |x_i|$$

It looks as if this norm is related to the H-infinity norm.

A.33 Norm of matrix

The measure of strength of a matrix can be determined in terms of its determinant or eigenvalues (e.g., the largest or the smallest eigenvalue). One measure is given as

$$\|A\| = \sup_{\|x\|=1} \{\|Ax\|\}$$

Often a singular value is used as a norm of a matrix.

A.34 Observability

This generally applies to state observability. It means that if the system (its representation) is (controllable and) observable, then given the input-output responses of the system, one must be able to determine/observe the states of the system (also given the model information, essentially its structure). Often certain assumptions on statistics of the noise processes are made.

A.35 Outliers

Often an outlier is considered a noisy data point that does not belong to normal (Gaussian) distribution. In a measurement if one encounters the noise processes that have very large variance and small variance also, the one with very large variance can be regarded as an outlier. The outliers need be handled very carefully; otherwise overall estimation results could be degraded. The methods to deal with the outliers should be an integral part of the estimation process. Outliers can be considered to belong to Gaussian distribution but with a very large variance. The proper use of the method would yield robust estimators. Depending upon the problem, outliers could also be considered to belong to other types of distribution, e.g., uniform, as well. Often, a simple approach to discard an outlier measurement is used. If the computed residual value from the predicted measurement is greater than three times the predicted standard deviation, then that measurement is ignored. This is an *ad hoc* method to make the filtering/estimation process robust, in the presence of outliers.

A.36 Parameter estimation error norm (PEEN)

$$\text{PEEN} = \frac{\text{norm}(\beta - \hat{\beta})}{\text{norm}(\beta)} \times 100$$

A.37 Pseudo inverse

A pseudo inverse for an $m \times n$ matrix A is given by

$$(A^T A)^{-1} A^T$$

For an $n \times n$ matrix, it degenerates to a conventional inverse. Also, singular value decomposition can be used to compute the pseudo inverse. We see from eq. (2.4), that the pseudo inverse naturally appears in the parameter estimator equation.

A.38 Root sum square error (RSSE)

Let x_t, y_t, z_t be the true trajectories and $\hat{x}$, $\hat{y}$, $\hat{z}$ be the estimated/predicted trajectories. Then

$$\text{RSSE}(t) = \sqrt{(x_t(t) - \hat{x}(t))^2 + (y_t(t) - \hat{y}(t))^2 + (z_t(t) - \hat{z}(t))^2}$$

This is valid also for the discrete-time signals.

$$\text{Percentage RSSE} = \frac{\text{RSSE}(t)}{\sqrt{x_t^2(t) + y_t^2(t) + z_t^2(t)}} \times 100$$

A.39 Root mean square error (RMSE)

$$\text{RMSE} = \frac{1}{N}\sqrt{\frac{(x_t(t) - \hat{x}(t))^2 + (y_t(t) - \hat{y}(t))^2 + (z_t(t) - \hat{z}(t))^2}{3}}$$

Percentage RMSE can also be defined.

A.40 Singular value decomposition (SVD)

A matrix $A(m \times n)$ can be factored into

$$A = USV^T$$

Here, U and V are orthogonal matrices with dim. (m, m) and (n, n) respectively. S is an (m, n) diagonal matrix. Its elements are real and non-negative and are called singular values, ρ_i, of the matrix A. The concept of singular values is used in control system analysis and design as well as in the determination of the model order of the system when significant SVs are retained to reduce the complexity of the identified model. Also, SVD is used in parameter/state estimation problems to obtain numerically stable algorithms.

A.41 Singular values (SV)

Singular values σ are defined for a matrix A as

$$\sigma_i(A) = \sqrt{\lambda_i\{A^T A\}} = \sqrt{\lambda_i\{AA^T\}}$$

Here λ_i are the eigenvalues of the matrix $A^T A$.

The maximum SV of a matrix A is called the spectral norm of A:

$$\sigma_{\max}(A) = \max_{x \neq 0} \frac{\|Ax\|_2}{\|x\|_2} = \|A\|_2$$

For a singular matrix A, one can use $\sigma_{\min}(A) = 0$.

Thus, for a vector, the Eucledian norm is

$$l_2 = \left(\sum_i |x_i|^2\right)^{1/2}$$

For a matrix A, $\sigma_{\max}(A)$ can be used.

A.42 Steepest descent method

The simplest form is explained below.

Let f be the function of a variable, say, parameter β, i.e., $f(\beta)$. We consider that $f(\beta)$ is a cost function with at least one minimum as shown in Fig. A.4.

Then we use the parameter estimation rule as

$$\frac{d\beta}{dt} = -\frac{\partial f(\beta)}{\partial \beta}$$

What this means is that the rate of change in the parameter (with respect to time) is in the negative direction of the gradient of the cost function with respect to the parameter.

We can discretise the above formulation as

$$\beta(i+1) = \beta(i) - \mu \frac{\partial f}{\partial \beta}$$

In the above expression, Δt is absorbed in the factor μ. We see from Fig. A.4 that at point p_2, the slope of f is positive and hence we get a new value of β (assuming $\mu = 1$) as

$$\beta = \beta_2 - (\text{positive value of the slope})$$

Hence, $\beta < \beta_2$ and β is approaching $\hat{\beta}$. Similarly, when the slope is negative, β will approach $\hat{\beta}$, and so on.

The method will have problems if there are multiple minima and there is high noise in the measurement data. Small values of μ will make the algorithm slow and large values might cause it to oscillate. Proper choice of μ should be arrived at by

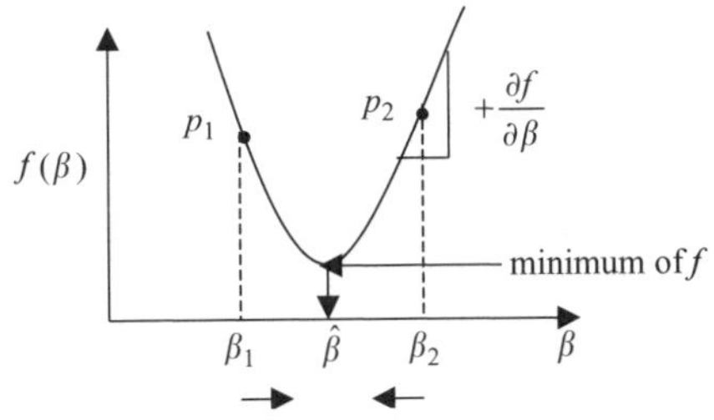

Figure A.4 Cost function

trials using the real data for the estimation purpose. The μ is called obviously the 'step size' or 'tuning parameter'.

The method is suitable also for a function of more than one variable. It is also known as the steepest ascent or 'hill climbing' method.

A.43 Transition matrix method

This method is used for solving the matrix Riccati equation (eq. (8.49)) [4].

Based on the development in Section 8.4, we have the following set of linear equations (for $a = Sb$):

$$\dot{b} = -f_{\hat{x}}^T b + 2H^T R^{-1} H a \qquad \text{(refer to eq. (8.54))}$$

$$\dot{a} = \tfrac{1}{2} Q^{-1} b + f_{\hat{x}} a \qquad \text{(refer to eq. (8.55))}$$

or, in a compact form, we have

$$\begin{bmatrix} \dot{b} \\ \dot{a} \end{bmatrix} = \begin{bmatrix} -f_{\hat{x}}^T & 2H^T R^{-1} H \\ \frac{1}{2} Q^{-1} & f_{\hat{x}} \end{bmatrix} \begin{bmatrix} b \\ a \end{bmatrix}$$

$\dot{X} = FX$ and its solution can be given as

$$X(t_0 + \Delta t) = \Phi(\Delta t) X(t_0)$$

Here, Φ is the transition matrix given as

$$\Phi(\Delta t) = e^{F\Delta t} = \begin{bmatrix} \phi_{bb} & \phi_{ba} \\ \phi_{ab} & \phi_{aa} \end{bmatrix}$$

Since the elements of matrix F are known, the solution X can be obtained which in turn gives b and a. Thus, S can be obtained as

$$S(t_0 + \Delta t) = [\phi_{ab}(\Delta t) + \phi_{aa}(\Delta t)\, S(t_0)][\phi_{bb}(\Delta t) + \phi_{ba}(\Delta t)\, S(t_0)]^{-1}$$

The above procedure can also be used to solve the continuous-time matrix Riccati equation for the covariance propagation in the continuous-time Kalman filter.

A.44 Variance of residuals

$$\sigma_r^2 = \frac{1}{N-1}\sum_{k=1}^{N}(r(k) - \underline{r})^2$$

Here, $\underline{r}$ is the mean of the residuals.

A.45 References

1 HSIA, T. C.: 'System identification – least squares methods' (Lexington Books, Lexington, Massachusetts, 1977)
2 SORENSON, H. W.: 'Parameter estimation – principles and problems' (Marcel Dekker, New York, 1980)
3 DRAKOS, N.: 'Untitled', Computer based learning unit. University of Leeds, 1996 (Internet site: rkb.home.cern.ch/rk6/AN16pp/mode165.html)
4 GELB, A. (Ed.): 'Applied optimal estimation' (M.I.T. Press, Cambridge, MA, 1974)
5 PAPOULIS, A.: 'Probability, random variables and stochastic processes' (McGraw Hill, Singapore, 1984, 2nd edn)
6 FORSYTHE, W.: 'Digital algorithm for prediction, differentiation and integration', *Trans. Inst. MC*, 1979, **1**, (1), pp. 46–52
7 KOSKO, B.: 'Neural networks and fuzzy systems – a dynamical systems approach to machine intelligence' (Prentice Hall, Englewood Cliffs, NJ, 1992)
8 KING, R. E.: 'Computational intelligence in control engineering' (Marcel Dekker, New York, 1999)
9 GREEN, M., and LIMEBEER, D. N.: 'Linear robust control' (Prentice-Hall, Englewood Cliffs, NJ, 1995)
10 HUSSAIN, A., and GANGIAH, K.: 'Optimization techniques' (The Macmillan Company of India, India, 1976)

Appendix B

Aircraft models for parameter estimation

B.1 Aircraft nomenclature

To understand aircraft dynamics and the equations of motion, it is essential to become familiar with the aircraft nomenclature. The universally accepted notations to describe the aircraft forces and moments, the translational and rotational motions and the flow angles at the aircraft are shown in Fig. B.1. The axis system is assumed fixed at the aircraft centre of gravity and moves along with it. It is called the body-axis system. The forces and moments acting on the aircraft can be resolved along the axes. The aircraft experiences inertial, gravitational, aerodynamic and propulsive forces. Of these, the aerodynamic forces X, Y and Z, and the moments L, M and N are of importance as these play the dominant role in deciding how the aircraft behaves.

Figure B.1 also shows the aircraft primary control surfaces along with the normally accepted sign conventions. All surface positions are angular in deflection. The aileron deflection causes the aircraft to roll about the X-axis, the rudder deflection causes the

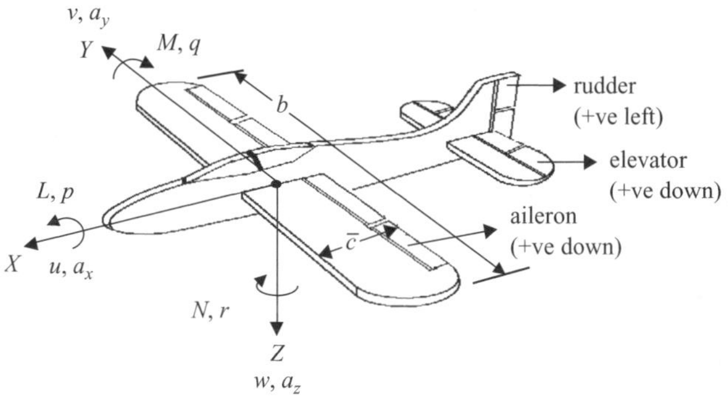

Figure B.1 Body-axis system

aircraft to yaw about the Z-axis and the elevator deflection causes it to pitch about the Y-axis.

The three Euler angles describing the aircraft pitch attitude, roll angle and heading angle are illustrated in Fig. B.2 [1].

The body-axis system notations are put together in Table B.1 below for better understanding.

As shown in Fig. B.3, the aircraft velocity can be resolved into u, v and w components along the X, Y and Z-axes. The total velocity V of the aircraft can be expressed as

$$V = \sqrt{u^2 + v^2 + w^2} \tag{B1.1}$$

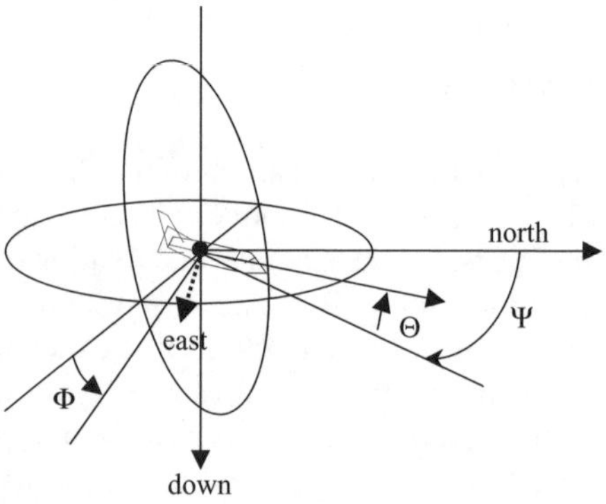

Figure B.2 Euler angles

Table B.1 Aircraft nomenclature

	X-axis Longitudinal axis Roll axis	Y-axis Lateral axis Pitch axis	Z-axis Vertical axis Yaw axis
Velocity components	u	v	w
Angular rates	Roll rate p	Pitch rate q	Yaw rate r
Euler angles	Roll angle ϕ	Pitch angle θ	Heading angle ψ
Accelerations	a_x	a_y	a_z
Aerodynamic forces	X	Y	Z
Aerodynamic moments	L	M	N
Control surface deflections	Elevator deflection δ_e	Aileron deflection δ_a	Rudder deflection δ_r
Moment of inertia	I_x	I_y	I_z

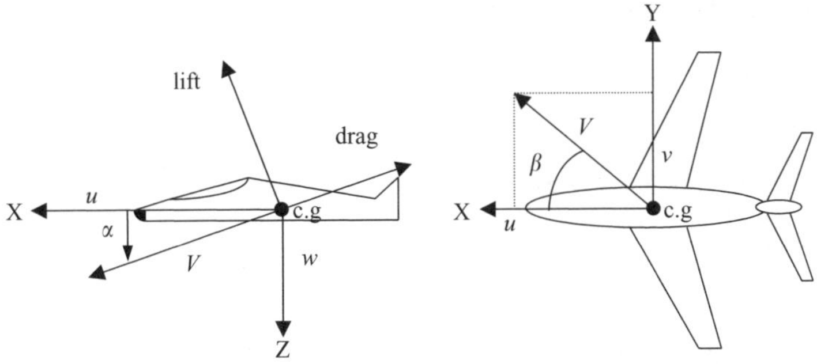

Figure B.3 Flow angles

The flow angles of the aircraft are defined in terms of angle-of-attack α and angle of sideslip β, which can be expressed in terms of the velocity components as

$$\begin{aligned} u &= V\cos\alpha\cos\beta \\ v &= V\sin\beta \\ w &= V\sin\alpha\cos\beta \end{aligned} \tag{B1.2}$$

or

$$\begin{aligned} \alpha &= \tan^{-1}\frac{w}{u} \\ \beta &= \sin^{-1}\frac{v}{V} \end{aligned} \tag{B1.3}$$

If S represents the reference wing area, $\bar{c}$ is the mean aerodynamic chord, b is the wingspan and $\bar{q}$ is the dynamic pressure $\frac{1}{2}\rho V^2$, and then the aerodynamic forces and moments can be written as

$$\begin{aligned} X &= C_X\bar{q}S \\ Y &= C_Y\bar{q}S \\ Z &= C_Z\bar{q}S \\ L &= C_l\bar{q}S\bar{c} \\ M &= C_m\bar{q}Sb \\ N &= C_n\bar{q}Sb \end{aligned} \tag{B1.4}$$

where the coefficients C_X, C_Y, C_Z, C_l, C_m and C_n are the non-dimensional body-axis force and moment coefficients. The forces acting on the aircraft are also expressed in terms of lift and drag. The lift force acts normal to the velocity vector V while the drag force acts in the direction opposite to V. The non-dimensional coefficients of lift and drag are denoted by C_L and C_D, and can be expressed in terms of the body-axis non-dimensional coefficients using the relations:

$$\begin{aligned} C_L &= -C_Z \cos\alpha + C_X \sin\alpha \\ C_D &= -C_X \cos\alpha - C_Z \sin\alpha \end{aligned} \tag{B1.5}$$

In a similar way, C_X and C_Z can be expressed in terms of C_L and C_D as

$$\begin{aligned} C_X &= C_L \sin\alpha - C_D \cos\alpha \\ C_Z &= -(C_L \cos\alpha + C_D \sin\alpha) \end{aligned} \tag{B1.6}$$

In flight mechanics, the normal practice is to express the non-dimensional force and moment coefficients in terms of aircraft stability and control derivatives. The objective of the aircraft parameter estimation methodology is to estimate these derivatives from flight data.

B.2 Aircraft non-dimensional stability and control derivatives

The process of expressing the non-dimensional force and moment coefficients in terms of stability and control derivatives was first introduced by Bryan [2]. The procedure is based on the assumption that the aerodynamic forces and moments can be expressed as functions of Mach number M, engine thrust F_T and other aircraft motion and control variables α, β, p, q, r, ϕ, θ, δ_e, δ_a and δ_r. Using Taylor series expansion, the non-dimensional coefficients can be represented as [3]:

$$\begin{aligned} C_D &= C_{D_0} + C_{D_\alpha}\alpha + C_{D_q}\frac{q\bar{c}}{2V} + C_{D_{\delta_e}}\delta_e + C_{D_M} M + C_{D_{F_T}} F_T \\ C_L &= C_{L_0} + C_{L_\alpha}\alpha + C_{L_q}\frac{q\bar{c}}{2V} + C_{L_{\delta_e}}\delta_e + C_{L_M} M + C_{L_{F_T}} F_T \\ C_m &= C_{m_0} + C_{m_\alpha}\alpha + C_{m_q}\frac{q\bar{c}}{2V} + C_{m_{\delta_e}}\delta_e + C_{m_M} M + C_{m_{F_T}} F_T \\ C_l &= C_{l_0} + C_{l_\beta}\beta + C_{l_p}\frac{pb}{2V} + C_{l_r}\frac{rb}{2V} + C_{l_{\delta_a}}\delta_a + C_{l_{\delta_r}}\delta_r \\ C_n &= C_{n_0} + C_{n_\beta}\beta + C_{n_p}\frac{pb}{2V} + C_{n_r}\frac{rb}{2V} + C_{n_{\delta_a}}\delta_a + C_{n_{\delta_r}}\delta_r \end{aligned} \tag{B2.1}$$

The body-axis force coefficients can also be expressed in the derivative form in a similar fashion:

$$\begin{aligned} C_X &= C_{X_0} + C_{X_\alpha}\alpha + C_{X_q}\frac{q\bar{c}}{2V} + C_{X_{\delta_e}}\delta_e + C_{X_M} M + C_{X_{F_T}} F_T \\ C_Y &= C_{Y_0} + C_{Y_\beta}\beta + C_{Y_p}\frac{pb}{2V} + C_{Y_r}\frac{rb}{2V} + C_{Y_{\delta_a}}\delta_a + C_{Y_{\delta_r}}\delta_r \\ C_Z &= C_{Z_0} + C_{Z_\alpha}\alpha + C_{Z_q}\frac{q\bar{c}}{2V} + C_{Z_{\delta_e}}\delta_e + C_{Z_M} M + C_{Z_{F_T}} F_T \end{aligned} \tag{B2.2}$$

Each force or moment derivative can be defined as the change in the force or moment due to unit change in the motion or control variable. For example, the stability derivative C_{L_α} is defined as:

$$C_{L_\alpha} = \frac{\partial C_L}{\partial \alpha} \tag{B2.3}$$

i.e., C_{L_α} is defined as the change in C_L for a unit change in α. Note that, while C_L is dimensionless, C_{L_α} has a dimension of '/rad'.

The above list of aircraft derivatives is by no means exhaustive. For example, the aerodynamic coefficients can also be expressed in terms of derivatives due to change in forward speed, e.g., C_{L_u}, C_{D_u}, C_{Z_u} and C_{m_u}. Use of higher order derivatives (e.g., $C_{X_{\alpha^2}}$, $C_{Z_{\alpha^2}}$ and $C_{m_{\alpha^2}}$) to account for nonlinear effects and $C_{L\dot{\alpha}}$ and $C_{m\dot{\alpha}}$ derivatives to account for unsteady aerodynamic effects is common. The choice of the derivatives to be included for representing the force or moment coefficients is problem specific.

Some more information on the aircraft stability and control derivatives is provided below [3, 4]:

a *Speed derivatives (C_{L_u}, C_{D_u} and C_{m_u})*
The drag, lift and pitching moment coefficients are affected by the change in forward speed. C_{L_u} affects the frequency of the slow varying longitudinal phugoid mode (discussed later). The change in C_{D_u} is particularly noticeable at high speeds. C_{m_u} is frequently neglected.

b *Angle-of-attack derivatives (C_{L_α}, C_{D_α} and C_{m_α})*
C_{L_α} is an important derivative that represents the lift-curve slope. The derivative C_{D_α} is often neglected in flight data analysis but can assume importance at low speeds, particularly during landing and take-off. C_{m_α} is the basic stability parameter. A negative value of C_{m_α} indicates that the aircraft is statically stable.

c *Pitch rate derivatives (C_{L_q}, C_{D_q} and C_{m_q})*
The aerodynamic forces on the aircraft wing and horizontal tail vary with change in pitch rate q. The contributions from C_{L_q} and C_{D_q} are usually not significant. However, the contribution to pitching moment from horizontal tail due to change in q is quite significant. The derivative C_{m_q} contributes to the damping in pitch. Usually more negative values of C_{m_q} signify increased damping.

d *Angle-of-attack rate derivatives ($C_{L\dot{\alpha}}$, $C_{D\dot{\alpha}}$ and $C_{m\dot{\alpha}}$)*
These derivatives can be used to model the unsteady effects caused by the lag-in-downwash on the horizontal tail (see Section B.18).

e *Sideslip derivatives (C_{Y_β}, C_{l_β} and C_{n_β})*
C_{Y_β} represents the side-force damping derivative ($C_{Y_\beta} < 0$). It contributes to the damping of Dutch-roll mode (discussed later). It is used to compute the contribution of the vertical tail to C_{l_β} and C_{n_β}. The derivative C_{l_β} represents the rolling moment created on the airplane due to sideslip (dihedral effect). For rolling stability, $C_{l_\beta} < 0$. The derivative C_{n_β} represents the directional or weathercock stability ($C_{n_\beta} > 0$ for aircraft possessing static directional stability). Both C_{l_β} and C_{n_β} affect the aircraft Dutch-roll mode and spiral mode.

f *Roll rate derivatives (C_{Y_p}, C_{l_p} and C_{n_p})*
C_{Y_p} has a small contribution and is often neglected. C_{l_p} (negative value) is the damping in roll parameter and determines roll subsidence. C_{n_p} is a cross derivative that influences the frequency of the Dutch-roll mode.

g *Yaw rate derivatives (C_{Y_r}, C_{l_r} and C_{n_r})*
C_{Y_r} is frequently neglected. C_{l_r} affects the aircraft spiral mode. C_{n_r} is the damping in yaw parameter that contributes to damping of the Dutch-roll mode in a major way.

h *Longitudinal control derivatives ($C_{L_{\delta e}}$, $C_{D_{\delta e}}$ and $C_{m_{\delta e}}$)*
Among the longitudinal control derivatives, $C_{m_{\delta e}}$ representing the elevator control effectiveness is the most important parameter.

i *Lateral control derivatives ($C_{Y_{\delta a}}$, $C_{l_{\delta a}}$ and $C_{n_{\delta a}}$)*
While $C_{Y_{\delta a}}$ is usually negligible, $C_{l_{\delta a}}$ and $C_{n_{\delta a}}$ are important derivatives that represent the aileron control effectiveness and the adverse yaw derivative, respectively. $C_{n_{\delta a}}$ is an important lateral-directional control derivative.

j *Directional control derivatives ($C_{Y_{\delta r}}$, $C_{l_{\delta r}}$ and $C_{n_{\delta r}}$)*
$C_{n_{\delta r}}$ is an important lateral-directional control derivative representing rudder effectiveness.

B.3 Aircraft dimensional stability and control derivatives

When the change in airspeed is not significant during the flight manoeuvre, the forces X, Y, Z and the moments L, M and N can be expanded in terms of the dimensional derivatives rather than non-dimensional derivatives for parameter estimation.

$$
\begin{aligned}
X &= X_u u + X_w w + X_q q + X_{\delta e}\delta_e \\
Y &= Y_v v + Y_p p + Y_q q + Y_r r + Y_{\delta_a}\delta_a + Y_{\delta_r}\delta_r \\
Z &= Z_u u + Z_w w + Z_q q + Z_{\delta e}\delta_e \\
L &= L_v v + L_p p + L_q q + L_r r + L_{\delta_a}\delta_a + L_{\delta_r}\delta_r \\
M &= M_u u + M_w w + M_q q + M_{\delta e}\delta_e \\
N &= N_v v + N_p p + N_q q + N_r r + N_{\delta_a}\delta_a + N_{\delta_r}\delta_r
\end{aligned}
\tag{B3.1}
$$

B.4 Aircraft equations of motion

The dynamics of aircraft flight are described by the equations of motion, which are developed from Newtonian mechanics. While in flight, the aircraft behaves like a dynamical system, which has various inputs (forces and moments) acting on it. For a given flight condition (represented by altitude, Mach no. and c.g. loading), a control input given by the pilot will cause the forces and moments to interact with the basic natural characteristics of the aircraft thereby generating certain responses, also called states. These responses contain the natural dynamical behaviour of the aircraft, which can be described by a set of equations.

An aircraft has six degrees of freedom motion in atmosphere. The use of the full set of equations of motion for aircraft data analysis, however, may not always turn out to be a beneficial proposition. Depending upon the problem definition, simplified equations can give results with less computational requirements and no loss in the accuracy of the estimated parameters.

Since most of the aircraft is symmetric about the X-Z plane, the six degrees of freedom equations of motion can be split into two separate groups – one characterising the longitudinal motion of the aircraft and the other pertaining to the lateral-directional motion. Thus, we assume that the longitudinal and lateral motions are not coupled. The other two major assumptions made in deriving the simplified aircraft equations of motion are: i) aircraft is a rigid body; and ii) deviations of the aircraft motion from its equilibrium are small. With these assumptions and following Newton's second law, the components of forces and moments acting on the aircraft can be expressed in terms of the rate of change of linear and angular momentum as follows [4]:

$$\begin{aligned}
X &= m(\dot{u} + qw - rv) \\
Y &= m(\dot{v} + ru - pw) \\
Z &= m(\dot{w} + pv - qu) \\
L &= I_x\dot{p} - I_{xz}\dot{r} + qr(I_z - I_y) - I_{xz}pq \\
M &= I_y\dot{q} + pr(I_x - I_z) + I_{xz}(p^2 - r^2) \\
N &= I_z\dot{r} - I_{xz}\dot{p} + pq(I_y - I_x) + I_{xz}qr
\end{aligned} \tag{B4.1}$$

Longitudinal equations of motion

The longitudinal motion consists of two oscillatory modes:

(i) Short period mode.
(ii) Long period (phugoid) mode.

Short period approximation (see Fig. B.4)

The short period motion is a well damped, high frequency mode of an aircraft. The variations in velocity are assumed small. Therefore, this mode can be represented by only two degrees of freedom motion that provides a solution to the pitch

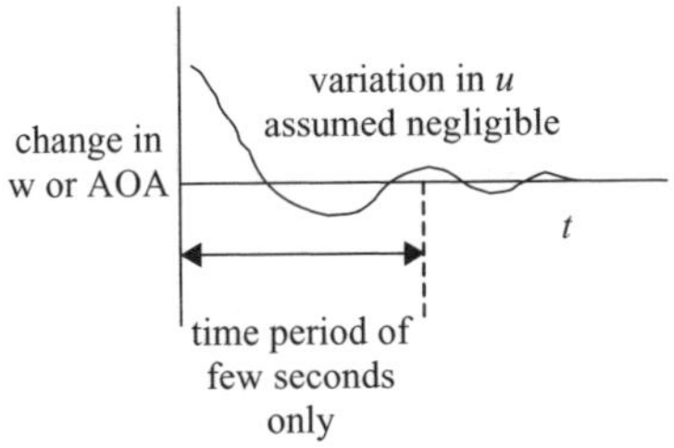

Figure B.4 Short period mode

moment and vertical force equations (the X-force equation need not be considered since there is no appreciable change in forward speed).

It is a normal practice to represent the aircraft equations as first order differential equations.

State equations

A simplified model of the aircraft longitudinal short period motion can then be written as:

$$\begin{aligned}\dot{w} &= Z_w w + (u_0 + Z_q)q + Z_{\delta_e}\delta_e \\ \dot{q} &= M_w w + M_q q + M_{\delta_e}\delta_e\end{aligned} \tag{B4.2}$$

Equation (B4.2) can be obtained by combining eqs (B3.1) and (B4.1) and using the definitions of the stability and control derivatives [4]:

$$\begin{aligned}Z_w &= \frac{1}{m}\frac{\partial Z}{\partial w}; \quad Z_q = \frac{1}{m}\frac{\partial Z}{\partial q}; \quad Z_{\delta_e} = \frac{1}{m}\frac{\partial Z}{\partial \delta_e} \\ M_w &= \frac{1}{I_y}\frac{\partial M}{\partial w}; \quad M_q = \frac{1}{I_y}\frac{\partial M}{\partial q}; \quad M_{\delta_e} = \frac{1}{I_y}\frac{\partial M}{\partial \delta_e}\end{aligned} \tag{B4.3}$$

Since $\alpha \approx w/u_0$, the above equations can also be written in terms of α instead of w:

$$\begin{aligned}\dot{\alpha} &= \frac{Z_\alpha}{u_0}\alpha + \left(1 + \frac{Z_q}{u_0}\right)q + \frac{Z_{\delta_e}}{u_0}\delta_e \\ \dot{q} &= M_\alpha \alpha + M_q q + M_{\delta_e}\delta_e\end{aligned} \tag{B4.4}$$

where u_0 is the forward speed under steady state condition and

$$Z_w = \frac{Z_\alpha}{u_0}; \quad M_w = \frac{M_\alpha}{u_0} \tag{B4.5}$$

Putting the short period two degrees of freedom model in state-space form $\dot{x} = Ax + Bu$, and neglecting Z_q:

$$\begin{bmatrix}\dot{\alpha} \\ \dot{q}\end{bmatrix} = \begin{bmatrix}\dfrac{Z_\alpha}{u_0} & 1 \\ M_\alpha & M_q\end{bmatrix}\begin{bmatrix}\alpha \\ q\end{bmatrix} + \begin{bmatrix}\dfrac{Z_{\delta_e}}{u_0} \\ M_{\delta_e}\end{bmatrix}\delta_e \tag{B4.6}$$

The characteristic equation of the form $(\lambda I - A)$ for the above system will be

$$\lambda^2 - \left(M_q + \frac{Z_\alpha}{u_0}\right)\lambda + \left(M_q\frac{Z_\alpha}{u_0} - M_\alpha\right) = 0 \tag{B4.7}$$

Solving for the eigenvalues of the characteristic equation yields the following frequency and damping ratio for the short period mode:

$$\text{Frequency} \qquad \omega_{n_{sp}} = \sqrt{\frac{Z_\alpha M_q}{u_0} - M_\alpha} \tag{B4.8}$$

Damping ratio $$\zeta_{sp} = -\frac{M_q + (Z_\alpha/u_0)}{2\omega_{n_{sp}}} \tag{B4.9}$$

Phugoid mode (long period mode; see Fig. B.5)

The Phugoid mode is a lightly damped mode with relatively low frequency oscillation. In this mode, α remains practically constant while there are noticeable changes in u, θ and altitude. An approximation to the phugoid mode can be made by omitting the pitching moment equation:

$$\begin{bmatrix} \dot{u} \\ \dot{\theta} \end{bmatrix} = \begin{bmatrix} X_u & -g \\ -\dfrac{Z_u}{u_0} & 0 \end{bmatrix} \begin{bmatrix} u \\ \theta \end{bmatrix} + \begin{bmatrix} X_{\delta_e} \\ 0 \end{bmatrix} \delta_e \tag{B4.10}$$

where g is the acceleration due to gravity.

Forming the characteristic equation, solving for eigenvalues yields the following expressions for the phugoid natural frequency, and damping ratio:

Frequency $$\omega_{n_{ph}} = \sqrt{-\frac{Z_u g}{u_0}} \tag{B4.11}$$

Damping ratio $$\zeta_{ph} = -\frac{X_u}{2\,\omega_{n_{ph}}} \tag{B4.12}$$

The aforementioned longitudinal approximations yield the simplest set of longitudinal equations of motion. However, these may not always yield correct results for all types of longitudinal manoeuvres. The following fourth order model is more likely to give better representation of the longitudinal motion of the aircraft in flight:

$$\begin{aligned} \dot{u} &= \frac{\bar{q}S}{m}C_X - qw - g\sin\theta \\ \dot{w} &= \frac{\bar{q}S}{m}C_Z + qu + g\cos\theta \\ \dot{q} &= \frac{\bar{q}S\bar{c}}{I_y}C_m \\ \dot{\theta} &= q \end{aligned} \tag{B4.13}$$

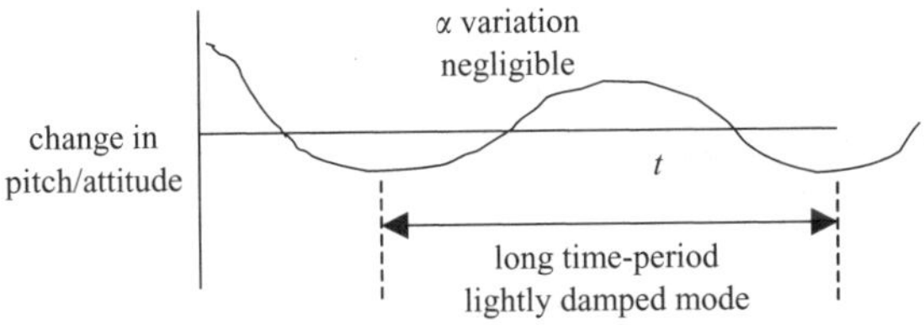

Figure B.5 Phugoid mode

where C_X, C_Z and C_m are the non-dimensional aerodynamic coefficients that can be expressed in terms of stability and control derivatives using Taylor series expansion.

Lateral equations of motion

The lateral motion is characterised by three modes:

(i) Spiral mode.
(ii) Roll subsidence.
(iii) Dutch-roll mode.

The lateral-directional state model consists of the side force, rolling and yawing moment equations. The following state-space model for lateral-directional motion yields satisfactory results for most applications.

$$\begin{bmatrix} \dot{\beta} \\ \dot{p} \\ \dot{r} \\ \dot{\phi} \end{bmatrix} = \begin{bmatrix} \dfrac{Y_\beta}{u_0} & \dfrac{Y_p}{u_0} & \dfrac{Y_r}{u_0} - 1 & \dfrac{g\cos\theta_0}{u_0} \\ L_\beta & L_p & L_r & 0 \\ N_\beta & N_p & N_r & 0 \\ 0 & 1 & 0 & 0 \end{bmatrix} \begin{bmatrix} \beta \\ p \\ r \\ \phi \end{bmatrix} + \begin{bmatrix} \dfrac{Y_{\delta_a}}{u_0} & \dfrac{Y_{\delta_r}}{u_0} \\ L_{\delta_a} & L_{\delta_r} \\ N_{\delta_a} & N_{\delta_r} \\ 0 & 0 \end{bmatrix} \begin{bmatrix} \delta_a \\ \delta_r \end{bmatrix} \tag{B4.14}$$

Solving for the eigenvalues from the lateral-directional characteristic equation will yield two real roots and a pair of complex roots.

Spiral mode

One of the real roots, having a small value (relatively long time-period) indicates the spiral mode. The root can have a negative or positive value, making the mode convergent or divergent. The mode is dominated by rolling and yawing motions. Sideslip is almost non-existent. The characteristic root λ for spiral mode is given by

$$\lambda = \frac{L_\beta N_r - L_r N_\beta}{L_\beta} \tag{B4.15}$$

Increasing N_β (yaw damping) will make the spiral mode more stable.

Roll mode

The dominant motion is roll. It is a highly damped mode with a relatively short time-period. The characteristic root λ for spiral mode is given by

$$\lambda = L_p \tag{B4.16}$$

where L_p is the roll damping derivative.

Dutch-roll mode

The Dutch-roll is a relatively lightly damped mode that consists of primarily the sideslip and yawing motions. Solving for the eigenvalues of the characteristic equation yields the following expressions for the natural frequency and damping ratio for this oscillatory mode:

$$\text{Frequency} \qquad \omega_{n_{DR}} = \sqrt{\frac{Y_\beta N_r - N_\beta Y_r + N_\beta u_0}{u_0}} \tag{B4.17}$$

$$\text{Damping ratio} \qquad \zeta_{DR} = -\left(\frac{Y_\beta + N_r u_0}{u_0}\right)\frac{1}{2\omega_{n_{DR}}} \tag{B4.18}$$

One can find several approximate forms of the equations of motion in literature. The following form of the lateral-directional equations of motion is more general and expressed using non-dimensional force and moment coefficients.

$$\begin{aligned}
\dot{\beta} &= \frac{\bar{q}S}{mV}C_Y + p\sin\alpha - r\cos\alpha + \frac{g}{V}\sin\phi\cos\theta \\
\dot{p} &= \frac{1}{I_x}[\dot{r}I_{xz} + \bar{q}SbC_l + qr(I_y - I_z) + pqI_{xz}] \\
\dot{r} &= \frac{1}{I_z}[\dot{p}I_{xz} + \bar{q}SbC_n + pq(I_x - I_y) - qrI_{xz}] \\
\dot{\phi} &= p + \tan\theta(q\sin\phi - r\cos\phi)
\end{aligned} \tag{B4.19}$$

The coefficients C_Y, C_l and C_n can be expressed in terms of stability and control derivatives using Taylor series expansion.

Aircraft six degrees of freedom equations of motion

With the advancement in parameter estimation methods and computing facilities, it has now become feasible to use the full set of six degrees of freedom aircraft equations of motion.

Aircraft six degrees of freedom motion in flight can be represented by the following set of state and observation equations.

State equations

$$\begin{aligned}
\dot{V} &= -\frac{\bar{q}S}{m}C_D + g(\cos\phi\cos\theta\sin\alpha\cos\beta + \sin\phi\cos\theta\sin\beta - \sin\theta\cos\alpha\cos\beta) \\
&\quad + \frac{F_T}{m}\cos(\alpha + \sigma_T)\cos\beta \\
\dot{\alpha} &= \frac{g}{V\cos\beta}(\cos\phi\cos\theta\cos\alpha + \sin\theta\sin\alpha) + q - \tan\beta(p\cos\alpha + r\sin\alpha) \\
&\quad - \frac{\bar{q}S}{mV\cos\beta}C_L - \frac{F_T}{mV\cos\beta}\sin(\alpha + \sigma_T) \\
\dot{\beta} &= \frac{g}{V}(\cos\beta\sin\phi\cos\theta + \sin\beta\cos\alpha\sin\theta - \sin\alpha\cos\phi\cos\theta\sin\beta) + p\sin\alpha \\
&\quad - r\cos\alpha + \frac{\bar{q}S}{mV}(C_Y\cos\beta + C_D\sin\beta) + \frac{Fe}{mV}cos(\alpha + \sigma_T)\sin\beta
\end{aligned}$$

$$\dot{p} = \frac{1}{I_x I_z - I_{zx}^2}\{\bar{q}\ Sb(I_z C_l + I_{zx} C_n) - qr\left(I_{zx}^2 + I_z^2 - I_y I_z\right)$$
$$+ pq I_{zx}(I_x - I_y + I_z)\}$$

$$\dot{q} = \frac{1}{I_y}\{\bar{q} S\bar{c} C_m - (p^2 - r^2) I_{xz} + pr(I_z - I_x) + F_T(l_{tx} \sin\sigma_t + l_{tz}\cos\sigma_T)\}$$

$$\dot{r} = \frac{1}{I_x I_z - I_{zx}^2}\{\bar{q} Sb(I_x C_n + I_{zx} C_l) - qr I_{zx}(I_x - I_y + I_z)$$
$$+ pq(I_{xz}^2 - I_x I_y + I_x^2)\} \qquad \text{(B4.20)}$$

$$\dot{\phi} = p + q \sin\phi \tan\theta + r \cos\phi \tan\theta$$

$$\dot{\theta} = q \cos\phi - r \sin\phi$$

$$\dot{\psi} = (q \sin\phi + r \cos\phi) \sec\theta$$

$$\dot{h} = u \sin\theta - v \cos\theta \sin\phi - w \cos\theta \cos\phi$$

Here, σ_T is the tilt angle of the engines and l_{tx} and l_{tz} represent the location of the engine relative to c.g. C_L, C_D and C_Y are the non-dimensional force coefficients, and C_l, C_m and C_n are the moment coefficients referred to the centre of gravity. The longitudinal flight variables are α, q and θ while the lateral-directional flight variables are β, p, r, ϕ and ψ. The aircraft velocity is V, and the engine thrust is F_T.

Observation model

$$\begin{aligned}
\alpha_m &= \alpha \\
\beta_m &= \beta \\
p_m &= p \\
q_m &= q \\
r_m &= r \\
\phi_m &= \phi \\
\theta_m &= \theta \\
a_{x_m} &= \frac{\bar{q}S}{m} C_X + \frac{F_T}{m}\cos\sigma_T \\
a_{y_m} &= \frac{\bar{q}S}{m} C_Y \\
a_{z_m} &= \frac{\bar{q}S}{m} C_Z - \frac{F_T}{m}\sin\sigma_T
\end{aligned} \qquad \text{(B4.21)}$$

The above equations pertain to rigid body dynamics and assume that all flight variables are measured at c.g. If the sensors are not mounted at c.g. (which is often the case), then corrections must be made to sensor measurements for the offset distance from c.g. before they can be used in the above equations (this aspect is treated separately in this appendix).

It is generally convenient to postulate the equations of motion in the polar coordinate form as given above, because it is easier to understand the effects of the changes in force and moments in terms of α, β and V. However, this formulation become singular at zero velocity where α, β are not defined. Under such conditions, one can formulate the equations in rectangular coordinates [1].

B.5 Aircraft parameter estimation

One of the important aspects of flight-testing of any aircraft is the estimation of its stability and control derivatives. Parameter estimation is an important tool for flight test engineers and data analysts to determine the aerodynamic characteristics of new and untested aircraft. The flight-estimated derivatives are useful in updating the flight simulator model, improving the flight control laws and evaluating handling qualities. In addition, the flight determined derivatives help in validation of the predicted derivatives. These predicted derivatives are often based on one or more of the following: i) wind tunnel; ii) DATCOM (Data Compendium) methods; and iii) some analytical methods.

The procedure for aircraft parameter estimation is well laid out. The aircraft dynamics are modelled by a set of differential equations (equations of motion already discussed). The external forces and moments acting on the aircraft are described in terms of aircraft stability and control derivatives, which are treated as unknown (mathematical model). Using specifically designed control inputs, responses of the test aircraft and the mathematical model are obtained and compared. Appropriate parameter estimation algorithms are applied to minimise the response error by iteratively adjusting the model parameters.

Thus, the key elements for aircraft parameter estimation are: manoeuvres, measurements, methods and models. A brief insight into the various aspects of these elements, also referred to as the Quad-M requirements of aircraft parameter estimation (Fig. B.6), is provided next [5].

B.6 Manoeuvres

The first major step in aircraft parameter estimation is the data acquisition. This primarily addresses the issue of obtaining measurements of the time histories of control surface deflections, air data (airspeed, sideslip and angle-of-attack), angular velocities, linear and angular accelerations, and attitude (Euler) angles. In addition to these variables, quantities defining flight conditions, aircraft configuration, instrumentation system, fuel consumption for estimation of aircraft, c.g. location, weight and inertias are also required. Detailed information of these requirements must be sought before commencing with the data analysis.

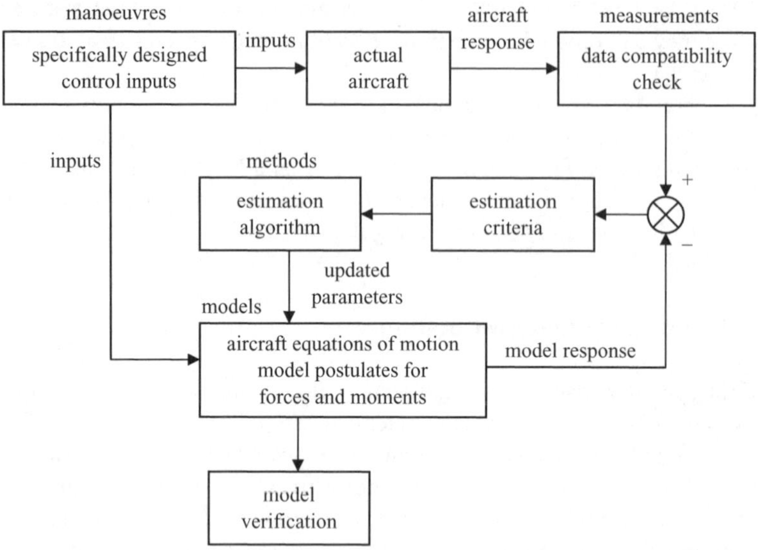

Figure B.6 Quad-M requirements of aircraft parameter estimation

A reliable estimation of the stability and control derivative from flight requires the aircraft modes to be excited properly. It will not be possible to estimate C_{m_α} and C_{m_q} if the longitudinal short period mode is not sufficiently excited. Specification of input forms is a critical factor because experience shows that the shape of the input signal has a significant influence on the accuracy of the estimated parameters. Some typical inputs (Fig. B.7) used to generate aircraft flight test data are listed below.

(i) *3211 input* This is a series of alternating step inputs, the duration of which satisfies the ratio $3:2:1:1$. It is applied to the aircraft control surface through the pilot's stick. This input signal has power spread over a wide frequency band. It can be effectively used to excite the aircraft modes of motion. When applied to ailerons, it excites the rolling motion that can be analysed to obtain derivatives for roll damping and aileron control effectiveness. At the end of the input, the controls are held constant for some time to permit the natural response of the aircraft to be recorded. Similar test signals can be used for rudder surface to determine yaw derivatives and rudder effectiveness. The aircraft short period longitudinal motion can be produced by applying the 3211 input to the elevator. The time unit Δt needs to be selected appropriately to generate sufficient excitation in the aircraft modes of motion.

(ii) *Pulse input* This control input signal has energy at low frequency and is not very suitable for parameter estimation purposes. Nonetheless, a longer duration pulse (of about 10 to 15 s) can be given to the elevator to excite the longitudinal phugoid motion of the aircraft. The aircraft response should be recorded for

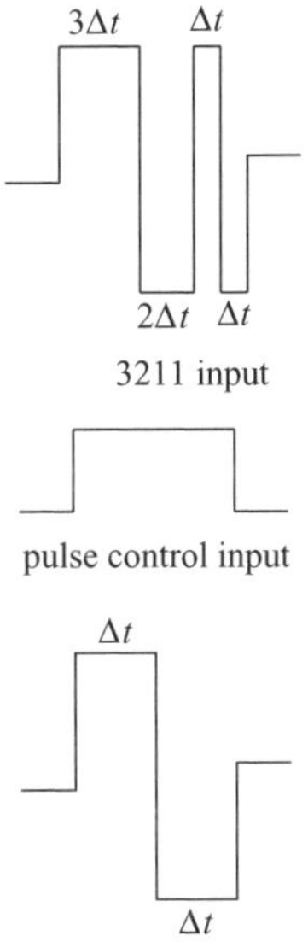

Figure B.7 Control inputs

a sufficient number of cycles before re-trimming. From this response, one can estimate speed related derivatives, and the phugoid damping and frequency.

(iii) *Doublet control input* This signal excites a band at higher frequency. It is used to excite longitudinal short period manoeuvres for estimating derivatives like $C_{m_\alpha}, C_{m_q}, C_{m_\delta}, \ldots$ and the Dutch-roll manoeuvres for estimating derivatives like $C_{l_\beta}, C_{n_\beta}, C_{n_r}, \ldots$ etc. If the natural frequency ω_n of the mode to be excited is known, then the approximate duration of the time unit Δt for a doublet can be determined from the expression $\Delta t = 1.5/\omega_n$

In nutshell, it is desirable to use inputs whose power spectral density is relatively wide band. In this context, the 3211 form of input is found to have power over a wide frequency range whilst doublet inputs tend to excite only a narrow band of frequencies. The pulse inputs have power at low frequencies and are therefore suitable for exciting low frequency modes of the system. A combination of various input forms is generally considered the best for proper excitation of the system response.

Some of the flight manoeuvres generally used to generate responses, which can be used for the estimation of aircraft stability and control derivatives, are listed below [6].

Longitudinal short period manoeuvre

Starting from a horizontal level trimmed flight at constant thrust, a doublet or 3211 multi step input is applied to the elevator. As far as possible, we try to avoid variations in the lateral-directional motion. The pulse width of the input signal is appropriately selected to excite the short period mode of the aircraft.

Phugoid manoeuvre
A longer duration pulse input signal is applied to the elevator keeping the thrust constant. The aircraft should be allowed to go through a minimum of one complete cycle of the phugoid before re-trimming.

Thrust input manoeuvre
The manoeuvre is used to determine the effect of a thrust variation on the aircraft motion. Starting from trimmed level flight, a doublet variation in thrust is applied and the flight data recorded.

Flaps input manoeuvre
This manoeuvre can be used to gather information for estimation of the flaps effectiveness derivatives. Data is generated by applying a doublet or 3211 input to the flaps. Other longitudinal controls and thrust are kept constant. Variations in the lateral-directional motion are kept small.

Doublet or 3211 aileron input manoeuvre
The purpose of this manoeuvre is to get information for estimation of the roll damping and aileron effectiveness. Starting from trimmed horizontal level flight, a doublet or 3211 input signal is applied to the aileron. The pulse width of the input signal should be appropriately selected to excite dominantly the aircraft rolling motion.

Doublet or 3211 rudder input manoeuvre
This manoeuvre is used to excite Dutch-roll motion to estimate yaw derivatives and rudder control effectiveness. Starting from trimmed level flight, a doublet or 3211 input signal is applied to the rudder keeping the thrust constant. Sufficient time is allowed for the oscillations to stabilise at the end of the input. The pulse width of the input signal is appropriately selected to match the Dutch-roll frequency.

Roll manoeuvre
The manoeuvre generates bank-to-bank motion that can be used to estimate roll derivatives. The roll manoeuvre is initiated with a pulse input to the aileron in one direction and after few seconds, the aircraft is brought back to the horizontal level position with an input to the aileron in reverse direction. The process is then repeated in the other direction. At the end of this manoeuvre, the heading angle should be approximately the same as at the beginning.

Roller coaster (pull-up push-over) manoeuvre
This manoeuvre is used to determine the aircraft drag polars. Starting from a trimmed level flight, the pitch stick (that moves the elevator) is first pulled to slowly increase the vertical acceleration from 1 g to 2 g (at the rate of approximately 0.1 g/s) and then return slowly to level flight in the same fashion. Next, the elevator stick is pushed

slowly, causing the vertical acceleration to change from 1 g to 0 g at a slow rate and then return slowly to trimmed level flight. Data is recorded at least for about 25 to 30 s in this slow response manoeuvre. This manoeuvre covers low angle-of-attack range.

Acceleration and deceleration manoeuvre
The purpose of this manoeuvre is to estimate the drag polars at high angles of attack and to study the effects of speed variation on the aerodynamic derivatives, if any. Starting from a trimmed horizontal level flight at the lowest speed, the manoeuvre is initiated by rapidly pushing the stick down, i.e., nose down. At constant thrust, this results in a continuous gain in the airspeed and loss of altitude. After reaching the maximum permissible airspeed, the control stick is pulled back causing the aircraft to pitch up. This results in deceleration and gain of altitude. The manoeuvre is terminated once the minimum airspeed is reached.

Experience with flight data analysis has shown that no single manoeuvre, no matter how carefully performed and analysed, can provide a definitive description of the aircraft motion over the envelope or even at a given flight condition in the envelope. Thus, it is always desirable to obtain data from several manoeuvres at a single flight condition or a series of manoeuvres as the flight condition changes. Often, two or more such manoeuvres are analysed to obtain one set of derivatives. This is more popularly known as multiple manoeuvre analysis.

B.7 Measurements

The accuracy of estimated derivatives depends on the quality of measured data. Measurements are always subjected to systematic and random errors. It is, therefore, essential to evaluate the quality of the measured data and rectify the measurements before commencing with parameter estimation. Such an evaluation can include consideration of factors like the frequency content of the input signals, sampling rates, signal amplitudes, signal-to-noise ratio, etc. A widely used procedure for data quality evaluation and correction is the kinematic consistency checking. Since the aircraft measurements are related by a set of differential equations, it is possible to check for consistency among the kinematic quantities. This is also true, in general, for other dynamical systems. The procedure is also popularly referred to as *flight path reconstruction* (especially for longitudinal kinematic consistency) [7]. For example, the measured roll and pitch attitudes should match with those reconstructed from the rate measurements. This process ensures that the data are consistent with the basic underlying kinematic models. Since the aircraft is flying, it must be according to the kinematics of the aircraft but the sensors could go wrong in generating the data or the instruments could go wrong in displaying the recorded data. In addition to data accuracy, the compatibility check also provides the error model, i.e., the estimates of the bias parameters and scale factors in the measured data. An accurate determination of the error parameters can help prevent problems at a later stage during actual estimation of the aerodynamic derivatives.

The following kinematic equations are used.

State equations

$$
\begin{aligned}
&\dot{u} = -(q - \Delta q)w + (r - \Delta r)v - g\sin\theta + (a_x - \Delta a_x), \qquad u(0) = u_0 \\
&\dot{v} = -(r - \Delta r)u + (p - \Delta p)w + g\cos\theta\sin\phi + (a_y - \Delta a_y), \qquad v(0) = v_0 \\
&\dot{w} = -(p - \Delta p)v + (q - \Delta q)u + g\cos\theta\cos\phi + (a_z - \Delta a_z), \qquad w(0) = w_0 \\
&\dot{\phi} = (p - \Delta p) + (q - \Delta q)\sin\phi\tan\theta + (r - \Delta r)\cos\phi\tan\theta, \qquad \phi(0) = \phi_0 \\
&\dot{\theta} = (q - \Delta q)\cos\phi - (r - \Delta r)\sin\phi, \qquad \theta(0) = \theta_0 \\
&\dot{\psi} = (q - \Delta q)\sin\phi\sec\theta + (r - \Delta r)\cos\phi\sec\theta, \qquad \psi(0) = \psi_0 \\
&\dot{h} = u\sin\theta - v\cos\theta\sin\phi - w\cos\theta\cos\phi, \qquad h(0) = h_0
\end{aligned}
\tag{B7.1}
$$

where Δa_x, Δa_y, Δa_z, Δp, Δq and Δr are the biases (in the state equations) to be estimated. The control inputs are a_x, a_y, a_z, p, q and r.

Observation equations

$$
\begin{aligned}
V_m &= \sqrt{u_n^2 + v_n^2 + w_n^2} \\
\alpha_m &= K_\alpha \tan^{-1}\left[\frac{w_n}{u_n}\right] + \Delta\alpha \\
\beta_m &= K_\beta \sin^{-1}\left[\frac{v_n}{\sqrt{u_n^2 + v_n^2 + w_n^2}}\right] + \Delta\beta \\
\phi_m &= K_\phi \phi + \Delta\phi \\
\theta_m &= K_\theta \theta + \Delta\theta \\
\psi_m &= K_\psi \psi \\
h_m &= h
\end{aligned}
\tag{B7.2}
$$

The velocity components u, v, w from the state equations are computed at c.g. whilst the flight variables α_m and β_m are measured at the nose boom. It is, therefore, necessary that u, v, w be computed at the nose boom (u_n, v_n, w_n) in order that the α computed from observation equations and that measured from the flight pertain to the same reference point (nose boom in this case). Alternatively, the measured α at the nose boom can be corrected for the c.g. offset. Both approaches are correct. The nose boom is the pitot location installed in front of the aircraft. The static and stagnation pressure measurements at the pitot location are used for obtaining V, α and β. The length of the boom is usually kept 2 to 3 times the fuselage diameter to avoid interference effects.

B.8 Correction for c.g. position

As mentioned above, all quantities in the state and observation equations should be defined w.r.t. c.g. Although the aircraft rates, and the roll and pitch attitudes are not affected by the c.g. location, the measurements of linear accelerations and velocity components are influenced by the distance between the c.g. and the sensor position. In most of the cases, the airspeed is measured at the pitot location installed in front of the aircraft. There is a separate α and β vane to record the angle-of-attack and sideslip angle (at the nose boom). To compare the model response with the measured response, the estimated model outputs of V, α and β obtained at c.g. should be transformed to the individual sensor location where the actual measurements are made.

Assuming the sensor locations in

x-direction (positive forward from c.g.):	x_n
y-direction (positive to the right of c.g.):	y_n
z-direction (positive downward of c.g.):	z_n

the speed components along the three axes at the sensor location are given by

$$\begin{aligned} u_n &= u - (r - \Delta r)y_n + (q - \Delta q)z_n \\ v_n &= v - (p - \Delta p)z_n + (r - \Delta r)x_n \\ w_n &= w - (q - \Delta q)x_n + (p - \Delta p)y_n \end{aligned} \tag{B8.1}$$

The V_n, α_n and β_n at sensor location are computed as

$$\begin{aligned} V_n &= \sqrt{u_n^2 + v_n^2 + w_n^2} \\ \alpha_n &= \tan^{-1}\left(\frac{w_n}{u_n}\right) \\ \beta_n &= \sin^{-1}\left(\frac{v_n}{V_n}\right) \end{aligned} \tag{B8.2}$$

Also, the linear accelerometers, in most of the cases, are not mounted exactly at the c.g. Knowing the c.g. location and the accelerometer offset distances x_a, y_a and z_a from the c.g., the accelerations a_x, a_y and a_z at the c.g. can be derived from the measured accelerations a_{xs}, a_{ys} and a_{zs} at the sensor location using the following relations:

$$\begin{aligned} a_x &= a_{xs} + (q^2 + r^2)x_a - (pq - \dot{r})y_a - (pr + \dot{q})z_a \\ a_y &= a_{ys} - (pq + \dot{r})x_a + (r^2 + p^2)y_a - (rq - \dot{p})z_a \\ a_z &= a_{zs} - (pr - \dot{q})x_a - (qr + \dot{p})y_a + (p^2 + q^2)z_a \end{aligned} \tag{B8.3}$$

Although the error parameters, consisting of scale factors and biases, can be estimated using any one of various parameter estimation techniques, i.e., equation error method, output error method or filter error method, for most of the applications reported in literature, the output error method has been found to be adequate for consistency checking.

B.9 Methods

The selection of the estimation technique is influenced by the complexity of the mathematical model, *a priori* knowledge about the system and information on the noise characteristics in measured data. The chosen estimation technique must provide the estimated values of the parameters along with their accuracies, usually in the form of standard errors or variances. The commonly used techniques for aircraft parameter estimation have been discussed in various chapters of this book. These include the equation error method, output error method (OEM) and filter error method. The other approach to aircraft parameter estimation is the one in which a nonlinear filter provides the estimates of the unknown parameters that are defined as additional state variables (EKF). The equation error method represents a linear estimation problem, whereas the remaining methods belong to a class of nonlinear estimation problem. The neural network (feedforward neural network and recurrent neural network) approach to aircraft parameter estimation has also been discussed in Chapters 10 and 11. The estimation before modelling and the model error estimation algorithms are also very popular for aircraft parameter estimation. Recently, frequency domain methods have also gained some impetus.

B.10 Models

We have already discussed the mathematical models to be used in aircraft parameter estimation. The characteristic motion of the aircraft is defined by the basic equations of motion derived from the Newtonian mechanics. They involve forces and moments, which include the aerodynamic, inertial, gravitational and propulsive forces. The forces and moments are approximated by stability and control derivatives using the Taylor's series expansion. Some simple sets of longitudinal and lateral-directional equations have already been discussed in this appendix. The complete set of six DOF equations of motion pertaining to the rigid body dynamics has also been described. Again, modelling of aerodynamic forces and moments raises the fundamental question of how complete the model should be. Although a more complete model can be justified for the correct description of the aircraft dynamics, it is not clear what should be the best relationship between the model complexity and measurement information. An attempt to identify too many parameters from a limited amount of data might fail or might yield estimates with reduced accuracy. The search for obtaining adequate aerodynamic models that can satisfactorily explain the various flow phenomena is still being vigorously pursued. Various techniques of model structure determination are discussed in Chapter 6. Modified forms of linear regression (SMLR method) for determining model structure are discussed in Chapter 7.

B.11 Model verification

Model verification is the last step in flight data analysis procedures and should be carried out no matter how sophisticated an estimation technique is applied. Several criteria help to verify the estimated model, namely: i) standard deviations (Cramer-Rao lower bounds) of the estimates; ii) correlation coefficients among the estimates; iii) fit error (determinant of the covariance matrix of residuals); iv) plausibility of estimates from physical understanding of the system under investigation or in comparison with other (analytical, wind tunnel etc.) predictions; and v) model predictive capability. The last of the criteria is the most widely used procedure for verification of the flight-estimated models. For verification, the model parameters are fixed to the estimated values and the model is driven by inputs that are different from those used in estimation. The model responses are then compared with the flight measurements to check upon the predictive capabilities of the estimated model.

B.12 Factors influencing accuracy of aerodynamic derivatives

Here, we briefly mention some factors, which, though seemingly unimportant, can often have a significant influence on the accuracy of the estimated aircraft stability and control derivatives.

The total aerodynamic force and moment coefficients are a function of the state and control variables. Therefore, any error in measuring the motion variables (e.g., use of incorrect calibration factors) will have a direct impact on the computation of total coefficients, which, in turn, will lead to estimation of incorrect derivatives. The choice of the axis system on which the measurements are based and the derivatives defined is also important. Before comparing the flight estimated derivatives with theoretical or wind tunnel estimates, one must ensure that all of them are converted to the same axis-system.

Another important factor is the dynamic pressure. The presence of the dynamic pressure term $\bar{q}$ in the equations of motion shows that any error in the measurement of $\bar{q}$ is likely to degrade the accuracy of the estimated parameters. Further, the fact that dimensional derivatives are directly multiplied by $\bar{q}$ (e.g., $M_\alpha = \bar{q}ScC_{m_\alpha}/I_y$) makes it essential to have $\bar{q}$ measurement as accurate as possible.

The dependence of one particular set of derivatives on another can also play an important role in influencing the accuracy of the identified derivatives. For example, a good estimate of the lift derivatives and an accurate α measurement are necessary for determining reliable drag derivatives. However, the reverse is not true, since the influence of drag derivatives in defining the lift force derivatives is small.

Beside the accuracy requirements in instrumentation, adequate knowledge about the mass and inertia characteristics is also important for accurate estimation of aircraft derivatives. The non-dimensional moment derivatives are directly influenced by the inertia calculations, while the force derivatives will be straightway affected by the errors in aircraft mass calculations. Information on the fuel consumption is useful to compute c.g. travel and actual mass of the aircraft at any time during the flight. For moment of inertia, manufacturer's data is mostly used.

The kinematic equations for data consistency check and the aircraft equations of motion for aerodynamic model estimation are formulated w.r.t. a fixed point. In the majority of the cases, this fixed point is assumed to be the aircraft centre of gravity.

Naturally, the motion variables to be used in the equations need to be measured at the c.g. However, the sensors are generally located at a convenient point, which, though not exactly at c.g., may lie close to it. For example, a flight log mounted on a boom in front of the aircraft nose is commonly used to measure airspeed V, angle-of-attack α and the sideslip angle β. Similarly, the accelerometers are also not located exactly at the c.g. Before commencing with consistency checks and parameter estimation, it is mandatory that the sensor measurements be corrected for offset from c.g. data correction for c.g. offset has already been discussed in this appendix.

B.13 Fudge factor

This is normally used along with Cramer-Rao bounds for aircraft parameter estimates. Actually, the uncertainty bound for parameter estimate is multiplied with a fudge factor to reflect correctly the uncertainty. When OEM is used for parameter estimation from data (often the flight test data of an aircraft), which are often affected by process noise (atmospheric turbulence), the uncertainty bounds do not correctly reflect the effect of this noise or uncertainty of the parameter estimates, since OEM does not, *per se*, handle process noise. A fudge factor of about 3 to 5 is often used in practice. It can be determined using an approach found in Reference 8. This fudge factor will also be useful for any general parameter estimation if the residuals have a finite (small) bandwidth.

B.14 Dryden model for turbulence

In Chapter 5, the longitudinal data simulation in the presence of turbulence (Example 5.1) is carried out using a Dryden model with an integral scale of turbulence $L = 1750$ ft and turbulence intensity $\sigma = 3$ m/s. The model generates moderate turbulence conditions whereby the forward speed, vertical speed and the pitch rate are modified to include the turbulence effects.

Consider the dynamic model of the form [9, 10]:

$$\begin{aligned}
\dot{y}_u &= \frac{[-y_u + x_u k_u \sqrt{\pi/\Delta t}]}{t_u} \\
\dot{y}_q &= -\left[\frac{\pi V_T}{4b}\right] y_q + w_{fturb} \\
\dot{y}_{w_2} &= y_{w_1} \\
\dot{y}_{w_1} &= -\frac{y_{w_2}}{t_w^2} - \frac{2y_{w_1}}{t_w} + x_w \sqrt{\frac{\pi}{\Delta t}}
\end{aligned} \qquad \text{(B14.1)}$$

where x_u and x_w are random numbers used to simulate the random nature of turbulence, and t_u, t_w, k_u and k_w are the time constants defined as follows:

$$t_u = \frac{L_u}{V_T}; \quad t_w = \frac{L_w}{V_T}; \quad k_u = \sqrt{\frac{2\sigma_u^2 t_u}{\pi}}; \quad k_w = \sqrt{\frac{2\sigma_w^2 t_w}{\pi}} \qquad \text{(B14.2)}$$

where

$$V_T = \sqrt{u^2 + w^2}; \quad \sigma_u = \sigma_w \quad \text{and} \quad L_u = L_w = 1750\,\text{ft} \tag{B14.3}$$

The dynamic model for turbulence is appended to the system state equations given in Example 5.1 and a fourth order Runge-Kutta integration is applied to obtain the longitudinal flight variables u, w, q and θ, and the turbulence variables y_u, y_q, y_{w2} and y_{w_1}. Following the procedure outlined [9, 10], the turbulence in forward velocity, vertical velocity and pitch rate, in the flight path axes, is given by

$$u_{fturb} = y_u; \quad w_{fturb} = \frac{k_w[(y_{w_2}/t_w) + \sqrt{3}y_{w_1}]}{t_w} \quad \text{and} \quad q_{fturb} = \left[\frac{\pi \dot{y}_q}{4b}\right] \tag{B14.4}$$

where b is the wingspan.

Since the flight variables u, w, q and θ are computed in the body-axis, the quantities u_{fturb}, w_{fturb} and q_{fturb} should be computed in the body-axis. The change over from flight path to body axes is carried out using the transformation [10]:

$$\begin{bmatrix} u_{turb} \\ w_{turb} \\ q_{turb} \end{bmatrix} = \begin{bmatrix} \cos\alpha & 0 & -\sin\alpha \\ 0 & 1 & 0 \\ \sin\alpha & 0 & \cos\alpha \end{bmatrix} \begin{bmatrix} u_{fturb} \\ w_{fturb} \\ q_{fturb} \end{bmatrix} \tag{B14.5}$$

In Chapter 5, the above Dryden model is used only for simulating the atmospheric turbulence and does not figure in the estimation of model parameters. The aircraft longitudinal response with turbulence can now be simulated using the equations:

$$\begin{aligned} u_m &= u - u_{turb} \\ w_m &= w - w_{turb} \\ q_m &= q - q_{turb} \\ \theta_m &= \theta \\ a_{xm} &= \frac{\bar{q}SC_x}{m} \\ a_{zm} &= \frac{\bar{q}SC_z}{m} \\ \dot{q}_m &= \frac{\bar{q}S\bar{c}C_m}{I_y} \end{aligned} \tag{B14.6}$$

Figure B.8 gives a complete picture of the process of simulating longitudinal aircraft motion in turbulence.

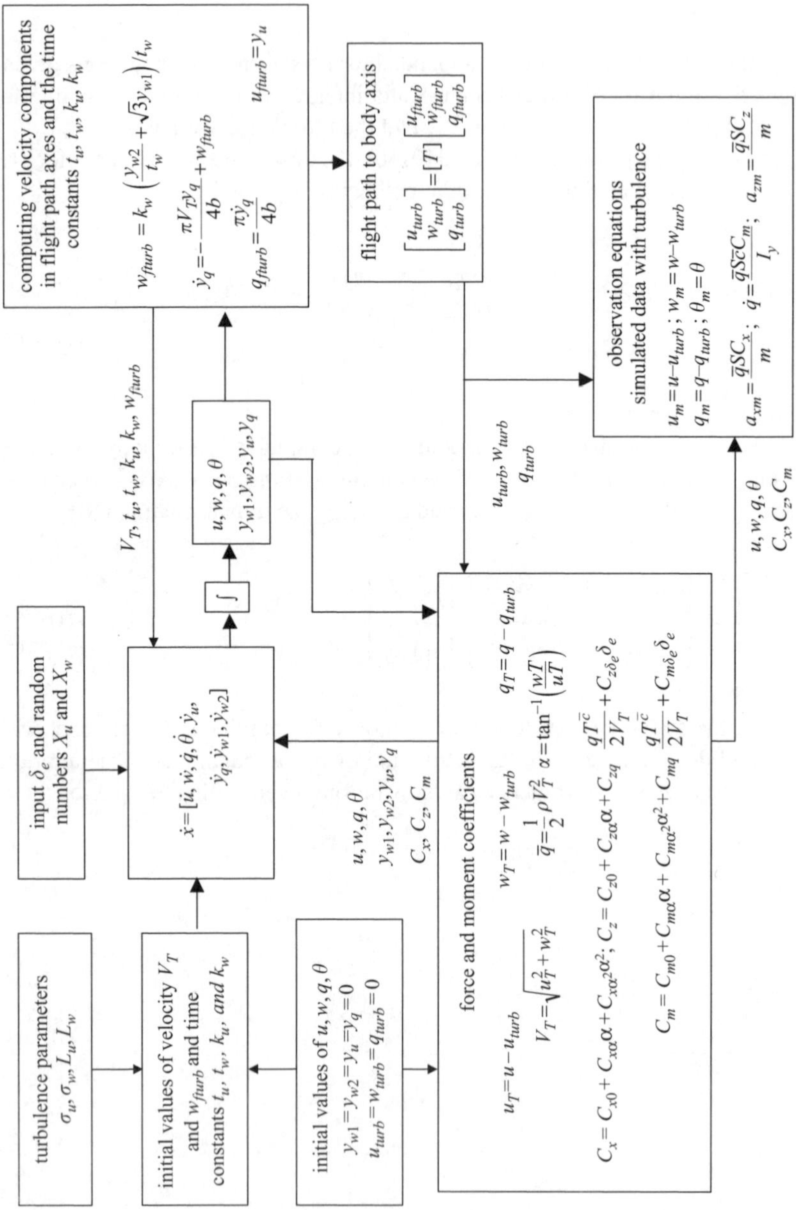

Figure B.8 *Simulation of aircraft longitudinal motion in turbulence*

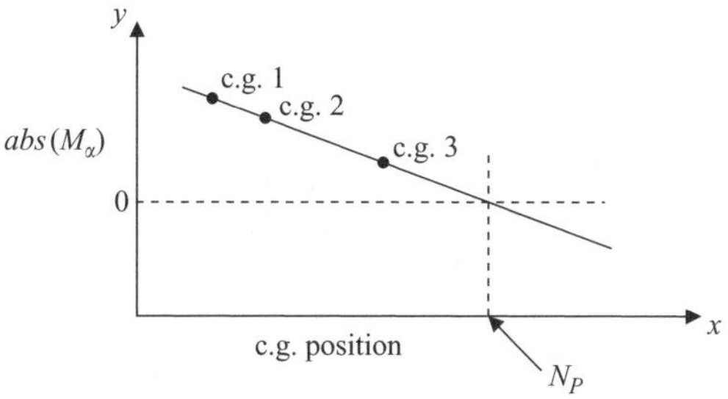

Figure B.9 Natural point estimation

B.15 Determination of aircraft neutral point from flight test data

The aircraft neutral point N_P is defined as the c.g. position for which the following condition is satisfied in straight and level flight of an aircraft [11]:

$$\frac{dC_m}{dC_L} = 0 \tag{B15.1}$$

In eq. (B15.1), C_m is the pitching moment coefficient and C_L is the lift coefficient. The distance between the neutral point N_P and the actual c.g. position is called the static margin. When this margin is zero, the aircraft has neutral stability. It has been established [11], that the neutral point is related to the short period static stability parameter M_α and natural frequency (see eq. (B4.8)). It means that we estimate $M_\alpha$values from short period manoeuvres of the aircraft (flying it for three different c.g. positions), plot it w.r.t. c.g., and extend this line to the x-axis. The point on the x-axis when this line passes through 'zero' on the y-axis is the neutral point (Fig. B.9). If M_w is estimated from short period manoeuvre, then M_α can be computed easily using eq. (B4.5).

B.16 Parameter estimation from large amplitude manoeuvres

Parameter estimation methods are generally applied to small manoeuvres about the trim flight conditions. The aircraft is perturbed slightly from its trim position by giving a control input to one or more of its control surfaces. Linear aerodynamic models are assumed for analysis of these small perturbation manoeuvres. However, it may not always be possible to trim an airplane at a certain angle-of-attack. For such situations, large amplitude manoeuvres and data partitioning techniques can be used to obtain aerodynamic derivatives over the angle-of-attack range covered by the large amplitude manoeuvre [12]. The method for analysing these manoeuvres consists of partitioning the data into several bins or subsets, each of which spans a smaller range of angle-of-attack. The principle behind partitioning is that in the range of angle-of-attack defined by each subspace, the variation in the aerodynamic force and moment coefficients due to the change in angle-of-attack can be neglected.

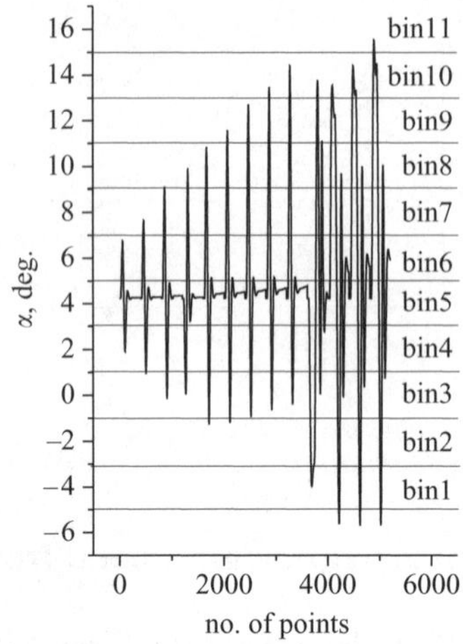

Figure B.10 Partitioning of data from large amplitude manoeuvres into bins

For example, the large amplitude manoeuvre data could be partitioned into several two deg. angle-of-attack subspaces as shown in Fig. B.10. Since time does not appear explicitly, the measured data points can be arranged in an arbitrary order. The normal practice is to estimate linear derivative models but, if necessary, a stepwise multiple linear regression approach (discussed in Chapter 7) can be used to determine a model structure with higher order terms (e.g., by including terms like α^2, αq, $\alpha\delta_e$) for better representation of the aircraft dynamics.

B.17 Parameter estimation with *a priori* information

When wind tunnel data or estimated parameter values from some previous flight data analysis are known, it seems reasonable to use *a priori* features in parameter estimation, thereby making use of all the information available to obtain estimates and ensuring that no change in the aircraft derivatives is made unless the flight data has sufficient information to warrant such a change.

The procedure used is to expand the cost function for the output error method defined in Chapter 3 (eq. (3.52)), to include a penalty for departure from the *a priori* value.

$$J = \frac{1}{2}\sum_{k=1}^{N}[z(k) - y(k)]^T R^{-1}[z(k) - y(k)] + \frac{N}{2}\ln|R| + \underbrace{(\theta_0 - \theta)^T K W^{-1}(\theta_0 - \theta)}_{\textit{inclusion of a priori values}}$$

The *a priori* values are defined by the parameter vector θ_0. It is to be noted that the fit error between the measured and model estimated response would marginally increase when *a priori* information is used, but it will reduce the scatter of the estimates and also the number of iterations to convergence. The matrix W helps to fix the relative weighting among the parameters and K is the overall gain factor.

$[W = \sigma_{ii}^2]$	Here, σ_{ii} represents the wind tunnel variance for each of the selected unknown parameters. W is considered a diagonal matrix.
K	Variation in K helps to change the overall weighting of the wind tunnel parameters to the flight estimated parameters. In general, one can use the value of K that doubles the fit error.

As mentioned earlier, the optimisation technique without the *a priori* feature would provide the best fit of the estimated response with flight response. However, addition of *a priori* values brings about only a slight change in the quality of fit. Thus, it can be safely concluded that the output error method with the *a priori* feature will provide a better chance to validate the predicted derivatives with flight-determined derivatives.

B.18 Unsteady aerodynamic effects

The process of expressing aerodynamic force and moment coefficients in terms of aircraft stability and control derivatives was discussed in Section B.2. In Section B.10, the fundamental question of how complete the model should be for parameter estimation was posed. For most of the cases (e.g., for developing high-fidelity simulators), we generally do not worry too much what derivatives are included in the estimation model, as long as the response predicted by the model gives an accurate representation of the aircraft behaviour in flight. On the other hand, if the model is to be used to understand the physics of a flow phenomenon, then the choice of stability and control derivatives to be included in the estimation model needs to be carefully considered. For example, the aircraft damping in pitch comes from the derivatives C_{m_q} and $C_{m_{\dot{\alpha}}}$. If the aim of parameter estimation is solely to have a model that can give an accurate match with flight response, we need not estimate C_{m_q} and $C_{m_{\dot{\alpha}}}$ separately. The estimation of C_{m_q} (which in fact will be the combination of both the derivatives) will suffice, as it will also include the effects arising from $C_{m_{\dot{\alpha}}}$. However, if the interest is in understanding the flow phenomenon that gives rise to $C_{m_{\dot{\alpha}}}$ (commonly known as the downwash lag effects in aircraft terminology), a separate estimation of C_{m_q} and $C_{m_{\dot{\alpha}}}$ would be mandatory. Such a model will be nonlinear in parameters and would require special treatment for estimation from flight data. One approach to induce aircraft excitation in the longitudinal axis to generate the data so that such separation is made possible, is to use pitch manoeuvre (short period) at different bank angles. The data from such manoeuvres provides necessary separation of the pitch rate q from the angle-of-attack rate $\dot{\alpha}$, thereby making it possible to estimate independently C_{m_q} and $C_{m_{\dot{\alpha}}}$ [13].

B.19 Drag polars

The drag polar is a curve that shows the graphical relationship between the aircraft lift coefficient C_L and drag coefficient C_D. The drag is least at $C_L = 0$ and increases in a parabolic fashion as C_L increases. Parameter estimation methods (see Chapter 9) can be used to determine C_L and C_D from flight data to obtain the aircraft drag polars. This helps in validation of the drag polars obtained from wind tunnel experiments.

B.20 References

1 MAINE, R. E., and ILIFF, K. W.: 'Application of parameter estimation to aircraft stability and control – the output error approach', NASA RP-1168, 1986

2 BRYAN, G. H.: 'Stability in aviation', (Macmillan, London, 1911)

3 NELSON, R. C.: 'Flight stability and automatic control' (McGraw-Hill International, Singapore, 1998, 2nd edn)

4 McRUER, D. T., ASHKENAS, I., and GRAHAM, D.: 'Aircraft dynamics and automatic control' (Princeton University Press, New Jersey, 1973)

5 HAMEL, P. G., and JATEGAONKAR, R.V.: 'Evolution of flight vehicle system identification', *Journal of Aircraft*, 1996, **33**, (1), pp. 9–28

6 JATEGAONKAR, R. V.: 'Determination of aerodynamic characteristics from ATTAS flight data gathering for ground-based simulator', DLR-FB 91-15, May 1991

7 MULDER, J. A., CHU, Q. P., SRIDHAR, J. K., BREEMAN, J. H., and LABAN, M.: 'Non-linear aircraft flight path reconstruction review and new advances', *Prog. in Aerospace Sciences*, 1999, **35**, pp. 673–726

8 MORELLI, E. A., and KLEIN, V.: 'Determining the accuracy of aerodynamic model parameters estimated from flight data', AIAA-95-3499, 1995

9 MADHURANATH, P.: 'Wind simulation and its integration into the ATTAS simulator', DFVLR, IB 111-86/21

10 MADHURANATH, P., and KHARE, A.: 'CLASS – closed loop aircraft flight simulation software', PD FC 9207, NAL Bangalore, October 1992

11 SRINATHKUMAR, S., PARAMESWARAN, V., and RAOL, J. R.: 'Flight test determination of neutral and maneuver point of aircraft', AIAA Atmoshperic Flight Mechanics Conference, Baltimore, USA, Aug. 7–9, 1995

12 PARAMESWARAN, V., GIRIJA, G., and RAOL, J. R.: 'Estimation of parameters from large amplitude maneuvers with partitioned data for aircraft', AIAA Atmospheric Flight Mechanics Conference, Austin, USA, Aug. 11–14, 2003

13 JATEGAONKAR, R. V., and GIRIJA, G.: 'Two complementary approaches to estimate downwash lag effects from flight data', *Journal of Aircraft*, 1991, **28**, (8), pp. 540–542

Appendix C
Solutions to exercises

Chapter 2

Solution 2.1

Let $z = H\beta + v$.

By pre-multiplying both sides by H^T, we obtain: $H^T z = H^T H\beta + H^T v$;

$$\hat{\beta} = (H^T H)^{-1} H^T z - (H^T H)^{-1} H^T v$$

We can postulate that measurement noise amplitude is low and not known (the latter is always true), to obtain

$$\hat{\beta} = (H^T H)^{-1} H^T z$$

This is exactly the same as eq. (2.4). We also see that the extra term is the same as in eq. (2.5).

Solution 2.2

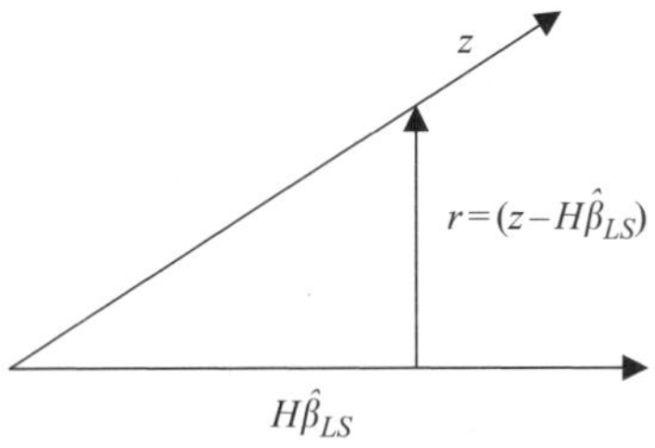

Figure C.1

Solution 2.3

The property tells us about the error made in the estimate of parameters. It also shows that if the measurement errors are large, this will reflect in the parameter estimation error directly if H is kept constant. Thus, in order to keep the estimation error low and have more confidence in the estimated parameters, the measurements must be more accurate. Use of accurate measurements will help. Pre-processing of the measurements might also help.

Solution 2.4

The responses are nonlinear. The point is that the dynamical system between S and V is linear, since it is described by a transfer function. In this case, V is an independent variable. However, the response of S is w.r.t. time and it is found to be nonlinear.

Solution 2.5

Let $\hat{z} = m\hat{x}$.

Then

$$(z - \hat{z}) = m(x - \hat{x}) + v$$

$$(z - \hat{z})(z - \hat{z})^T = (m(x - \hat{x}) + v)(m(x - \hat{x})^T + v^T)$$

$$\text{cov}(\tilde{z}) = E\{(z - \hat{z})(z - \hat{z})^T\} = E\{m^2(x - \hat{x})(x - \hat{x})^T + vv^T\}$$

by neglecting the cross covariance between $(x - \hat{x})$ and v, thereby assuming that $\tilde{x}$ and v are uncorrelated.

$$\text{cov}(\tilde{z}) = m^2 \,\text{cov}(\tilde{x}) + R$$

where R is the covariance matrix of v.

Solution 2.6

Using eqs (2.6) and (2.12), we get

$$P_{GLS} = (\underline{H}^T\underline{H})^{-1}\underline{H}^T R_e\underline{H}(\underline{H}^T\underline{H})^{-1}$$

with

$$\underline{H} = H'; \quad \underline{v} = v'$$

and

$$R_e = \text{cov}(\underline{v}\,\underline{v}^T) = S^T RS$$

$$P_{GLS} = (\underline{H}^T\underline{H})^{-1}\underline{H}^T S^T RS\underline{H}(\underline{H}^T\underline{H})^{-1}$$

Further simplification is possible.

Solution 2.7

If H is invertible, then we get $K = H^{-1}$. However, in general, it is a non-square matrix and hence not invertible. We can expand $K = H^{-1}RH^{-T}H^TR^{-1}$ of eq. (2.15) to

$$K = H^{-1}RR^{-1} = H^{-1}$$

provided H is invertible which is not the case. Hence, the major point of eq. (2.15) is that the pseudo inverse of H is used, which is (assuming $R = I$):

$$(H^TH)^{-1}H^T$$

Solution 2.8

(i) Forward difference method $\qquad \dfrac{\partial h(x)}{\partial \beta} = \dfrac{h(x + \Delta\beta) - h(x)}{\Delta\beta}$

(ii) Backward difference method $\qquad \dfrac{\partial h(x)}{\partial \beta} = \dfrac{h(x) - h(x - \Delta\beta)}{\Delta\beta}$

(iii) Central difference method $\qquad \dfrac{\partial h(x)}{\partial \beta} = \dfrac{h(x + \Delta\beta) - h(x - \Delta\beta)}{2\Delta\beta}$

The $\Delta\beta$ can be chosen as $\Delta\beta = \varepsilon\beta$ where $\varepsilon = 10^{-6}$.

If β is too small, then $\Delta\beta = \varepsilon$.

Solution 2.9

$$z = H\beta + X_v\beta_v + e$$

$$z = [H|X_v]\begin{bmatrix}\beta \\ \beta_v\end{bmatrix} + e$$

Then

$$\begin{bmatrix}\hat{\beta} \\ \hat{\beta}_v\end{bmatrix} = \left((H|X_v)^T(H|X_v)\right)^{-1}(H|X_v)^Tz = \left[\begin{bmatrix}H^T \\ X_v^T\end{bmatrix}(H|X_v)\right]^{-1}(H|X_v)^Tz$$

$$= \begin{bmatrix}H^TH & H^TX_v \\ X_v^TH & X_v^TX_v\end{bmatrix}^{-1}(H|X_v)^Tz$$

Solution 2.10

One can pass the white noise input to the linear-lumped parameter dynamical system or low pass filter. The output process will be the correlated signal with a band-limited spectrum, since the noise at high frequencies will be filtered out.

Solution 2.11

Let

$$y(t) = e^{\sigma t}$$

When

$$t = 0; \quad y(0) = 1$$

Let

$$y(t_d) = 2$$

then

$$2 = e^{\sigma t_d} \quad \Rightarrow \quad \ln 2 = \sigma t_d$$

or

$$t_d = \frac{\ln 2}{\sigma} = \frac{0.693}{\sigma}$$

Chapter 3

Solution 3.1

Let

$$x_1 = y; \quad \dot{x}_1 = \dot{y} = x_2$$

Then

$$\ddot{y} = \dot{x}_2$$

and we have

$$m\dot{x}_2 + dx_2 + Kx_1 = w(t)$$

Thus,

$$\dot{x}_1 = x_2$$
$$\dot{x}_2 = -\frac{d}{m}x_2 - \frac{K}{m}x_1 + \left(\frac{1}{m}\right) w(t)$$

Putting in matrix form, we get

$$\begin{bmatrix} \dot{x}_1 \\ \dot{x}_2 \end{bmatrix} = \begin{bmatrix} 0 & 1 \\ -\frac{K}{m} & -\frac{d}{m} \end{bmatrix} \begin{bmatrix} x_1 \\ x_2 \end{bmatrix} + \begin{bmatrix} 0 \\ \frac{1}{m} \end{bmatrix} w(t)$$

$$\dot{x} = Ax + Bu$$

We finally have

$$\frac{\partial \dot{x}}{\partial K} = \begin{bmatrix} 0 & 0 \\ -\frac{1}{m} & 0 \end{bmatrix} x(t) + A\frac{\partial x}{\partial K}$$

and

$$\frac{\partial \dot{x}}{\partial d} = \begin{bmatrix} 0 & 0 \\ 0 & -\dfrac{1}{m} \end{bmatrix} x(t) + A\frac{\partial x}{\partial d}$$

Solution 3.2

Both the methods are batch-iterative and equally applicable to nonlinear systems. The GLSDC involves a weighting matrix, which is not explicit in OEM, rather matrix R appears. Sensitivity computations are also needed in both the methods. GLSDC is essentially not based on the ML principle, but perhaps could give equally good estimates.

Solution 3.3

Let

$$\dot{x} = A(\beta_2)x(\beta_1, \beta_2) + B(\beta_2)u \quad \text{and} \quad y = C(\beta_2)x(\beta_1, \beta_2) + D(\beta_2)u$$

Then, we have

$$\frac{\partial \dot{x}}{\partial \beta_1} = A\frac{\partial x(\beta_1, \beta_2)}{\partial \beta_1}$$

$$\frac{\partial \dot{x}}{\partial \beta_2} = A\frac{\partial x(\beta_1, \beta_2)}{\partial \beta_2} + \frac{\partial A}{\partial \beta_2}x(\beta_1, \beta_2) + \frac{\partial \beta}{\partial \beta_2}u$$

$$\frac{\partial y}{\partial \beta_1} = C\frac{\partial x(\beta_1, \beta_2)}{\partial \beta_1} \quad \text{and finally} \quad \frac{\partial y}{\partial \beta_2} = C\frac{\partial x(\beta_1, \beta_2)}{\partial \beta_2} + \frac{\partial C}{\partial \beta_2}x + \frac{\partial D}{\partial \beta_2}u$$

Solution 3.4

$$\frac{\partial Y}{\partial \beta} = \begin{bmatrix} x_1 & x_2 & 0 & 0 & 0 & 0 \\ 0 & 0 & x_1 & x_2 & 0 & 0 \\ 0 & 0 & 0 & 0 & x_1 & x_2 \end{bmatrix}_{3\times 6}$$

Assuming $R = I$, we get

$$\nabla^2_\beta(J) = \sum_{k=1}^{N} \left(\frac{\partial Y}{\partial \beta}\right)^T R^{-1} \left(\frac{\partial Y}{\partial \beta}\right)$$

$$= \sum_{k=1}^{N} \begin{bmatrix} x_1 & 0 & 0 \\ x_2 & 0 & 0 \\ 0 & x_1 & 0 \\ 0 & x_2 & 0 \\ 0 & 0 & x_1 \\ 0 & 0 & x_2 \end{bmatrix} \begin{bmatrix} x_1 & x_2 & 0 & 0 & 0 & 0 \\ 0 & 0 & x_1 & x_2 & 0 & 0 \\ 0 & 0 & 0 & 0 & x_1 & x_2 \end{bmatrix}$$

$$
= \begin{bmatrix}
\sum x_1^2 & \sum x_1x_2 & 0 & 0 & 0 & 0 \\
\sum x_1x_2 & \sum x_2^2 & 0 & 0 & 0 & 0 \\
0 & 0 & \sum x_1^2 & \sum x_1x_2 & 0 & 0 \\
0 & 0 & \sum x_1x_2 & \sum x_2^2 & 0 & 0 \\
0 & 0 & 0 & 0 & \sum x_1^2 & \sum x_1x_2 \\
0 & 0 & 0 & 0 & \sum x_1x_2 & \sum x_2^2
\end{bmatrix}
$$

Comparing the elements of the above equation for the second gradient with the elements of eq. (10.51), we see that they have a similar structure and signify some correlation like computations in information matrices.

Solution 3.5

We see that if the bias is zero, then the variance in the parameter estimate is greater than $I_m^{-1}(\beta)$. When the estimate is biased, this bound will be greater.

Solution 3.6

We see that in the ML method, parameter β is obtained by maximising the likelihood function eq. (3.33), which is also equivalent to minimising the negative log likelihood function of eq. (3.34). Comparing eq. (2.2) with eq. (3.34), we infer that the LS estimate is a special case of ML for Gaussian assumption and linear system.

Solution 3.7

Both the expressions give respective covariance matrices for the parameter estimation error. In eq. (3.56), the sensitivities $\partial y/\partial\Theta$ are to be evaluated at each data point. Looking at eq. (2.1), we see that $H = \partial z/\partial\beta$ is also a sensitivity matrix. Practically, the inverse of these two matrices gives the information matrices for the respective estimators. The major difference is the route used to arrive at these formulae. MLE has a more probabilistic basis and is more general than LS.

Chapter 4

Solution 4.1

Let $\hat{z} = \hat{y}$; then

$$
\begin{aligned}
\operatorname{cov}(z - \hat{z}) &= \operatorname{cov}(y + v - \hat{y}) \\
E\{(z - \hat{z})(z - \hat{z})^T\} &= E\{(y + v - \hat{y})(y + v - \hat{y})^T\} \\
&= E\{(y - \hat{y})(y - \hat{y})^T\} + E\{vv^T\}
\end{aligned}
$$

Here, we assume that the measurement residuals $(y - \hat{y})$ and measurement noise v are uncorrelated. Then, we get

$$\text{cov}(z - \hat{z}) = \text{cov}(y - \hat{y}) + R$$

Solution 4.2

$$\phi = e^{A\Delta t} = I + A\Delta t + \frac{A^2 \Delta t^2}{2!}$$

$$\phi = \begin{bmatrix} 1 & 0 \\ 0 & 1 \end{bmatrix} + \begin{bmatrix} 0 & \Delta t \\ 0 & -a\Delta t \end{bmatrix} + \begin{bmatrix} 0 & -\dfrac{a\Delta t^2}{2} \\ 0 & \dfrac{a^2 \Delta t^2}{2} \end{bmatrix} = \begin{bmatrix} 1 & \dfrac{\Delta t - a\Delta t^2}{2} \\ 0 & \dfrac{1 - a\Delta t + a^2 \Delta t^2}{2} \end{bmatrix}$$

$$\phi = \begin{bmatrix} 1 & \Delta t \\ 0 & 1 - a\Delta t \end{bmatrix}$$

Solution 4.3

Since w is unknown,

$$\tilde{x}(k+1) = \phi \hat{x}(k) + bu$$

$$\tilde{\sigma}_x^2 = \phi \hat{\sigma}_x^2 \phi^T + g^2 \sigma_w^2$$

Since u is a deterministic input, it does not appear in the covariance equation of the state error. The measurement update equations are

$$r(k+1) = z(k+1) - c\tilde{x}(k+1)$$

$$K = \frac{\tilde{\sigma}_x^2 c}{(c^2 \tilde{\sigma}_x^2 + \sigma_v^2)}$$

$$\hat{\sigma}_x^2 = (1 - Kc)\tilde{\sigma}_x^2$$

Solution 4.4

We have

$$\begin{bmatrix} \dot{x}_1 \\ \dot{x}_2 \end{bmatrix} = \begin{bmatrix} a_{11} & a_{12} \\ a_{21} & a_{22} \end{bmatrix} \begin{bmatrix} x_1 \\ x_2 \end{bmatrix} + \begin{bmatrix} w_1 \\ w_2 \end{bmatrix}$$

Since a_{ij} are unknown parameters, we consider them as extra states:

$$\dot{x}_1 = a_{11}x_1 + a_{12}x_2 + w_1$$
$$\dot{x}_2 = a_{21}x_1 + a_{22}x_2 + w_2$$
$$\dot{x}_3 = 0$$
$$\dot{x}_4 = 0$$
$$\dot{x}_5 = 0$$
$$\dot{x}_6 = 0$$

with $x_3 = a_{11}$, $x_4 = a_{12}$, $x_5 = a_{21}$ and $x_6 = a_{22}$.

We finally get

$$\begin{aligned}
\dot{x}_1 &= x_1x_3 + x_2x_4 + w_1 \\
\dot{x}_2 &= x_1x_5 + x_2x_6 + w_2 \\
\dot{x}_3 &= 0 \\
\dot{x}_4 &= 0 \\
\dot{x}_5 &= 0 \\
\dot{x}_6 &= 0
\end{aligned}$$

Then $\dot{x} = f(x) + w$, where f is a nonlinear vector valued function.

Solution 4.5

Let the linear model be given by

$$\begin{aligned}
\dot{x} &= A_1x + Gw_1 \\
z &= Hx + v
\end{aligned}$$

By putting the equations for x and v together, we get

$$\begin{aligned}
\dot{x} &= A_1x + Gw_1 \\
\dot{v} &= A_2v + w_2
\end{aligned}$$

We define joint vector $\left[\begin{smallmatrix} x \\ v \end{smallmatrix}\right]$ to get

$$\begin{bmatrix} \dot{x} \\ \dot{v} \end{bmatrix} = \begin{bmatrix} A_1 & 0 \\ 0 & A_2 \end{bmatrix} \begin{bmatrix} x \\ v \end{bmatrix} + \begin{bmatrix} G & 0 \\ 0 & 1 \end{bmatrix} \begin{bmatrix} w_1 \\ w_2 \end{bmatrix}$$

and

$$z = \begin{bmatrix} H & I \end{bmatrix} \begin{bmatrix} x \\ v \end{bmatrix}$$

We see that the vector v, which is correlated noise, is now augmented to the state vector x and hence, there is no measurement noise term in the measurement equation. This amounts to the situation that the measurement noise in the composite equation is zero, leading to $R^{-1} \to \infty$, and hence the Kalman gain will be ill-conditioned. Thus, this formulation is not directly suitable in KF.

Solution 4.6

The residual error is the general term arising from, say, $z - \hat{z}$ (see Chapter 2).

Prediction error
Consider $\tilde{x}(k+1) = \phi\hat{x}(k)$. Then, $z(k+1) - H\tilde{x}(k+1)$ is the prediction error, since $\hat{z} = H\tilde{x}(k+1)$ is the predicted measurement based on the estimate $\tilde{x}$.

Filtering error

Assume that we have already obtained the estimate of the state after incorporating the measurement data:

$$\hat{x}(k+1) = \tilde{x}(k+1) + K(z(k+1) - H\tilde{x}(k+1))$$

Then, the following quantity can be considered as a filtering error:

$$z(k+1) - H\hat{x}(k+1)$$

since the error is obtained after using $\hat{x}(k+1)$, the filtered state estimate.

Solution 4.7

The main reason is that the measurement data occurring at arbitrary intervals can be easily incorporated in the Kalman filtering algorithm.

Solution 4.8

The quantity S is the theoretical (prediction) covariance of the residuals, whereas the $\mathrm{cov}(rr^T)$ is the actual computed covariance of the residuals. For proper tuning of KF, both should match. In fact the computed residuals should lie within the theoretical bounds predicted by S.

Solution 4.9

Let

$$x(k+1) = \phi x(k) + gw(t)$$
$$z(k) = cx(k) + v(k)$$

Then

$$\tilde{p} = \phi\hat{p}\phi^T + g^2\sigma_w^2$$
$$\hat{p} = (1 - Kc)\tilde{p}$$

Also

$$K = \tilde{p}c\left(c^2\tilde{p} + \sigma_v^2\right)^{-1} = \frac{\tilde{p}c}{\tilde{p}c^2 + \sigma_v^2}$$

and hence

$$\hat{p} = \left(1 - \frac{\tilde{p}c^2}{\tilde{p}c^2 + \sigma_v^2}\right)\tilde{p} = \frac{\tilde{p}\sigma_v^2}{c^2\tilde{p} + \sigma_v^2} = \frac{\tilde{p}}{1 + (c^2\tilde{p}/\sigma_v^2)}$$

If σ_v^2 is low, then $\hat{p}$ is low, meaning thereby, we have more confidence in the estimates. We can also rearrange $\hat{p}$ as

$$\hat{p} = \frac{\sigma_v^2}{c^2 + (\sigma_v^2/\tilde{p})}$$

then if $\tilde{p}$ is low, then $\hat{p}$ is low. If the observation model is strong, then $\hat{p}$ is also low.

Solution 4.10

$$\sigma_x^2 = E\{(x - E\{x\})^2\} = E\{x^2 - 2xE\{x\} + (E\{x\})^2\}$$
$$= E\{x^2\} + (E\{x\})^2 - 2E\{x\}E\{x\}$$
$$\sigma_x^2 = E\{x^2\} - (E\{x\})^2$$

Solution 4.11

Std. $= \sqrt{\sigma_x^2} = \sigma_x =$ RMS if the random variable has zero mean.

Solution 4.12

$$P = UDU^T$$

Now, we can split D into its square root as

$$P = UD^{1/2}D^{1/2}U^T = (UD^{1/2})(UD^{1/2})^T$$
$$P = RR^T$$

So, the propagation of U, D factors of covariance matrix P does not involve the square-rooting operation, but it is the square-root type, by the expression of P above.

Solution 4.13

$$\hat{P} = (I - KH)P(I - KH)^T + KRK^T$$
$$\hat{P} = (I - PH^TS^{-1}H)P(I - PH^TS^{-1}H)^T + PH^TS^{-1}RS^{-T}HP^T$$
$$= (P - PH^TS^{-1}HP)(I - PH^TS^{-1}H)^T + PH^TS^{-1}RS^{-T}HP^T$$
$$= (P - PH^TS^{-1}HP) - PH^TS^{-T}HP^T + PH^TS^{-1}HPH^TS^{-T}HP^T + PH^TS^{-1}RS^{-T}HP^T$$
$$= P - PH^TS^{-1}HP - PH^TS^{-T}HP^T + PH^TS^{-1}HPH^TS^{-T}HP^T + PH^TS^{-1}RS^{-T}HP^T$$

Since, P is symmetric

$$\hat{P} = P - PH^TS^{-1}HP - PH^TS^{-T}HP + PH^TS^{-1}HPH^TS^{-T}HP + PH^TS^{-1}RS^{-T}HP$$
$$= P - 2PH^TS^{-1}HP + PH^TS^{-1}(HPH^T + R)S^{-T}HP$$
$$= P - 2PH^TS^{-1}HP + PH^TS^{-T}HP = P - PH^TS^{-1}HP = (I - KH)P$$

Solution 4.14

The residual is given as $r(k) = z(k) - H\tilde{x}(k)$, where $\tilde{x}(k)$ is the time propagated estimates of KF. We see that $z(k)$ is the current measurement and the term $H\tilde{x}(k)$ is the effect of past or old information derived from the past measurements. Thus, the term $r(k)$ generates new information and, hence, it is called the 'innovations' process.

Solution 4.15

Let

$$\underline{x} = \frac{1}{N}\sum_{k=1}^{N} x(k) = \frac{1}{N}\left[\sum_{k=1}^{N-1} x(k) + x(N)\right]$$
$$= \frac{1}{N}\left[\frac{(N-1)}{(N-1)}\sum_{k=1}^{N-1} x(k) + x(N)\right]$$
$$\underline{x} = \frac{1}{N}[(N-1)\underline{x}(N-1) + x(N)]$$

Thus

$$\underline{x(k)} = \frac{1}{k}[(k-1)\underline{x}(k-1) + x(k)]$$

Similarly, for variance of x, we get

$$\sigma_x^2(k) = \frac{1}{k}\left[(k-1)\sigma_x^2(k-1) + x^2(k)\right]$$

Chapter 5

Solution 5.1

Let $\phi = e^{F\Delta t}$ and hence $\phi^{-1} = e^{-F\Delta t} = 1 - F\Delta t$.

Then, we obtain

$$P - \phi^{-1}P(\phi^T)^{-1} = P - (I - F\Delta t)P(I - F^T\Delta t)$$
$$= FP\Delta t + PF^T\Delta t + FPF^T\Delta t^2$$

Neglecting Δt^2 for small values of Δt, we get

$$P - \phi^{-1}P(\phi^T)^{-1} = (FP + PF^T)\Delta t$$

Solution 5.2

Since P is the covariance matrix and obtained as squared-elements/cross products of the components of the variable x, it should be at least the semi-positive definite matrix. This will be ensured if $\tilde{P}$ is semi-positive definite and the eigenvalues of KH are also equal to or less than 1; otherwise, due to the negative sign in the bracket term, $\hat{P}$ will not retain this property.

Chapter 6

Solution 6.1

Let

$$\mathrm{LS}(1) = \frac{b_0}{1 + a_1 z^{-1}}$$

Then, by long division, we get

$$\mathrm{AR} = b_0 + a_1 z^{-1} + a_1^2 z^{-2} + a_1^3 z^{-3} + a_1^4 z^{-4} + \cdots$$

$$\mathrm{AR} = b_0 + b_1 z^{-1} + b_2 z^{-2} + b_3 z^{-3} + b_4 z^{-4} + \cdots + b_n z^{-n}$$

with $b_1 = a_1, b_2 = a_1^2, b_3 = a_1^3$, etc.

This is a long AR model of an order higher than original model with order 1.

Solution 6.2

Let the 1st order AR model be

$$y(k) = \frac{e(k)}{1 + a_1 q^{-1}}$$

We can replace q by z [2], and z as the complex frequency $z = \sigma + j\omega$ to get

$$y(k) = \frac{e(k)}{1 + a_1 z^{-1}}$$

Then

$$\frac{y(z)}{e(z)} = \frac{z}{a_1 + z} = \frac{\sigma + j\omega}{a_1 + \sigma + j\omega}$$

Often we obtain T.F. on unit circle and presume the presence of only the $j\omega$ term:

$$\frac{y}{e}(\omega) = \frac{j\omega}{a_1 + j\omega} = \frac{(a_1 - j\omega)j\omega}{(a_1 + j\omega)(a_1 - j\omega)} = \frac{\omega^2 + a_1 j\omega}{a_1^2 + \omega^2}$$

Then magnitude of T.F. is

$$\mathrm{mag}(\omega) = \frac{\sqrt{\omega^4 + (a_1\omega)^2}}{a_1^2 + \omega^2} \quad \text{and} \quad \text{phase}\,\theta(\omega) = \tan^{-1}\frac{a_1\omega}{\omega^2} = \tan^{-1}\frac{a_1}{\omega}$$

The plot of $\mathrm{mag}(\omega)$ and phase $\theta(\omega)$ versus ω gives the discrete Bode diagram.

Solution 6.3

The first order LS model (without the error part) is

$$y(k) = \frac{b_0}{1 + a_1 q^{-1}} u(k)$$

Next, we get

$$\frac{y(k)}{u(k)} = \frac{b_0}{1 + a_1 z^{-1}} = \frac{b_0 z}{z + a_1} = \frac{b_0(1 + \tau s)}{a_1 + 1 + \tau s} \Rightarrow \frac{y(s)}{u(s)}$$

$$\frac{y(s)}{u(s)} = \frac{b_0 + b_0 \tau s}{1 + a_1 + \tau s} = \frac{b_0 \tau((1/\tau) + s)}{\tau(((1 + a_1)/\tau) + s)} = \frac{b_0(s + (1/\tau))}{s + (1 + a_1)/\tau}$$

Solution 6.4

$$\frac{y(s)}{u(s)} = \frac{b_0((2 + \tau s)/(2 - \tau s))}{a_1 + (2 + \tau s)/(2 - \tau s)} = \frac{b_0(2 + \tau s)}{2 + \tau s + a_1(2 - \tau s)}$$

$$= \frac{b_0(2 + \tau s)}{2(1 + a_1) + (1 - a_1)\tau s} = \frac{b_0 \tau((2/\tau) + s)}{(1 - a_1)\tau(s + 2(1 + a_1)/(1 - a_1)\tau)}$$

$$\frac{y(s)}{u(s)} = \left.\frac{(b_0/(1 - a_1))(s + (2/\tau))}{s + (2/\tau)((1 + a_1)/(1 - a_1))}\right|_{\text{for } s = j\omega}$$

It is called a bilinear transformation.

Solution 6.5

Magnitude $(e^{\tau s}) = \text{mag}(e^{j\omega\tau}) = \text{mag}(\cos \omega\tau + \sin \omega\tau) = 1$.

$$\text{Phase } (e^{j\omega\tau}) = \theta = \omega\tau$$

$$\text{mag}\left(\frac{2 + \tau s}{2 - \tau s}\right) = 1$$

This transformation is preferable to the one in Exercise 6.3 because the magnitude of the transformation is preserved, it being '1'.

Solution 6.6

We have, based on

(i) $\quad s = \dfrac{1 - q^{-1}}{\tau}$ and

(ii) $\quad s = \dfrac{2}{\tau}\dfrac{1 - q^{-1}}{1 + q^{-1}}$

We see a marked difference between the two s-domain operators, obtained using the above transformations.

Solution 6.7

Since the first term is the same, the major difference will be due to the second term. For $N = 100$, $\ln(N) = 4.6$ and this factor is greater than factor '2' in eq. (6.26), and hence, this part of the B statistic will rise faster and will put a greater penalty on the number of coefficients for given N.

Solution 6.8

$$\begin{aligned}
&(2+\tau s)z^{-1} = 2-\tau s \\
&2z^{-1} + \tau s z^{-1} = 2-\tau s \\
&\tau s + \tau s z^{-1} = 2 - 2z^{-1} \\
&\tau s(1+z^{-1}) = 2(1-z^{-1}) \\
&s = \frac{2}{\tau}\frac{1-z^{-1}}{1+z^{-1}}
\end{aligned}$$

Solution 6.9

$$z = e^{\tau(\sigma+j\omega)} = e^{\tau\sigma}e^{j\omega\tau}$$

$$|z| = e^{\tau\sigma} \quad \text{and} \quad \angle z = \theta = \omega\tau$$

Thus, we have

$$\sigma = \frac{1}{\tau}\ln|z| \quad \text{and} \quad \omega = \frac{\angle z}{\tau}$$

Using these expressions, we can determine the roots in the s-domain given the roots in the z-domain (discrete pulse transfer function domain).

Chapter 7

Solution 7.1

$$\dot{x} = \frac{x(t+\tau) - x(t)}{\tau}$$

$$\ddot{x} = \frac{1}{\tau^2}(x(t+2\tau) - 2x(t+\tau) + x(t))$$

The above equation follows from

$$\ddot{x}(t) = \frac{1}{\tau}\left[\frac{1}{\tau}(x(t+2\tau) - x(t+\tau)) - \frac{1}{\tau}(x(t+\tau) - x(t))\right]$$

Thus, we have

$$\frac{m}{\tau^2}[x(t+2\tau) - 2x(t+\tau) + x(t)] + \frac{d}{\tau}[x(t+\tau) - x(t)] + Kx = u$$

or

$$mx(t+2\tau) + (-2m + \tau d)x(t+\tau) + (m - \tau d + \tau^2 K)x(t) = \tau^2 u$$

or

$$mx_{k+2} + (-2m + \tau d)x_{k+1} + (m - \tau d + \tau^2 K)x_k = \tau^2 u_k$$

Solution 7.2

Method 1

$$\dot{y} = \dot{A}_2 x + A_2 \dot{x}$$
$$\dot{\hat{y}} = \dot{A}_2 \hat{x} + A_2(A\hat{x} + Bu)$$
$$\dot{\hat{y}} = (\dot{A}_2 + A_2 A)\hat{x} + A_2 Bu$$

Method 2

$$\hat{y}(k+1) - \hat{y}(k) = A_2(\hat{x}(k+1) - \hat{x}(k))$$

$$\frac{\hat{y}(k+1) - \hat{y}(k)}{\Delta t} = \frac{A_2}{\Delta t}(\hat{x}(k+1) - \hat{x}(k))$$

We obtain the right hand side term from

$$\frac{\hat{x}(k+1) - \hat{x}(k)}{\Delta t} = A\hat{x}(k) + Bu$$

Thus, we get

$$\frac{\hat{y}(k+1) - \hat{y}(k)}{\Delta t} = A_2 A\hat{x}(k) + A_2 Bu$$

As $\Delta t \to 0$, we get

$$\dot{\hat{y}} = A_2 A\hat{x}(k) + A_2 Bu$$

So, we have two results

(i) $\dot{\hat{y}} = \dot{A}_2 \hat{x} + A_2 A\hat{x} + A_2 Bu$

(ii) $\dot{\hat{y}} = A_2 A\hat{x} + A_2 Bu$

We see that Method 1 is more accurate if A_2 is a time varying matrix.

Solution 7.3

We see from eq. (7.13) that

$$\sigma_s^2 = \hat{\sigma}_x^2 + \frac{\hat{\sigma}_x^2\phi}{\tilde{\sigma}_x^2}(\sigma_s^2 - \tilde{\sigma}_x^2)\frac{\hat{\sigma}_x^2\phi}{\tilde{\sigma}_x^2} = \sigma_x^2 + \phi(\sigma_s^2 - \tilde{\sigma}_x^2)\phi$$

Then

$$\sigma_s^2 - \phi\phi\sigma_s^2 = (1 - \phi\phi)\sigma_x^2$$

$$(1 - \phi\phi)\sigma_s^2 = (1 - \phi\phi)\sigma_x^2$$

Thus, $\sigma_s^2 = \sigma_x^2$

Solution 7.4

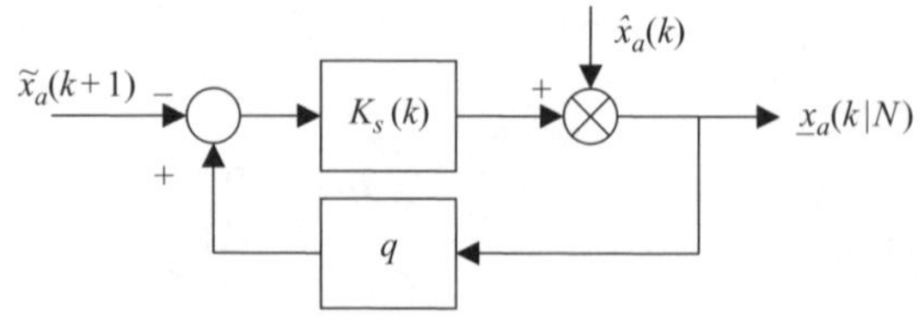

Figure C.2

where $\underline{x}_a(k|N) = q^{-1}\underline{x}_a(k+1|N)$

Solution 7.5

We have $I_m = P^{-1}$ and hence

$$I_{ff} = \left(\sigma_f^2\right)^{-1} \quad \text{and} \quad I_{ff} = \left(\sigma_b^2\right)^{-1} \text{ thus giving}$$

$$I_{fs} = \left(\sigma_f^2\right)^{-1} + \left(\sigma_b^2\right)^{-1} = I_{ff} + I_{fb}$$

Thus, we see that the smoother gives or utilises enhanced information.

Chapter 8

Solution 8.1

No. The reason is that d is the deterministic discrepancy (in the model). It is a time-history, which is estimated by the IE method. As such, it is not a random variable. We can regard Q^{-1}, perhaps, as some form of information matrix, deriving a hint from the fact that in GLS, W is used and if $W = R^{-1}$, we get the so-called Markov estimates. And since R^{-1} can be regarded as some form of information matrix (R being the covariance matrix), Q^{-1} may be called an information matrix. It is a very important tuning parameter for the algorithm.

Solution 8.2

The idea is to have correct estimates of the state as the integration of eq. (8.4), and simultaneously the correct representation of model error estimation d. In order that both these things happen, eqs (8.3) and (8.4) should be satisfied. The estimate should evolve according to eq. (8.3) and eq. (8.4) should be satisfied in order to get proper tuning by Q to obtain a good estimate of d. In eq. (8.2), the second term is also to be minimised thereby saying that only accurate d needs to be obtained by choosing the appropriate penalty by Q. Too much or too less d will not obtain the correct estimate of x.

Solution 8.3

Use of R^{-1} normalises the cost function, since $E\{(y - \hat{y})(y - \hat{y})^T\}$ is a covariance matrix of residuals and R is the measurement noise covariance matrix. Then $E\{(y - \hat{y})^T R^{-1}(y - \hat{y})\}$ will be a normalised sum of squares of residuals.

Solution 8.4

In KF, a similar situation occurs, and it is called 'covariance matching'. The computed covariance from the measurement residuals is supposed to be within the theoretical bounds (which are specified by the diagonal elements of the covariance matrix of innovations), computed by the filter itself as $S = HPH^T + R$.

Solution 8.5

In order to determine the additional model from d, the least squares method will be used and the residuals arising from the term will be treated as measurement noise.

Solution 8.6

Continuously replace computed S by $(S + S^T)/2$ before updating S.

Solution 8.7

Following eq. (8.2), we obtain the cost function as

$$J = \sum_{k=0}^{N} (z(k) - \hat{x}(k))^2(\sigma^2)^{-1} + \int_{t_0}^{t_f} d^2 Q\, dt$$

The Hamiltonian is

$$H = \Psi(x(t), u(t), t) + \lambda^T(t) f(x(t), u(t), t)$$

$$H = d^2 Q + \lambda^T d$$

Solution 8.8

The term $\phi(x(t_f), t_f)$ will be replaced by the following term [1]:

$$\sum_{k=0}^{N} \phi_k(x(t_k), t_k)$$

This will signify the inclusion of penalty terms at times between t_0 and t_f.

Solution 8.9

We have

$$\frac{\partial H}{\partial x} = -\lambda^T(t)\frac{\partial f}{\partial x} + \frac{\partial \psi}{\partial x}$$

From Pontryagin's necessary condition, we have

$$\frac{\partial H}{\partial x} = \dot{\lambda}^T$$

and hence

$$\dot{\lambda}^T = -\lambda^T(t)\left(\frac{\partial f}{\partial x}\right) + \left(\frac{\partial \psi}{\partial x}\right)$$

which can be rewritten as

$$\dot{\lambda} = -\left(\frac{\partial f}{\partial x}\right)^T \lambda^T(t) + \left(\frac{\partial \psi}{\partial x}\right)^T$$

$\dot{\lambda}(t) = A\lambda(t) + u(t)$ with appropriate equivalence.

It must be noted that since f_x and ψ_x are matrices evaluated at estimated state $\hat{x}$, we see that the co-state equation has a similar structure as the state equation.

Chapter 9

Solution 9.1

Let

$$\dot{x} = Ax + Bu$$

Then

$$\dot{x} = Ax + B(Kx + L\dot{x} + \delta) = Ax + BKx + BL\dot{x} + B\delta$$

$$(I - BL)\dot{x} = Ax + BKx + B\delta$$

Hence

$$\dot{x} = (I - BL)^{-1}[(A + BK)x + B\delta]$$

Solution 9.2

From the expression for the integrating feedback, we have

$$\dot{u} = -Fu + Kx + \delta$$

$$\dot{u} = Kx - Fu + \delta$$

Putting the state equation $\dot{x} = Ax + Bu$ and the above equation together, we get

$$\dot{x} = [A \quad B]\begin{bmatrix} x \\ u \end{bmatrix} + [0]\delta$$

$$\dot{u} = [K \quad -F]\begin{bmatrix} x \\ u \end{bmatrix} + 1.\delta$$

We get

$$\begin{bmatrix} \dot{x} \\ \dot{u} \end{bmatrix} = \begin{bmatrix} A & B \\ K & -F \end{bmatrix} \begin{bmatrix} x \\ u \end{bmatrix} + \begin{bmatrix} 0 \\ 1 \end{bmatrix} \delta$$

Solution 9.3

$$\dot{x} = Ax + Bu + w$$

Also, we have

$$Kx - x = 0 \quad \Rightarrow \quad (K - I)x = 0$$

Adding the above two equations, we get

$$\dot{x} + 0 = Ax + Bu + w + (K - I)x$$
$$\dot{x} = (A + (K - I))x + Bu + w$$

We can multiply $(K - I)$ by an arbitrary matrix B_a to get

$$\dot{x} = [A + B_a(K - I)]x + Bu + w$$

Solution 9.4

Let

$$\begin{bmatrix} Y \\ a \end{bmatrix} = \begin{bmatrix} X \\ C_{OE} \end{bmatrix} \beta$$

be represented as

$$Z = H\beta; \quad H^T = \begin{bmatrix} X^T & C_{OE}^T \end{bmatrix}$$

The observability matrix is

$$O_b = [H^T | A^T H^T | \cdots | (A^T)^{n-1} H^T]$$
$$= [[X^T \quad C_{OE}^T] | A^T [X^T \quad C_{OE}^T] | \cdots | (A^T)^{n-1} [X^T \quad C_{OE}^T]]$$

In order that the system is observable, the O_b should have rank n (dimension of β).

Solution 9.5

In the LS estimator, we have $\hat{\beta}_{LS} = (X^T X)^{-1} X^T Y$ and the term $(X^T X)^{-1}$ signifies the uncertainty, or the variance of the estimator.

Actually

$$\text{cov}(\beta - \hat{\beta}) = \sigma_r^2 (X^T X)^{-1}$$

This means that $(X^T X)$ can be regarded as the information matrix. From eq. (9.47), we see that the information matrix of the new (ME) estimator is enhanced by the term $C_{OE}^T W^{-1} C_{OE}$ and hence the variance of the estimator is reduced. This is intuitively appealing, since the *a priori* information on certain parameters will reduce uncertainty in the estimates.

Solution 9.6

We have from the first equation

$$\frac{\partial x(k+1)}{\partial \beta} = \phi \frac{\partial x(k)}{\partial \beta} + \frac{\partial \phi}{\partial \beta} x(k) + \psi \frac{\partial B}{\partial \beta} u(k) + \psi B \frac{\partial u(k)}{\partial \beta} + \frac{\partial \psi}{\partial \beta} Bu(k)$$

and

$$\frac{\partial y(k)}{\partial \beta} = H \frac{\partial x(k)}{\partial \beta} + \frac{\partial H}{\partial \beta} x(k) + \frac{\partial D}{\partial \beta} u(k) + D \frac{\partial u(k)}{\partial \beta}$$

Solution 9.7

$$\phi = e^{A\Delta t} = I + A\Delta t + A^2 \frac{\Delta t^2}{2!} + \cdots$$

$$\psi = \int_0^{\Delta t} e^{A\tau} d\tau \approx I\Delta t + A\frac{\Delta t^2}{2!} + A^2 \frac{\Delta t^3}{3!} + \cdots$$

Solution 9.8

The eigenvalues are $\lambda_1 = -1$ and $\lambda_2 = 2$.

The new system matrix should be $\bar{A} = A - I\delta$, and in order that $\bar{A}$ has stable eigenvalues, we have

$$\bar{A} = \begin{bmatrix} -1 & 0 \\ 0 & 2 \end{bmatrix} - \begin{bmatrix} \delta & 0 \\ 0 & \delta \end{bmatrix} = \begin{bmatrix} -1-\delta & 0 \\ 0 & 2-\delta \end{bmatrix}$$

$$\lambda_1 = -1 - \delta \quad \text{and} \quad \lambda_2 = 2 - \delta = -2 \text{ (say)}$$

This gives $\delta = 4$ and $\lambda_1 = -5$.

Thus, the new matrix with stable eigenvalues will be

$$\bar{A} = \begin{bmatrix} -5 & 0 \\ 0 & -2 \end{bmatrix}$$

Solution 9.9

$$\phi_A = I + \begin{bmatrix} -\Delta t & 0 \\ 0 & 2\Delta t \end{bmatrix} = \begin{bmatrix} 1-\Delta t & 0 \\ 0 & 1+2\Delta t \end{bmatrix}$$

$$\phi_{\bar{A}} = I + \begin{bmatrix} -5\Delta t & 0 \\ 0 & -2\Delta t \end{bmatrix} = \begin{bmatrix} 1-5\Delta t & 0 \\ 0 & 1-2\Delta t \end{bmatrix}$$

Since we have $\bar{A} = A - I\delta$;

$$\phi_{\bar{A}} = e^{\bar{A}\Delta t} = e^{(A-I\delta)\Delta t} = I + (A - I\delta)\Delta t = I + A\Delta t - I\delta\Delta t$$
$$= \phi_A - I\delta\Delta t$$

Thus $\phi_{\bar{A}} = \phi_A - I\delta\Delta t$ and the equivalent $\delta_{eq} = \delta\Delta t$.

Solution 9.10

$$A_d = \begin{bmatrix} -1 & 0 \\ 0 & 4 \end{bmatrix}; \quad A_{od} = \begin{bmatrix} 0 & -2 \\ -3 & 0 \end{bmatrix}$$

We see that A_d still has one eigenvalue at $\lambda = 4$; an unstable solution.

Solution 9.11

$$A_s = \begin{bmatrix} -1 & -2 \\ 3 & 0 \end{bmatrix} \quad \text{and} \quad A_{us} = \begin{bmatrix} 0 & 0 \\ 0 & 4 \end{bmatrix}$$

Solution 9.12

Since Δt is a constant, the above expression gives the autocorrelation of the process $r(k)$ for τ, and the time lag is '1' unit (of Δt). Thus, we have

$$R_{rr}(\tau = 1) = \frac{\Delta t}{N-1} \sum_{k=1}^{N} r(k)r(k-1)$$

Since r is a white process, $R_{rr}(\tau = 1) \to 0$ or within the bound $\pm 1.97/\sqrt{N}$.

Solution 9.13

Using the three expressions of Example 9.6, we have

$$\dot{w} = (Z_w + Z_{\delta_e}K)w + (u_0 + Z_q)q + Z_{\delta_e}\delta_p$$
$$\dot{q} = (M_w + M_{\delta_e}K)w + M_q q + M_{\delta_e}\delta_p$$

Thus, if $M_w = 0.2$, we can make

$$M_w + M_{\delta_e}K = -0.4$$

and choose

$$K = \frac{-0.4 - M_w}{M_{\delta_e}} = \frac{-0.4 - 0.2}{M_{\delta_e}} = \frac{-0.6}{M_{\delta_e}}$$

And since $M_{\delta_e} = -12.8$, we get

$$K = \frac{-0.6}{-12.8} = \frac{0.6}{12.8}$$

Solution 9.14

We have from Fig. 9.7

$$y(s) = G(s)u(s) \quad \Rightarrow \quad u(s) = \delta(s) - H(s)y(s)$$
$$= \delta(s) - H(s)G(s)u(s)$$

and hence we have $u(s) + H(s)G(s)u(s) = \delta(s)$ and finally

$$\frac{u(s)}{\delta(s)} = \frac{1}{1 + G(s)H(s)} = \text{the sensitivity function}$$

Solution 9.15

Since input u (the closed loop system error) is affected by the output noise v due to the feedback, u and v are correlated. However, since the $\hat{u}$ is an estimate of u, hopefully, drastically reducing the effect of noise, $\hat{u}$ and v are considered uncorrelated.

Chapter 10

Solution 10.1

We use

$$\frac{dW_2}{dt} = -\frac{\partial E(W_2)}{\partial W_2} = (z - u_2)\frac{\partial u_2}{\partial W_2} = (z - u_2)\frac{\partial f(y_2)}{\partial W_2}$$
$$= f'(y_2) \cdot (z - u_2) \cdot u_1^T; \qquad \text{since } \frac{\partial y_2}{\partial W_2} = u_1^T$$

Using the discretisation rule, we get

$$\frac{W_2(i+1) - W_2(i)}{\Delta t} = f'(y_2).(z - u_2) \cdot u_1^T$$
$$W_2(i+1) = W_2(i) + \Delta t e_{2b} u_1^T = W_2(i) + \mu e_{2b} u_1^T$$

by defining $e_{2b} = f'(y_2)(z - u_2)$.

Δt can be absorbed in μ, the learning rate parameter.

Solution 10.2

$$\frac{dW_1}{dt} = -\frac{\partial E}{\partial W_1} = (z - u_2)\frac{\partial u_2}{\partial W_1} = (z - u_2) f'(y_2)\frac{\partial y_2}{\partial W_1}$$
$$= (z - u_2) f'(y_2) W_2^T \frac{\partial u_1}{\partial W_1} = (z - u_2) f'(y_2) W_2^T f'(y_1) u_0^T$$

$$\frac{dW_1}{dt} = e_{1b} u_0^T$$

Defining

$$e_{1b} = f'(y_1)W_2^T e_{2b}$$

Finally we get

$$W_1(i+1) = W_1(i) + \mu e_{1b} u_0^T; \quad \Delta t \text{ is absorbed in } \mu.$$

Solution 10.3

In the computational algorithm, one can do the following:

If $z_i = 1$
then $z_i = z_i - \varepsilon$
else end

Here, ε is a small positive number.

Solution 10.4

In eq. (10.12), the μ term has $e_{1b}u_0^T$ whereas in eq. (10.21), the μ term has $e_{1b}K_1^T$ as the factors. Here $K_1^T = (f_1 + u_0 P_1 u_0)^{-T} u_0^T P_1^T$, thereby having additional quantities as $(f_1 + u_0 P_1 u_0)^{-T}$ and P_1^T. These factors will have varying range and for the same problem, the range of values of μ in the learning rules will be different.

Solution 10.5

The KF equations are

$$K = \tilde{P}H^T(H\tilde{P}H^T + R)^{-1} \quad \text{and} \quad \hat{P} = (I - KH)\tilde{P}$$

Equations (10.15) and (10.16) are:

$$K_2 = P_2 u_1 (f_2 + u_1 P_2 u_1)^{-1}$$

or

$$K_2 = P_2 u_1 (u_1 P_2 u_1 + f_2)^{-1}$$

and

$$P_2 = \frac{(I - K_2 u_1)P_2}{f_2}$$

We see that $H^T = u_1$; $R \to f_2$. This means that $R = I$, and the forgetting factor appears instead. In principle, this FFNN learning rule is derived from the application of the KF principle to obtain weight update rules [11].

Solution 10.6

$$\frac{W_1(i+1) - W_1(i)}{\Delta t} = \frac{\mu e_{1b} u_0^T}{\Delta t} + \Omega \frac{(W_1(i) - W_1(i-1))}{\Delta t}$$

We can absorb Δt into μ, and then as $\Delta t \to 0$, we get

$$\dot{W}_1|_{t=i+1} = \mu e_{1b} u_0^T + \Omega \dot{W}_1\big|_{t=i}$$

Solution 10.7

We see from eq. (10.51) that the elements are the sum of the products of $\dot{x}_i, x_i, u_i$, etc. These are approximate computations of various correlations like quantities between $\dot{x}$, x_0 and u. W can be viewed as the information providing matrix.

Solution 10.8

$$\beta_i = \rho \left[\frac{1 - e^{-\lambda x_i}}{1 + e^{-\lambda x_i}}\right]$$

$$\beta_i (1 + e^{-\lambda x_i}) = \rho - \rho e^{-\lambda x_i}$$

$$\beta_i + \beta_i e^{-\lambda x_i} = \rho - \rho e^{-\lambda x_i}$$

$$(\beta_i + \rho) e^{-\lambda x_i} = \rho - \beta_i$$

$$e^{-\lambda x_i} = \left(\frac{\rho - \beta_i}{\rho + \beta_i}\right)$$

$$-\lambda x_i = \ln\left(\frac{\rho - \beta_i}{\rho + \beta_i}\right)$$

$$x_i = -\frac{1}{\lambda} \ln\left(\frac{\rho - \beta_i}{\rho + \beta_i}\right)$$

Solution 10.9

$$\frac{\partial f}{\partial x_i} = f' = f(x_i)[1 - f(x_i)]$$

This function $f(x_i)$ is infinitely differentiable.

Since

$$f(x) = (1 + e^{-x})^{-1}$$

$$f'(x) = (-1) \frac{-e^{-x}}{(1 + e^{-x})^2} = \frac{e^{-x}}{(1 + e^{-x})^2} = \frac{1}{1 + e^{-x}} \left(1 - \frac{1}{1 + e^{-x}}\right)$$

$$= f(x)(1 - f(x))$$

Solution 10.10

We can consider that weights W are to be estimated during the training of the FFNN and that these can be considered as the states of the KF to be estimated.

Then we have

$$W(k+1) = W(k) + w(k)$$

as the state model and

$$z(k) = f(W(k), u_2(k)) + v(k)$$

Here, function f is defined by the FFNN propagation. The weight vector W will contain weights as well as biases of the network. Then the W can be estimated using the EKF described in Chapter 4.

Solution 10.11

Let RNN-S dynamics be given as

$$\dot{x}_i(t) = \sum_{j=1}^{n} w_{ij} x_j(t) + b_i; \qquad i = 1, \ldots, n$$

and

$$\dot{x} = Ax + Bu$$

Here

$$A \Rightarrow \sum_{j=1}^{n} w_{ij} \qquad \text{and} \qquad B = 1, \quad u = b_i$$

which are known quantities. Interestingly, both the states have a similar meaning: *internal states of the system*.

In addition, $z = Hx$ and $\beta_j(t) = f(x_j(t))$ Here, β is the output state of RNN whereas in the linear system, z is the output. For nonlinear measurement model, we will have: $z = h(x)$ and we see striking similarity of h with f. Here, h could be any nonlinearity whereas f has a specific characteristic like sigmoid nonlinearity.

Solution 10.12

$$\frac{\partial E}{\partial \beta} = -\rho \sum_{k=1}^{N} \tanh(\lambda(\dot{x}(k) - Ax(k)))x^T(k)$$

Here, β contains the elements of A.

Solution 10.13

$$\text{Rule 1:} \quad \beta_1 = -\mu \int \frac{\partial E}{\partial \beta_1} dt$$

$$\text{Rule 2:} \quad \beta_1 = f \int \frac{-\partial E}{\partial \beta_1} dt = f(\beta) \quad \text{where } \beta = \int \frac{-\partial E}{\partial \beta_1} dt \text{ and hence } \frac{d\beta}{dt} = -\frac{\partial E}{\partial \beta_1}$$

$$\text{Rule 3:} \quad \frac{d\beta_1}{dt} = f'(\beta)\frac{d\beta}{dt} \quad \Rightarrow \quad \frac{d\beta}{dt} = \frac{1}{f'(\beta)}\frac{d\beta_1}{dt} = -\frac{\mu}{f'(\beta)}\frac{\partial E}{\partial \beta_1}$$

The detailed development can be found in Reference 18.

Solution 10.14

Step 1: $e(k) = \dot{x}(k) - Ax(k)$ assuming some initial values of A

Step 2: nonlinearity effect : $e'(k) = f(e(k))$

Step 3: $$\frac{\partial E}{\partial \beta (= A)} = \sum_{k=1}^{N} e'(k)(-x(k))^T$$

Step 4: adaptive block : $\dfrac{d\beta}{dt} = -\mu \dfrac{\partial E}{\partial \beta}$ μ is as a tuning or learning parameter.

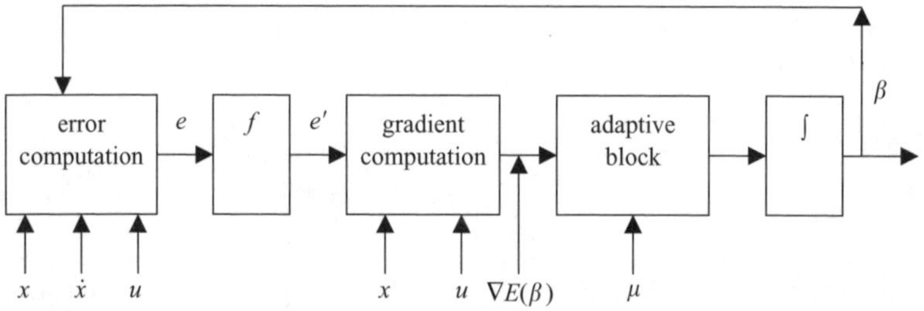

Figure C.3

Solution 10.15

During the training, the weights might vary drastically and the training algorithm might oscillate and not converge. The term with the momentum factor is related to the rate of change of weights at successive iterations: $(W(i) - W(i-1))/\Delta t$, where Δt could be absorbed in the momentum factor. Thus, the approximation of the derivative of the weight vector is used to control the weights. This is similar to using anticipatory action in the control system, somewhat equivalent to derivative control action.

Chapter 11

Solution 11.1

$$X^T X = (A^T - jB^T)(A + jB) = A^T A - jB^T A + jA^T B + B^T B$$
$$= A^T A + B^T B + j(A^T B - B^T A)$$

$$\text{Real } (X^T X) = (A^T A + B^T B)$$

Solution 11.2

Let

$$X^{-1} = C + jD$$

Then, we have

$$XX^{-1} = (A + jB)(C + jD) = I + jO$$

Simplifying, we get

$$AC + jBC + jAD - BD = I + jO$$

By collecting comparative terms, we get

$$AC - BD = I$$

$$BC + AD = O$$

$$\begin{bmatrix} A & -B \\ B & A \end{bmatrix} \begin{bmatrix} C \\ D \end{bmatrix} = \begin{bmatrix} I \\ O \end{bmatrix}$$

$$\begin{bmatrix} C \\ D \end{bmatrix} = \begin{bmatrix} A & -B \\ B & A \end{bmatrix}^{-1} \begin{bmatrix} I \\ O \end{bmatrix}$$

The above expression involves only real operations.

Solution 11.3

(Here 'T' is replaced by the prime sign for simplicity.)

$$\begin{aligned} \hat{\beta} &= [\mathrm{Re}\{(A' - jB')(A + jB)\}]^{-1}[\mathrm{Re}\,\{(A' - jB')(C + jD)\}] \\ &= [\mathrm{Re}\,(A'A - jB'A + jA'B + B'B)]^{-1} \\ &\quad \times [\mathrm{Re}\,(A'C - jB'C + jA'D + B'D)] \\ &= (A'A + B'B)^{-1}(A'C + B'D) \end{aligned}$$

Index